CARL SAGAN
&
IMMANUEL
VELIKOVSKY

Other Titles From New Falcon Publications

Undoing Yourself With Energized Meditation
Secrets of Western Tantra
The Tree of Lies
    By Christopher S. Hyatt, Ph.D.
The Enochian World of Aleister Crowley
    By Aleister Crowley, L. M. DuQuette, and C. S. Hyatt
Aleister Crowley's Illustrated Goetia: Sexual Evocation
    By C. S. Hyatt, L. M. DuQuette and D. Wilson
Pacts With The Devil
    By S. Jason Black and Christopher S. Hyatt, Ph.D.
Sufism, Islam and Jungian Psychology
Buddhism and Jungian Psychology
Catholicism and Jungian Psychology
    By J. Marvin Spiegelman, Ph.D.
Cosmic Trigger: The Final Secret of the Illuminati
Cosmic Trigger II: Down to Earth
Cosmic Trigger III
Prometheus Rising
The New Inquisition
    By Robert Anton Wilson
Equinox of the Gods
Eight Lectures on Yoga
Gems From the Equinox
Temple of Solomon the King
    By Aleister Crowley
Neuropolitique
Info-Psychology
Game of Life
    By Timothy Leary, Ph.D.
Zen Without Zen Masters
A Handful of Zen
    By Camden Benares
The Complete Golden Dawn System of Magic
What You Should Know About The Golden Dawn
Golden Dawn Tapes—Series I, II, and III
    By Israel Regardie

And to get your free catalog of *all* of our titles, write to:
New Falcon Publications (Catalog Dept.)
1739 East Broadway Road, Suite 1-277
Tempe, Arizona 85282 U.S.A

CARL SAGAN
&
IMMANUEL VELIKOVSKY

By

Charles Ginenthal

1995
NEW FALCON PUBLICATIONS
TEMPE, ARIZONA U.S.A.

International Standard Book Number: 1-56184-075-0

First Falcon Edition 1995

Cover art by S. Jason Black

Special thanks to Ralph Dravin for his editing

The paper used in this publication meets the minimum require-ments of the American National Standard for Permanence of Paper for Printed Library Materials Z39.48-1984

Address all inquiries to:
New Falcon Publications
1739 East Broadway Road Suite 1-277
Tempe, Arizona 85282 U.S.A.
(or)
1605 East Charleston Blvd.
Las Vegas, NV 89104 U.S.A.

TO JOAN

ACKNOWLEDGMENTS

There are many individuals whose work helped make this book possible. The staff of the journal *Kronos* played a very large role in presenting some of the data contained in this book. I especially owe thanks to Clark Whelton for his encouragement, to Dr. Earl Milton, Dr. C.J. Ransom and Dr. George R. Talbott for a careful examination of the scientific materials, and to Dr. Lynn E. Rose, Frederick B. Jueneman and David N. Talbott for their fine advice. And to Lynn Schneider Ginenthal who patiently typed this book, my special thanks.

TABLE OF CONTENTS

PART II THE SCIENTIFIC EVIDENCE

PREFACE

'Dogma differs from hypothesis by the refusal of its adherents even to consider the aspects of its validity. Legitimate disagreement or controversy creates dogma when arguments are no longer listened to. Although usually belonging to the realm of theoretical models where direct experiment (or observation) is not possible dogmatism may sometimes induce its followers to misquotation or misrepresentation of the most indisputable facts, even to statements made in print by their opponents.'

E.J. Opik, "About Dogma in Science..."
Annual Review of Astronomy and Astrophysics, (1977). Vol. 15, p. 1

'Yet it remained for the scientific community to launch the most vicious and unreasoning attack on both the ideas and the author of *Worlds in Collision.'*

Fred Warshofsky, *Doomsday the Science of Catastrophe,* (1977), p. 42

'It is extremely difficult for a really radical idea to get a hearing, much less a fair hearing. And if the originator of the radical idea does not have normal credentials, getting a (fair) hearing can be virtually impossible.'

David M. Raup, *The Nemesis Affair,* (1986), p. 203

Dr. Carl Sagan, a professor of astronomy from Cornell University, a well known public personality and writer of popular books of science, in 1974 at a symposium of the American Association for the Advancement of Science (AAAS) delivered a paper, "An Analysis of *Worlds in Collision*". This paper was later edited and presented in a book, *Scientists Confront Velikovsky,* published by Cornell University Press. The paper was further edited and presented in Sagan's book *Broca's Brain,* under the title "Venus and Dr. Velikovsky". Sagan's paper is a critique of Immanuel Velikovsky's book *Worlds in Collision.*

Having read Velikovsky, I also read Sagan's paper; I thereafter discovered that a group of scientists and scholars had written critiques of Sagan's analysis. After reading these criticisms I began a search of the literature and over a period of time I became convinced that Sagan's critique lacked substance. Most surprising was the number of statements made by Sagan that proved to be clearly untrue. Further reading reinforced this discovery of the glaringly unscientific and unscholarly quality of Sagan's paper. What was much worse, was that it was difficult to imagine that even Sagan was unaware of the misrepresentation of evidence presented as scholarly criticism by him and offered to the public.

Thereafter, I encountered a colleague who, learning that I was interested in the thesis of Dr. Velikovsky, informed me that in *Broca's Brain* was an essay by Professor Sagan that demolished Velikovsky and his thesis. When he informed me that he had not read any of Velikovsky's books nor any criticisms of Sagan's article I asked, "How can you make a proper judgment if you haven't read both sides of the issue." To my astonishment he replied, "I don't have to read both sides to know which side is right!" His closed-minded attitude made discussion futile and I let the remark pass. Several days later I received a letter in which he presented citations from Sagan's paper and posed, "What possible arguments could be raised on Velikovsky's behalf?"

In response I composed a long letter which dealt with merely one of Sagan's criticisms. This posted I awaited his response—none came. A few weeks later at a monthly conference, we ran into each other. In a very friendly manner he approached me, smiling broadly, he shook my hand. "What did you think of my reply to your letter?" I asked. He admired the scholarship of my reply to Sagan and admitted frankly, "There are two sides to this Velikovsky business." This I followed up by asking if there were any other aspects of Sagan's criticism which he wished to clarify. He shook his head 'no' and I dropped the matter. However, I noted that he seemed shocked by the evidence of the rebuttal presented.

It was at that moment that the realization struck that Carl Sagan's criticisms had been uncritically read by a wide audience. This was soon discovered to be the case among friends and relatives. Seemingly, they had all read Sagan's side, but not Velikovsky's. With little or no scientific background with which to judge, they had accepted Sagan's word on all matters. It was then that I conceived the idea for this book. It is hoped that reading the other side will permit laymen to clarify the issues.

I must admit that doing the research for this book over about an eight-year period has brought to my attention much more than I had imagined regarding Sagan's critique. It has been a deeply saddening experience to discover again and again the crassness of Sagan's work on Velikovsky. It has also been a deeply shocking experience to learn the political nature of the way science operates. Even if Velikovsky's theories are completely wrong, no one deserves to be maligned as he has been. The deceit exposed in the following pages is an outrage to decency.

PART I

INTRODUCTION

Discussing the reception of the scientific community to Darwin's *Origin of Species,* Thomas Huxley wrote, 'It was badly received by the generation to which it was first addressed, and the angry outpouring of angry nonsenses to which it gave rise is sad to think upon. But the present generation will probably behave just as badly if another Darwin should arise, and inflict upon them what the generality of mankind most hate— the necessity of revising their convictions.'

Thomas Huxley cited by Daniel J. Boorstin,
The Discoverers, (1985), p. 476

'Only later, when the Quarterly Review article appeared and his friends had persuaded him that [St. George] Mivart's criticisms were not only unjust but also influential, did Darwin have second thoughts about him. He himself observed that Mivart had twice neglected to complete quotations from the *Descent,* but now he was told that the omitted words were essential to the argument, upon which [Darwin]...sorrowfully concluded that, 'though he [Mivart] means to be honourable, he is so bigoted that he cannot act fairly.'

Charles Darwin, cited by Gertrude Himmelfarb,
Darwin and the Darwinian Revolution, (1968), pp. 359-60

'Science writers, if they do it well, both inform and entertain, but the task of informing is primary. They must under no circumstances, misinform. If they do, their work is worthless and even harmful—all the more worthless and harmful if it is entertaining and attracts readers.'

Isaac Asimov, *The Planets,* (1985), p. 20

AN IMPROBABLE TALE

Some forty years ago, Immanuel Velikovsky triggered a venomous scientific controversy when he claimed that, within the past few thousand years, errant planets have nearly destroyed life on Earth.

Though his book, *Worlds in Collision,* was highly successful commercially (becoming a number-one best seller in 1950), Velikovsky was quickly repudiated *en masse* by accredited astronomers, physicists, historians and other specialists. His claims, it was said, violated the self-evident principles of physics and astronomy—and most everything we had come to know about our Earth and the solar system.

In sweeping terms, Velikovksy appeared to cast aside the most treasured assumptions of the scientific age. And worst of all, he drew the better part of his testimony from early mythical and religious texts—an outlandish reservoir of "evidence" in the opinion of most physical scientists.

Yet it is a fact that Velikovsky's case was persuasive enough to convince a good many critical readers, while an even larger number urged open-mindedness and a fair consideration of Velikovsky's unusual thesis.

Analyzing ancient sources from around the world, Velikovsky noted a consistent story of interplanetary conflagrations. It seems that early man recalled the planetary gods as fearsome powers, armed with thunderbolts and missiles, whose battles threw the world into confusion. Velikovsky believed that these stories were based on memories of actual events, a time when the planets moved on erratic courses, waging battles in the sky and menacing our own planet. Near collisions disturbed the terrestrial axis, removed the Earth from its established path and produced worldwide catastrophes.

Velikovsky's research led him to an extraordinary theory about the planet Venus. He claimed that Venus was born explosively from the planet Jupiter only a few thousand years ago, taking on the appearance of a comet, and moving on an Earth-threatening orbit around the Sun. On at least two occasions, he said, the Earth passed through the trailing debris of the comet-like Venus, showers of stone and fire descending on terrestrial inhabitants and leveling civilizations the world over.

It was the first of these two disasters, in Velikovsky's reconstruction, which ended the Egyptian Middle Kingdom and provided the catastrophic backdrop to the Hebrew Exodus.

To build his case Velikovsky undertook a global survey of mythical and historical records, supplemented by archaeological testimony. He claimed that ancient observations of planetary motions make no sense, at least in terms of the heavens we know today. Ancient sundials and water clocks, he reported, likewise testify to an altered celestial order, while the star chart on an Egyptian tomb represents a confused, "upside down" sky. From opposite sides of the world, Velikovsky produced surprising, dovetailing reports describing celestial upheavals and apparent devastations from Earth-changing encounters.

All told, it was an extraordinary and exciting thesis, but less than convincing to astronomers and physicists reading summaries in *Colliers* magazine and *Reader's Digest*. Not just Velikovsky's conclusions, but the approach itself was patently misplaced, at least in the eyes of those who had long ago dismissed ancient astrology, religion, and myth as valueless to modern science. That Velikovsky gathered his evidence from such untrustworthy sources, then used it to rewrite the history of the solar system, seemed to violate every canon of acceptable methodology.

In the years since publication of *Worlds in Collision,* the "Velikovsky Affair" has been the subject of continuing discussion in popular articles and books. But no one entering the fray has brought more attention to the issue than the prominent astronomer, author and television personality, Carl Sagan. In at least four books published in the past fifteen years, Sagan has presented a detailed position on the Velikovsky question, concluding that there is no admissible scientific evidence to support the latter's claims.

Sagan's widely-read treatments of the issue were preceded by a face-to-face encounter between Sagan and Velikovsky at a 1974 special symposium of the American Association for the Advancement of Science (AAAS) in San Francisco.

There are a number of reasons for addressing the Velikovsky issue squarely. It is, to begin with, an excellent case study in the history of unconventional ideas and the way such ideas are handled by the guardians of orthodox science. More importantly, many issues raised by Velikovsky simply have not gone away.

AN INTERDISCIPLINARY SCHOLAR

That Velikovsky, neither an astronomer nor a physicist, proposed to re-write the recent history of the solar system was for many specialists an exercise in futility. How could one untrained in celestial mechanics speak intelligently of wandering planetary orbs and violent exchanges between planets?

Velikovsky derived his thesis from a systematic exploration of or the study of history and the roots of ancient myth and symbolism. And whatever one may think of such an enterprise, it is clear that for this Velikovsky lacked neither the formal training nor the credentials, as the briefest of biographies will show:

Born in Vitebsk, Russia in 1895, he learned several languages as a child, and graduated with a gold medal from the Medvdenikov Gym- nasium in Moscow in 1913.

Barred from entering Moscow University by clauses restricting Jews, he began premedical studies in Scotland, returning to Russia during World War I to study law and ancient history at Moscow's Free University. In 1915, he was finally admitted to Moscow University, receiving his medical degree in 1921.

Shortly thereafter, Velikovsky moved to Berlin, where he founded and edited an international series of monographs by outstanding Jewish scholars, *Scripta Universitatis,* for which Albert Einstein edited the mathematics–physics section and became acquainted with Velikovsky. (His friendship with Einstein would continue until the latter's death in 1955. In Einstein's later years, there were many long evenings of discussion with Velikovsky in Princeton, and, as Velikovsky's supporters often remind us, *Worlds in Collision* was the one book which lay open on Einstein's desk at the time of his death.)

In 1923 Velikovsky married a young violinist, Elisheva Kramer and the following year moved to Jerusalem, to practice medicine. Later, after a stay in Vienna, where he studied under Freud's first pupil Wilhelm Stekel, Velikovsky moved to Tel Aviv, beginning another series of monographs, *Scripta Academica Hierosolymitana,* conceived as the cornerstone of an academy of science in Jerusalem. In 1930 he published the first paper to suggest that epileptics are characterized by pathological encephalograms.

In 1939 Velikovsky came to the United States to research a com- mentary on Freud's work *Moses and Monotheism.* While reflecting on Freud's thesis, Velikovsky had conceived the possibility that the Pharaoh Akhnaton, the real hero of Freud's book, was the legendary Oedipus, an idea later developed in Velikovsky's book *Oedipus and Akhnaton.* It was this research that began to carry him further and further into apparent incongruities of ancient history.

TALES OF UPHEAVAL

In April, 1940, Velikovsky was first struck by the idea that a great natural catastrophe might have taken place at the time of the Israelites' Exodus from Egypt—a time when, according to the Biblical account, plagues occurred, the Sea of Passage parted, Mount Sinai erupted, and the pillar of smoke and fire moved in the sky.

Velikovsky wondered: Does any Egyptian record of a similar catastrophe exist? He found the answer in an obscure papyrus stored in Leiden, Holland—the lamentations of an Egyptian sage, Ipuwer. As in the *Exodus* account, the complaints of the Egyptian sage spoke of rivers turning to blood and the destruction of the land. "Plague is throughout the land. Blood is everywhere," bewailed Ipuwer. "Men shrink from tasting, human beings thirst after water... That is our water! That is our happiness! What shall we do in respect thereof? All is ruin... The towns are destroyed... Upper Egypt has become waste... The residence is overturned in a minute."

The Ipuwer document, Velikovsky became convinced, described the very natural catastrophe recorded in the Hebrew *Exodus*. On this conviction, then, he began to reconstruct piece by piece the fragments of ancient Middle Eastern history, taking the catastrophe as a starting point from which to synchronize the chronologies of Egypt and Israel. The result was a series of volumes, beginning with *Ages in Chaos,* published in 1952.

The cause of the catastrophe which he believed to have terminated the Middle Kingdom remained unexplained. But one afternoon in October, 1940, Velikovsky noticed an interesting passage in *The Book of Joshua.* In connection with the flight of the Canaanites in the valley of Beth-horon, a destructive shower of meteorites is said to have occurred—this before the sun "stood still" in the sky. Was this a coincidence, or were the ancients recording a cosmic disturbance that must have shaken the entire Earth and might have been related to the upheavals approximately 50 years earlier during the Exodus? From a survey of other sources around the world, Velikovsky concluded that two global cataclysms had indeed overtaken the Earth, and that the agent of these disturbances was the now distant and settled planet Venus. Moreover, in its destructive role, Venus seems to have been depicted more like a comet than a planet.

Velikovsky noted, from one land to another, certain unique but repeated associations of Venus with well-known cometary images. Among the Mexicans, Velikovsky found, Venus was called a "star that smoked," the very phrase which Mexican astronomy used to describe a comet. On the other side of the world, the Hindu Vedas depict Venus "like a fire accompanied by smoke." "Fire is hanging down from the planet Venus," states the Hebrew *Talmud.* To the Egyptians, Venus, as Sekhmet, was "a circling star which scatters its flames and fire."

The Aztecs called Venus the "heart" of Quetzalcoatl—whose name means "the plumed serpent," and whose feathers are acknowledged to signify "flames of fire." The serpent or dragon is one of the most universal glyphs for the "comet" in the ancient world.

Other serpent or dragon figures that Velikovsky identifies with the Venus-comet include the Greek Typhon, Egyptian Set, Babylonian Tiamat, Hindu Vrtra—all of whom, in highly vivid accounts, raged in the sky and brought overwhelming destruction to the world.

The Greek word comet comes from *coma,* meaning "hair." Ancient astronomers referred to comets as stars with "hair" or with a "beard." But Venus, too, apparently possessed a comet-like tail: one of the Mexican names for Venus was "the mane." The Peruvian chaska, the word for Venus, means "wavy-haired." The Arabs called Venus the "one with hair."

Most compelling is the convergence of the above "comet" images in the instance of the Babylonian goddess Ishtar, one of the most famous goddesses in the ancient world and recognized by all authorities as a figure of Venus. Ishtar is the "bright torch of heaven"; she is "clothed with fire"; and she is the "fearful dragon," while her planet—Venus—is called "the one with hair" and the "bearded" planet.

Unless one refuses on principle to entertain such thinking, it is im-possible to review the evidence gathered by Velikovsky without at least suspecting that Venus did indeed once possess a comet-like tail, and that the planet may have contributed to some extraordinary celestial events.

THE COMETARY 'NEWCOMER'
Searching through the early records of man, Velikovsky looked for indications as to how—and why—a cometary Venus may have arisen, or entered the solar system. He knew that the ancients, aware of a link between the circuit of heavenly bodies and the ruin of previous civilizations, diligently watched the planetary motions. Their traditions recalled that when former epochs dissolved, the new "age" was marked by different celestial motions. Early astronomers and seers looked for any change which might signal approaching destruction and the end of an age.

Velikovsky noticed that prior to the second millennium B.C. ancient Hindu records spoke of four visible planets, excluding Venus. Babylonians, meticulous in their observations, likewise excluded Venus in their earliest list of the planets.

To these considerations, Velikovsky added the interesting fact that Venus was designated "the Newcomer." Could it be that the now peaceful planet originated as a cometary "protoplanet," and only settled into its present orbit within the past few thousand years?

Figuring crucially into Velikovsky's argument is the well-known story of the Greek goddess Athena (identified by Velikovsky with Venus). In the account of Homer, Athena is "born" from the head of Zeus, the planet Jupiter. It was apparently this story that first led Velikovsky to surmise that a cometary Venus may have exploded from Jupiter during a period of Jovian instability—a possibility that soon grew into a firm conviction. Here, then, is Velikovsky's scenario:

Some time before 1500 B.C. a brilliant, fiery object burst forth from the largest planet in the solar system, entering in cometary fashion upon a long, elliptical orbit around the sun. (Venus, a Chinese astro-nomical text recalls, spanned the heavens, rivaling the sun in brightness. "The brilliant light of Venus," records an ancient rabbinical source, "blazes from one end of the cosmos to the other.")

For an indeterminate period the Venus comet moved on its elongat-ed path, intersecting the orbit of the Earth. Then, around 1500 B.C. occurred a disastrous close approach. As Venus arched away from its perihelion, the Earth entered the outer reaches of its cometary tail. A rusty ferrous dust filtered down upon the globe, imparting a reddish hue to land and sea, and turning the water to "blood." As the Earth's path carried it more deeply into the comet's tail, the rain of particles grew steadily more coarse and perilous. Soon a great hail of gravel pelted the Earth. "There was hail, and fire mingled with hail, grievous, such as there was none like it in all the land of Egypt since it became a nation," states the author of *Exodus*.

Fleeing from the torrent of meteorites, men abandoned their livestock to the holocaust. Fields of grain, the life substance of great civilizations, perished. Cried the Egyptian Ipuwer: "No fruits, no herbs are found. That has perished which yesterday was seen. The land is left to its weariness like the cutting of flax." Such things happened, say the Mexican Annals of Cuauhtitlan, when the sky "rained, not water, but fire and red-hot stones."

As our planet plunged still deeper into the comet's tail, hydrocarbon gases enveloped the Earth, exploding in bursts of fire in the sky. Unignited trains of petroleum poured onto the planet, sinking into the surface and floating on the seas. From Siberia to the Caucasus to the Arabian desert, great spills of naphtha burned for years, their billows of smoke lending a dark shroud for mankind's struggle to survive.

In this celestial encounter, according to Velikovsky, the axis of the Earth was displaced, leaving half of the globe in prolonged darkness for several days, as global windstorms, Earthquakes and continental sweeps of mud and sea devastated the entire surface of our planet.

Above, the glistening comet shone like a dragon through the tempest of dust and smoke, as Venus and its writhing serpentine tail exchanged gigantic thunderbolts. The world's myths, Velikovsky tells us, have memorialized this conflagration as the combat of a light god and dragon of darkness. The Babylonians told of the celestial warrior Marduk striking the dragon Tiamat with bolts of fire. Egyptian chroniclers saw Isis and the serpent-dragon Set in deadly combat, while the Hindus described the great god Vishnu battling the "crooked serpent." Zeus, in the Greek account, struggled with the coiled viper Typhon.

The battle in the sky raged for weeks, with the cometary apparition taking on the appearance of a column of smoke by day, a pillar of fire by night. Through a series of close approaches the comet's tail en-closed the Earth in a shadow of death, a thick gloomy haze that lasted for many years.

In the age to follow, the sun rose in the east, where formerly it set. Now, the quarters of the world were displaced, and the seasons no longer came in their appointed times. "The winter is come as summer, the months are reversed, and the hours are disordered," reads an Egyptian papyrus. An inscription from before the tumult says that the sun "riseth in the west," while numerous records tell of Earth "turning over."

In the wake of these events, Venus continued on its threatening course around the Sun, and—some 50 years after the Exodus—again drew near. Under Joshua, the Israelites had entered the Promised Land. As the Canaanites fled from before the hand of Joshua in the valley Beth-horon, the daughter of Jupiter unleashed a second storm. "The Lord cast down great stones from heaven upon them unto Azekah, and they died," reports the *Book of Joshua*. Once more, the terrestrial axis tilted and the Earth shook. Above Beth-horon the sun stood still for hours, while on the other side of the Earth chroniclers recorded a prolonged night, lit only by the burning landscape. A destruction of this kind, according to Mexican sources, occurred about 50 years after an earlier world-destroying catastrophe.

Now the priests and astrologers began to fear a renewal of cosmic upheavals on 50 year cycles. With bloody orgies and incantation nations enjoined the dreaded queen of the planets to remain far from them. "How long will thou tarry, O lady of heaven and Earth?" inquired the Babylonians, while the Zoroastrian priests declared, "We sacrifice unto Tistrya, the bright and glorious star, whose rising is watched by the chiefs of deep understanding."

In both hemispheres, Velikovsky says, men fixed their gaze anxious-ly on the comet as, for centuries, it continued its circuit, crossing the orbits of both Earth and Mars. Before the middle of the eighth century B.C., astrologers observed irregularities in its wandering. Viewed from Babylonia, Venus rose, disappeared in the west for over nine months, then reappeared in the east. Dipping below the eastern horizon, it was not seen for over two months, until it shone in the west. The following year Venus vanished in the west for eleven years before reappearing in the east.

There is, of course, more to Velikovsky's scenario in *Worlds in Collision:* eventually Venus dislodged the planet Mars from its orbit, initiating a new period of instability and disaster, continuing from 747 to 686 B.C. For our purpose, however, the above summary should be sufficient to give the flavor of Velikovsky's thesis, and to make clear why the scientific community as a whole found the story to be—a story, but not science.

And let us not understate the difficulty. In many ways Velikovsky's account could not fail to tax one's credulity, no matter how far one might carry an open-mindedness on the underlying idea. When, in the account of the Exodus, the vengeful pharaoh pursues the fleeing Israelites across the Sea of Passage, the waters have already been divided by the tug of the celestial combatants. As told by Velikovsky, the entire band of Israelites had not yet crossed to the far side when a giant electrical bolt flew between the two planets. Instantly the waters collapsed. The pharaoh, his soldiers and chariots, and those Israelites who still remained between the divided water, were cast furiously into the air and consumed in a seething whirlpool. Though the timing wasn't perfect, it was certainly very good.

Then there is the seemingly miraculous case of descending manna, a mysterious life-giving substance which Velikovsky believes to have precipitated in the heavy atmosphere—possibly derived from Venus' hydrocarbons through bacterial action, he said. When heated, this "bread of heaven" dissolved, but when cooled, it precipitated into grains which could be preserved for long periods or ground between stones. Without this convenient turn of nature—too convenient, in the eyes of skeptics—the human race might have perished altogether.

Details such as these, when separated from the more fundamental thesis, simply provided the critics with easy targets for ridicule, which some extended to anything and everything about *Worlds in Collision*.

A CASE OF PROFESSIONAL HYSTERIA

Even before *Worlds in Collision* had reached the bookstore it was enveloped in controversy.

In 1950, after more than a dozen publishing houses had rejected Velikovsky's manuscript, it was accepted by Macmillan. Having announced the forthcoming release of the book, Macmillan was soon caught in what appeared to be an organized boycott, initiated by the well-known astronomer Harlow Shapley, then director of the Harvard College Observatory. In a personal letter to the publisher, Shapley sought to block the book's release, threatening to "cut off" his rela-tions with Macmillan. Letters from other authors of Macmillan books followed, along with threats from professors who could not imagine using the company's textbooks any longer if the publisher were to discredit itself in the rumored fashion.

Though the book had already been reviewed by several critics at Macmillan's request, and though it was now on press, the company hastily submitted the manuscript to three additional reviewers. These, too, recommended publication by a two-to-one vote.

So, in April, 1950, Macmillan decided to go ahead with publication of the already controversial book.

Despite the immediate furor, one of those who saw merit in Velikovsky's ideas was Gordon Atwater, chairman and curator of the Hayden Planetarium of the American Museum of Natural History. In a preface to a 1950 article by Fulton Oursler in *Reader's Digest* Atwater contended that, in light of the Velikovsky thesis, "the underpinnings of modern science can now be re-examined." In fact, Atwater himself planned to mount a star show at the planetarium illustrating the new possibilities opened up by *Worlds in Collision*. And in *This Week* magazine, a cover story by Atwater called for an open mind on Velikovsky's theory.

But the day before the article appeared, and in a move that seemed to set the tenor of the events to follow, Atwater was, without explana-tion, dismissed from the museum. Under growing pressure to abandon *Worlds in Collision*, Macmillan fired the editor who contracted the book, then, eight weeks after its publication, transferred its rights to Doubleday—a move unparalleled in publishing history: the book had already become number one on the *New York Times* non-fiction best-seller list.

The many bizarre responses by professional scholars—before and after publication of *Worlds in Collision*—have been fully detailed elsewhere. They include horrendous misrepresentations of the thesis by well respected astronomers and others who had never seen the book; repeated refusals by scientific journals to grant Velikovsky an opportunity to reply to his critics; and refusals to retract factually erroneous and even farcical "summaries" of his views.

For two decades following the appearance of *Worlds in Collision* Velikovsky was, with rare exceptions *persona non grata* on college and university campuses and his work treated as a joke by established publications.

This was to change somewhat toward the end of the sixties, however. By this time the space age was well underway, with volumes of extraterrestrial data flowing into Earth's computers. Stunning pictures, rock samples, measurements of every kind. The profiles of the planets were shifting with each subsequent revelation, and it was clear that many surprises on balance weighed in Velikovsky's favor. The unexpected, massive clouds of Venus, the planet's strange retrograde rotation and its surpassing temperature, the stark figures of the tortured planet Mars, verification by the Moon landings of radioactive hot spots and remanent magnetism predicted by Velikovsky; the growing recognition of electromagnetism in celestial mechanics—these and other discoveries may not have produced the pristine verdicts proclaimed by some of Velikovsky's loyalists, but were enough to encourage a number of scholars to take a new look at Velikovsky's thesis.

In 1972 a group out of Portland, Oregon began publishing a ten-issue series *Immanuel Velikovsky Reconsidered,* presenting a wide range of scholarly opinions on Velikovsky, with many contributors calling for a wholesale re-evaluation of his work in view of new data. The first issue published produced quite a stir, both in this country and abroad. In the following months, most of the country's general scientific publications addressed the Velikovsky question—some calling for more openness and tolerance of unpopular views, others wondering aloud how to preserve the integrity of science from intellectual con artists.

This was the beginning of some new and fascinating episodes, culminating in a widely publicized symposium on Velikovsky in 1974, sponsored by the American Association for the Advancement of Science.

LOOKING FOR VELIKOVSKY'S COMET

Since publication of *Worlds in Collision* in 1950, many aspects of Velikovsky's thesis have been debated by various scientific spokesmen who have assured us that certain ironclad principles of astronomy and the Earth sciences refute all of the book's primary claims. But can it honestly be said that the sum of the discussion so far has provided a definitive answer to the issues first raised by Velikovsky 40 years ago?

What is the evidence and how does it relate to Velikovsky's hypothesis? The question of the evidence is, of course, related to Sagan's criticism. For some, Sagan's criticisms of Velikovsky are sufficient to put the views he offers out of the realm of science. For example, Anthony R. Aveni's article "A Marshaling of Arguments", presented in *Science,* (Jan. 20, 1978), pp. 288–89, states, "Carl Sagan's paper...is amusing, acrid, and totally devastating...his essay alone is sufficient to reduce the Velikovsky theory to anile fancy... Velikovsky is flatly and totally disproven... As far as Velikovskyanism is concerned, it is dead and buried. The final nail has been driven. It is now hoped that we can move on to more exciting things." When letters were sent to Aveni critical of his review presenting evidence contrary to that presented by Sagan, Aveni sent a letter in response. "My review says that I'm tired of listening. I've spent too much time listening, and all of it isn't worth listening to—and that is an objective statement."

As pointed out earlier, E.J. Opik stated, "Dogma differs from hypothesis by the refusal of its adherents even to consider the aspects of its validity. Legitimate disagreement or controversy creates dogma when arguments are no longer listened to." In science, evidence dominates all other forms of argument. Therefore, Aveni's attitude may well be a personal standard for science. Only evidence should determine the nature of a scientific debate.

In the following pages, this author has gathered evidence from the scientific sources and cited them verbatim on each of Sagan's criticisms. It is only the evidence that will be of paramount importance in evaluating Sagan's critique.

WHAT IS SCIENCE?
In his introductory remarks Sagan offers his views of science,

> "Scientists, like other human beings, have their hopes and fears, their passions and despondencies—and their strong emotions may sometimes interrupt the course of clear thinking and sound practice... The history of science is full of cases where previously accepted theories and hypotheses have been entirely overthrown, to be replaced by new ideas that more adequately explain the data. While there is an understandable psychological inertia—usually lasting about one generation—such revolutions in scientific thought are widely accepted as a necessary and desirable element of scientific progress."[1]

There is, indeed, a clear distinction to be made between the psychological and sociological behavior of individual scientists, on the one hand and the requirement of truthfulness and responsible behavior of scientists in their symposia and journals on the other. Therefore, to determine whether or not *Science* and in particular, the AAAS symposium held on Velikovsky reflects science governed by passion or science governed by reason, we must investigate the AAAS symposium held on Velikovsky and the scientific journalistic treatment of Velikovsky.

Sagan states further that,

> "The most fundamental axioms and conclusions may be challenged. The prevailing hypotheses must survive confrontation with observation. Appeals to authority are impermissible. The...reasoned argument must be set out for all to see."[2]

Not only do these requirements demand that Velikovsky adhere to the rational scientific position but that Sagan in his criticisms fulfill these same ideals. If as Sagan suggests reason has come to rule passion in the case of Velikovsky then criteria of fairness and justice will be observed. If passion rules reason then dishonesty and injustice will be observed. Sagan adds,

> "Indeed, the reasoned criticism of a prevailing belief is a service to the proponents of that belief; if they are incapable of defending it they are well advised to abandon it. This self-questioning and error correcting aspect of the scientific method is its most striking property, and sets it off from many other areas of human endeavor such as politics and theology"[3] [or] "where credulity is the rule."[4]

For anyone to defend his views he must have access to the journals that raise criticisms of his thesis. The question arises: Was Velikovsky permitted full access to the scientific journals to defend his hypothesis and also to the AAAS publication for this debate? Furthermore, was Velikovsky given sufficient space to answer all attacks on his evidence?

As a case in point, the *Bulletin of Atomic Scientists* for April, 1964 saw fit to publish an "abusive" article by Howard Margolis.

> "The editor of the *Bulletin*, Dr. Eugene Rabinowitch, in a letter to Professor Alfred de Grazia [who as] editor of the *American Behavioral Scientist* [had protested the 'abusive' article] offered Velikovsky an opportunity to reply with an article 'not more abusive' than that of Margolis, or, instead to have some of his views presented in the *Bulletin* by some scientist of repute. Then Professor Harry H. Hess [Chairman of the of Geology at Princeton and President of the American Geological Society] submitted Velikovsky's article 'Venus—A Youthful Planet' to Dr. Rabinowitch. The latter then returned it with the statement that he did not read Velikovsky's book, nor the article."[5]

How can science be a self-correcting mechanism if it refuses to read or permit a reasoned response in the organs of scientific literature? Although the deplorable, irrational behavior of the *Bulletin of Atomic Scientists* took place in 1964 was there a different attitude—one more just and rational—governing the AAAS symposium on Velikovsky held ten years later? Was the symposium convened in San Francisco, at which Sagan presented his paper, a meeting to honestly discuss and debate Velikovsky's thesis or was it actually organized to ridicule and humiliate Velikovsky?

Professor of Philosophy, Lynn E. Rose of SUNY Buffalo published the following letter sent to Velikovsky, in which he states,

"...I urge you [Velikovsky] not to waste any more time with AAAS people or with their volume that was supposed to be a report of the AAAS sessions on your work held in San Francisco in 1974. The behavior of the AAAS people has been deplorable from the start. Their intention never was to examine or to debate your work; all along their intention was to find a way to ridicule and to belittle your work before the public.

"The AAAS people set up the program so that four panelists would speak against your theories and you alone would be allowed to speak in your defense... Not a single scientist working with you was allowed to participate in the panel discussion. This violated the AAAS promise that there would be as many panelists speaking *for* your theories as there were panelists speaking *against* your theories.

"All the panelists, including yourself, were to be given 'equal time.' Each of the four negative panelists then proceeded to enumerate alleged errors on your part and alleged evidence against your theories. Clearly, the intention was that these 'equal time' arrangements would permit them to introduce so many points that you would not have enough time to answer them all.

"This same strategy is being used by the AAAS people, in the arrangement for their proposed volume on the San Francisco sessions. They wish to retain the four-to-one odds, and have still not allowed anyone in addition to yourself to argue in support of your theories. They wish to keep all the arrangements for the volume in their own hands, and to prevent any balanced and serious examination of your work. They wish to provide far more space for negative comments from your opponents than for positive comments from you. And they wish to allow the four negative participants to include additional remarks that you will not have the opportunity to answer. It is possible that they will not even show you those additional remarks until the volume has already gone to press. It is also possible that, after you have spent so much time preparing material for their volume, they may suddenly decide not to publish it at all, thus leaving you with little to show for your time and efforts..."

"When a volume really is devoted to serious examination and criticism of a man's work, the format and atmosphere are light-years away from what the AAAS people are doing. I have in mind, for example, the Library of Living Philosophers series edited by Paul A. Schilpp. That series includes publications of Einstein, on Russell, and on many others. Each such volume includes a long bibliography of the man's writings, and a long preliminary essay by him in the form of an intellectual autobiography. There are a number of critical articles included in such a volume, but the man whose work is at issue is given as much time and space as he needs to reply to each criticism. The entire approach is serious and fair; there is debate and argument, but not abuse and slander. And the volume is presented to the reading public as if it were an *honor* and a form of *recognition* for the man who is its subject. What a far cry from the way the AAAS people are treating you!..."[6]

There is a difference between the behavior of the editors of the *Bulletin of Atomic Scientists* and the AAAS scientists. However, the difference is of degree. While the *Bulletin* acted crassly and openly to suppress Velikovsky's work, the AAAS scientists acted subtly and cunningly to give the appearance to the public of holding an open forum which was all the while a public relations gimmick to accomplish the same ends. We shall return to this irrational aspect of the AAAS scientists further on. Sagan proceeds,

"The idea of science as a method rather than as a body of knowledge is not widely appreciated outside of science, or indeed in some corridors inside of science. For this reason I and some of my colleagues in the American Association for the Advancement of Science have advocated a regular set of discussions at the annual AAAS meeting of hypotheses that are on the borderline of science and that have attracted substantial public interest. The idea is not to attempt to settle such issues definitively, but rather to illustrate the process of reasoned disputation to show how scientists approach a problem that does not lend itself to crisp examination, or is unorthodox in it interdisciplinary nature or otherwise evokes strong emotions."[7]

Commendably, Sagan asks for "reasoned disputation" as the proper approach to Velikovsky's "unorthodox" and "interdisciplinary" material especially materials "that have attracted substantial public interest."

RELIGION, ASTROLOGY, SUPERSTITION

Sagan continues,

"Vigorous criticism of new ideas is a commonplace in science. While the style of the critique may vary with the character of the critic, overly polite criticism benefits neither the proponents of new ideas nor the scientific enterprise. Any substantive objection is permissible and encouraged; the only exception being *ad hominem* attacks on the personality or motives of the author are excluded."[8]

This statement though laudable is, however, belied by Sagan impugning the motives of Velikovsky wherein he states,

"...how is it that *Worlds in Collision* has been so popular? Here I can only guess. For one thing, *it is an attempted validation of religion.* The old Biblical stories are literally true, Velikovsky tells us, if only we interpret them in the right way... *Velikovsky attempts to rescue not only religion but also astrology;* the outcomes of wars, the fates of whole peoples, are determined by the positions of the planets."[9] [emphasis added]

This undisguised slur on Velikovsky's motives by Sagan was strongly responded to by Velikovsky when he stated,

> "Sagan next presents 'Velikovsky's Principal Hypothesis' and he purports faithfully to tell what it is... Sagan says, 'at the moment Moses strikes his staff upon the rock, the Red Sea parts...' Later, 'after the death of Moses...the same comet comes screeching back for another grazing collision with the Earth. At the moment when Joshua says, 'Sun, stand thou still upon Gibeon; and thou Moon, in the Valley of Agalon...the Earth obligingly ceases its rotation.' He [Sagan] later says that I, 'attempt to rescue old time religion.' To tell of Velikovsky's principal hypothesis in this vein is nothing but purposely misleading."[10]

Velikovsky was justifiably incensed because, in *Worlds in Collision*, just the opposite information was presented,

> "The sea was torn apart. The people attributed this act to the intervention of their leader; he lifted his staff over the waters and they divided. Of course, there is no person who can do this and no staff with which it can be done. Likewise in the case of Joshua who commanded the sun and the Moon to halt their movements."[11]

To attribute to Velikovsky information which he never presented in his book is essentially an *ad hominem* attack on Velikovsky's personality and motives. The earlier laudable statements of Sagan are contradicted by his own words.

What is Velikovsky's view of religion, astrology and superstition? In *Earth in Upheaval*, Velikovsky presented his opinion regarding Darwin and The Church on evolution.

> "Darwin's theory represented progress as compared with the teaching of the Church. The Church assumed a world without change since the Beginning. Darwin introduced the principle of slow but steady change in one direction, from one age to another, from one eon to another. In comparison with the Church's teaching of immutability, Darwin's theory of slow evolution through natural selection or the survival of the fittest was an advance."[12]

Sagan's view that Velikovsky's hypothesis is "an attempted validation of religion" does not correspond with this statement. If Velikovsky wished to validate religion, his position should have been just the opposite. Sagan also claims that "Velikovsky attempts to rescue... astrology." Astrology is a pseudoscience which holds that our destiny is determined by where the planets and the Sun and Moon are in the twelve signs of the zodiac. Velikovsky does say that when a planet on a cometary orbit nearly collided with the Earth, whole nations were destroyed. This is not astrology. Astrology holds that certain days are unlucky while others are lucky. In particular, the thirteenth day of the month is astrologically unlucky. Here is what Velikovsky has to say regarding the thirteenth day of the month,

> "In the calendar of the Western Hemisphere on the thirteenth day of the month, called *olin,* 'motion' or 'Earthquake,' a new sun is said to have initiated another world age..." [The Earth experienced a global catastrophe.]
>
> "Here we have *en passant,* the answer to the open question concerning the origin of the superstition which regards the number 13, and especially the thirteenth day, as unlucky and inauspicious. It is still the belief of many superstitious persons, unchanged through thousands of years and even expressed in the same terms: 'The thirteenth day is a very bad day. You shall not do anything on this day.'" [13]

Again Sagan's claim is not supported by Velikovsky's statements. It is difficult to conclude that Velikovsky, who calls "superstitious" people who believe that the thirteenth day of the month unlucky, is in any way validating astrology.

Lastly, Sagan's remarks regarding Moses and Joshua suggest that Velikovsky accepts supernatural causes for events. In *Ages of Chaos,* Velikovsky tells us,

> "The biblical story of the last plague [of *Exodus*] has a distinctly supernatural quality in that all the firstborn and only the firstborn were killed on the night of the plagues. An earthquake that destroys only the firstborn is inconceivable, because events can never attain that degree of coincidence. No credit should be given to such a record.
>
> "Either the story of the last plague, in its canonized form, is a fiction, or it conceals a corruption of the text." [14]

In this case it is also rather clear that Velikovsky rejected the idea that there is a supernatural cause of events.

In the first three major works of Velikovsky: *Worlds in Collision, Earth in Upheaval* and *Ages in Chaos* are concise statements that indicate Sagan's impugning Velikovsky's motives and evidence are thoroughly misinformed. When Velikovsky called Darwin's theory an advance over the teachings of the Church, he was not rescuing religion; when he called people who believe the thirteenth day of the month unlucky, "superstitious," he was not defending astrology; and when he held that the biblical story of the last plague of the Exodus, in which only the firstborn are killed, "supernatural," "inconceivable" and "no credit should be given to such a record," he was attacking supernatural interpretation of events.

Eric Larrabee remarks that Velikovsky's thesis, "in no way involved the supernatural, even by implication. Either Velikovsky's thesis could be proven scientifically or it would fall to pieces. Far from seeking to confirm fundamentalist beliefs (as he was accused of doing), he offered them the most fundamental challenge of all, which was to provide a natural interpretation of 'miraculous' events rather than merely to dis - miss them as legendary."[15] In fact, at the symposium at which Sagan presented his paper, one of his colleagues, Dr. Derral Mulholland argued that "Velikovsky's challenge is not one to be decided on the basis of belief or unbelief. He does not say 'trust me,' he says 'this conclusion is suggested by the observations'...that involve testable ideas. He is not a mystic."[16] Thus, Sagan's smear of Velikovsky's motives is even denied by Mulholland.

One of the implications of Sagan's criticism is that Velikovsky's work validates and supports the entire *Bible*. Robert Anton Wilson, in *The New Inquisition*, (Tempe, Arizona 1991), p. 70 explains:

"Dr. Velikovsky examined the myths of the ancients and speculated that they might contain a few facts—sombunall [some but not all] in our terms ...Facts that could still be deduced by comparing various myth systems and noting what they have in common. For instance, there are over 120 flood legends in addition to the one in the *Old Testament*. They came from every part of the world—Asia, Africa, Australia, Russia, Scandinavia, Ireland, North America, South America, Polynesia. Throw out the local details and you have one constant: the idea that there was once a flood. So maybe there was? And maybe a comet created it."

On page 72 Wilson goes on to say:

"To proceed from 'Something like Noah's flood once happened' to 'The whole *Bible* is true' is not very logical, and I can't find anything like it in any book of Velikovsky's that I've read; and it would be just as...illogical to proceed from 'Something like the Polynesian flood story once happened.' to 'The whole Polynesian mythology is true,' and Velikovsky does not say that either as far as I have read him."

Sagan states in *Broca's Brain*, p. 84 that "Catastrophism began largely in the minds of those geologists who accepted a literal interpretation of the *Book of Genesis*, and in particular, the account of the Noahic flood." How accurate is this statement?

Velikovsky has employed the *Bible* and folklore and legends of ancient people to show that ancient man witnessed global catastrophes. His approach is similar to that of Georges Cuvier, the founder and father of the science of paleontology—the study of fossils. Stephen Jay Gould, the well-known Harvard biologist and historian of science says this about Cuvier's method of employing the *Bible* and folklore and legends of ancient people to prove that there was a universal flood in ancient times.

"Cuvier's methodology may have been naive, but one can only admire his trust in nature and his zeal for building a world by direct and patient observation, rather than by fiat, or unconstrained feats of imagination. His rejection of received doctrine as a source of necessary truth is, perhaps, most apparent in the section of the *Discours preliminaire* that might seem, superficially, to tout the *Bible* as infallible—his defense of Noah's flood. He does argue for a world-wide flood some five thousand years ago, and he does cite the *Bible* as support. But his thirty-page discussion is a literary and ethnographic compendium of all traditions from Chaldean to Chinese. And we soon realize that Cuvier has subtly reversed the usual apologetic tradition. He does not invoke geology and non-Christian thought as window dressing for 'how do I know, the *Bible* tells me so.' Rather, he uses the *Bible* as a single source among many of equal merit as he searches for clues to unravel the Earth's history. Noah's tale is but one local and highly imperfect rendering of the last major paroxysm."[17] Gould has remarked "...it seems unjust that catastrophists, [like Cuvier] who almost followed a caricature of objectivity and fidelity to nature, should be saddled with a charge that they abandoned the real world for their Bibles."[18]

The same year as Gould's statement regarding Cuvier was published, Velikovsky wrote in *Stargazers and Gravediggers*, (NY 1983) p. 284,

"In the astronomer's view there can be no greater effrontery than the questioning of their truths, and nothing enrages them more than to challenge such a perfect science by recourse, *horribile dictu,* to the Scriptures as a historical document. That *Worlds in Collision* contains much folklore, or 'old wives' tales,' was not so ludicrous as the fact that it brought the *Old Testament* back into the debate. The citation of passages from the Vedas, the Koran, and Mexican holy books was not so insulting as quotation from the Hebrew *Bible*. It is irrelevant that this book is among the most ancient of written literary documents in existence. As the theologian believes with blind faith that the Scriptures contain only truth, that their authorship is from God, and therefore, that every verse in them can be quoted as an irresistible argument, so the astronomer believes that where a passage is reproduced from the Scriptures, there must be a blunder, a softening of the brain tissue, or an attempt to hoax the credulous, as if the Scriptures were written by the devil.

"To my way of thinking, these books of the *Old Testament* are of human origin; though inspired, they are not infallible and must be handled in a scientific manner as other literary documents of great antiquity. Yet I must admit that I had a share of satisfaction upon discovering that the so-called miracles of the Hebrew *Bible* were physical phenomena, and like the disturbance...[seen by] other peoples of great antiquity in different parts of the world, they are also found preserved in the ancient literature of other nations."

Like Cuvier, Velikovsky "uses the *Bible* as a single source among many of equal merit as he searches for clues to unravel the Earth's history." The charge brought by Sagan against Velikovsky's aims and motives is precisely the same as that he used to describe the early catastrophists and is described by Gould of the attack upon Cuvier.

Velikovsky had become reacquainted with Professor Albert Einstein while both lived at Princeton, and Einstein read *Worlds in Collision,* which he often discussed with Velikovsky. What was Einstein's opinion? He stated, "not once and not twice [but] also in the presence of his secretary: 'The scientists make a grave mistake in not studying your book (*Worlds in Collision*) because of the exceedingly important material it contains.'" [19] Was Albert Einstein so naive as to believe that Velikovsky was presenting his book, *Worlds in Collision,* to validate religion, astrology and the supernatural? The week of Einstein's death he was rereading *Worlds in Collision* because evidence from Jupiter had confirmed one of Velikovsky's predictions.

R.F. Shaw writes in *Nature* (June 13, 1985, p. 536) "Critics have made much of Velikovsky's alleged appeal to the ignorant and also to his supposed religious motivation, *something never documented and which I do not find in his books.*" [emphasis added]

Thus, Sagan's claim that Velikovsky eschewed scientific evidence to support his theory is without substance. Stephen J. Gould's comment in *Times Arrow Times Cycle,* (Cambridge, MA 1987), p. 113, applies to Sagan's accusation of Velikovsky:

> "What a vulgar misrepresentation! Cuvier, perhaps the finest intellectual in the nineteenth century science was a child of the French Enlightenment who viewed dogmatic theology as anathema in science. He was a great empiricist who believed in the literal interpretation of geological phenomena... His earth, though subject to intermittent paroxysm was as ancient as Lyell's."

The reader shall see that in the fourth problem there is much geological evidence that supports Velikovsky's view for a recent catastrophe to the Earth.

HOW SCIENCE OPERATES

When Sagan upholds the objective scientific model of debate it seems strangely at odds with his statements. Why then did Sagan resort to such tactics? Here, Velikovsky's words may indicate causes.

> "As my opponent for the fourth tournament, the astronomical establishment selected Sagan. To answer his nearly 90 pages and nearly 30,000 words (1976 version), I am left with barely one-tenth of that amount, though an answer usually requires more space than an accusation, especially those that are bland and unsupported: I must first state what the charge was, then state what the truth is, what I really wrote, etc., and then present the evidence for what I said...therefore, I am in the position of standing against the entire establishment, though greatly limited as to

space and time, and blindfolded as to any additional counterarguments my opponents may bring, before I see the printed book... I am not abandoning the project and will do my best under the circumstances, to the limits of what decency can tolerate." [20]

Therefore when Sagan remarks, "The objective of such criticism [namely his own or that of the AAAS scientists] is not to suppress but rather to encourage the advance of new ideas," it is cynically amusing since it has been shown that the AAAS scientists used none of Sagan's criteria in dealing with Velikovsky. Sagan continues, "those [papers] that survive a firm skeptical scrutiny have a fighting chance of being right or at least useful." [21] How can a response which is censored by being limited in presenting a full answer have a chance of being fairly evaluated? Such a tactic is devised strictly to suppress rather than to encourage the advance of new ideas.

Sagan states, "My own view is that no matter how unorthodox the reasoning process or how unpalatable the conclusions, there is no excuse for any attempt to suppress new ideas—least of all by scien-tists." [22] If this is so, why didn't Sagan or any of the AAAS scientists *demand that Velikovsky be given sufficient time and space in the publication to answer all attacks?* Why did he and they take part in a blatantly one sided debate where the scholar under attack was so unfairly treated?

Frederic B. Jueneman, Director/Research for Innovative Concepts Associates of San Jose, chemist, and columnist discussed the AAAS symposium.

"Jueneman called [Ivan] King [one of the symposium's organizers] to inquire about the symposium and the events which led to it. According to Jueneman, King stated that the intent was to take another look at Velikovsky's work since there was renewed interest in it. He also said that the participants would be from the 'hard' sciences, which do not include sociology.

"Jueneman asked if it might be a move to stem criticism of the AAAS for the actions of its members in the Velikovsky affair. King replied that to some extent it was, but that only individual members of the AAAS were involved in the excesses against Velikovsky, not the AAAS itself...

"Soon it became apparent that the organizers of the symposium had no intention of pursuing a scientific discussion. King later said, 'None of us in the scientific establishment believes that a debate about Velikovsky's views of the Solar System would be remotely justified at a serious scientific meeting.' ...It is clear, however, that the meeting was arranged, as Jueneman said, to be a contemporary court of inquisition, and that the discussion was designed to convince the public that they should ignore the increasing number of scientists who were taking the time to analyze Velikovsky's work. Since the organizers admitted that they did not consider the meeting a scientific one, perhaps that is how they justified, to themselves, the misleading and sometimes false statements used to support their position."[23]

Actually the full statement by Ivan King is as follows:

"What disturbs the scientists is the persistence of these [Velikovsky's] views, in spite of all the efforts that scientists have spent on educating the public. It is in this context that the AAAS undertakes the Velikovsky symposium. Although the symposium necessarily includes a presentation of opposing views, we do not consider this to be the primary purpose of the symposium. None of us in the scientific establishment believes that a debate about Velikovsky's views...would be remotely justified at a serious scientific meeting."

Mark Washburn in his book, *Mars at Last,* (NY 1977), p. 95, states,

"There is something to be said for Velikovsky's side of it, however. To continue the structure-of-science metaphor a little longer, Velikovsky argued that the scientific establishment had constructed its own castle, complete with moat, drawbridge and battlements. If you didn't belong to the club, you weren't welcome. There was no room for the radical theorist who had new ideas about how the structure should be built.

"There was enough truth in Velikovsky's charges to make the scientific establishment uncomfortable. It was a difficult situation. If they debated Velikovsky's theories in the same manner as they would the theories of a reputable scientist, they would be lending legitimacy to a man who had perverted the principles of science... But if they refused to debate Velikovsky, it would seem that they were afraid of him."

Based on King's and Washburn's remarks, the scientific establishment set up the AAAS symposium on Velikovsky, but not to debate Velikovsky's theories in the same honest and respectful manner as they would the theories of members of their club. To do so would imply that Velikovsky's work was scientific. Washburn and King are telling us that Velikovsky's work was not discussed in the same way as that of other scientists, that is, the rules of the debate were no longer to be carried out in an honest and respectful manner. Objectivity had been thrown out the window. The aim of the meeting was to discredit, not evaluate, Velikovsky's work. What appears to be obvious at the outset, is that the ugly clannish passions of the scientific establishment had come to rule reasoned debate. George Orwell in his book *1984* called this "double speak," which for Orwell meant "double talk." The debate was not a debate. The outsider was to be destroyed. And, as Sagan said, "overly polite criticism" was not to be employed.

Therefore, the meaning of Sagan's statement, "I was very pleased that the AAAS held a discussion on *Worlds in Collision,* in which Velikovsky took part"[24] seems clear. Sagan took part in a meeting in which the organizer said, "None of us in the scientific establishment [including, of course, Carl Sagan] believes that a serious debate about Velikovsky's views...would be remotely justified at a serious scientific meeting."

PEER REVIEW

Sagan discusses how scientific papers are properly dealt with in science journals. He tells us that "Most scientists are accustomed to receiving...referees' criticisms every time they submit a paper to a scientific journal. Almost always the criticisms are helpful. Often a paper revised to take these critiques into account is subsequently accepted for publications."[25] In total, Sagan suggests that a scientific hypothesis offered to the scientific community be subject to review by peers—other scientists—that it be published in recognized science journals and that the submitter comply with valid criticisms.

The question arises: Does Sagan himself always follow this time honored procedure? In recent years, Carl Sagan has become the leading exponent of a very controversial theory termed, "Nuclear Winter." This hypothesis offers an explanation for the death of the dinosaurs. If a meteor about six kilometers in diameter struck the Earth 65 million years ago, Sagan claims that the dust thrown into the atmosphere and the smoke from forest fires would be so great as to have blocked sufficient sunlight from reaching the Earth and thus caused a global freeze which he calls "nuclear winter." Sagan further claims an atomic war would produce the same effect. However, in the "News and Comment" section of *Science,* an organ of the AAAS, Sagan's use of scientific procedure is subjected to criticism.

"A study by the National Center for Atmospheric Research suggests most of the world would experience a mild nuclear winter, not a deep freeze...[however] the best known presenter of the original theory, Carl Sagan of Cornell, claims there is 'nothing new' to make him alter his description of nuclear winter or the conclusions drawn from it... Sagan's refusal to acknowledge merit in the NCAR's (Nat. Cent. for Atmos. Res.) analysis—known as 'nuclear autumn'—sends some people up the wall. One wall climber is George Rathjens, professor of political science at M.I.T... 'Is this another case of Lysenkoism?' he asks, referring to an erroneous genetic theory forced on Soviet scientists in the late 1940's... Rathjens answers himself: 'I am afraid there's a certain amount of truth in that. The claim that the original nuclear winter model is unimpeached,' he adds, is 'the greatest fraud we've seen in a long time'... [this has led to other criticisms of Sagan's theory]. One such attack by Russell Seitz, a fellow at Harvard's Center for International Affairs, appeared recently in *The National Interest,* a Washington D.C. quarterly, and the *Wall Street Journal.* Seitz, who is not a diploma-holding scientist gibes at TTAPS's [Sagan and his co-authors] for mixing of physics and advertising. Seitz notes that Sagan published the nuclear winter thesis in *Parade* magazine a month before it appeared in *Science*. He writes: 'The peer review process at *Parade* presumably consists in the contributing editor conversing with the writer, perhaps while shaving—Sagan is both.' Anyone who wants to verify the data on which the conclusions were based, according to Seitz, has to set off on a 'paper chase' [Sagan's conclusions] rested on data published... *Science* article, 'details may be found in (15).' Reference 15 states in full: 'R.P. Turco, O.B. Toon, T.P. Ackerman, J.B. Pollack, C. Sagan in preparation.' It refers to a paper that has never been published in a peer-reviewed (or any other) journal. Rathjens also grumbles about the hard to get data. The entire thesis, he says, is 'a house of cards built on reference 15.'" [26]

Nor did Sagan's first Nuclear Winter article in *Science* benefit from the standard review process.

Did Velikovsky play by the rules of peer review? Before publication of *Worlds in Collision,* Velikovsky reported in *Stargazers and Gravediggers,* (NY 1983), p. 87, "The book was given to the [peer review] censors... [Velikovsky] was not informed of what was going on... As [he] heard...at a much later date, in 1952, two of the three censors were for the publication of the book, and one was against."

Thus, it is quite clear that Velikovsky's book *Worlds in Collision* was evaluated by the peer review process that Sagan requires. On this matter of peer review, it appears that Velikovsky's book passed the review while Carl Sagan's paper on Nuclear Winter essentially bypassed the review process. The only suggestion that seems to offer itself is that Sagan should follow his own advice.

When Sagan states, "...the reasoned criticism of a prevailing belief is a service to the proponents of that belief; if they are incapable of defending it, they are well advised to abandon it. This self-questioning and error correcting aspect of the scientific method is its most striking property," [27] what is observed is that Sagan neither subscribes to nor follows the ideals he so readily professes. Hence it is suggested that Sagan follow his own advice. It is further suggested that the AAAS scientists ignored not only the high ideals to which Sagan alludes, but that they ignored the simple canons of ordinary decency.

THE ORIGIN OF CRATERS

Sagan states,

"There is nothing absurd in the possibility of cosmic collisions," and, "Collisions and catastrophism are part and parcel of modern astronomy," and "The cratered surfaces of Mercury, Mars, Phobos, Deimos and the Moon bear eloquent testimony to the fact that there have been abundant collisions during the history of the solar system." [28]

Stated in this manner, Sagan's assertion conveys as a fact that craters are exclusively a collision event. This, however, has long been in question and is so in the present. Velikovsky claims that some craters are the result of impacts, but also that many craters on the Moon and elsewhere are the result of close passage of a large celestial body which produced more tidal forces (gravitational pull) on one hemisphere of the Moon than on the other which caused volcanism. He maintains the Moon has a long history of such heatings. From Sagan's statement one might readily conclude that science has settled this issue.

In their book, *The Cosmic Serpent*, Bill Napier and Victor Clube, two British astronomers tell us,

> "In 1667 Robert Hooke had dropped bullets into a stiff clay, creating little impact craters. However, he had also boiled a mixture of powered alabaster and water, and the bursting bubbles had also formed craters. A lively controversy over crater cosmology, engendered by these simple experiments has swung back and forth for 300 years, and only within the last decade or so has something like a consensus been reached."[29]

Consensus, of course, is not evidence nor science. It is only a general agreement among scientists without conclusive proof.

What is some of the evidence? In the *Annual Review of Astronomy and Astrophysics* for 1974 (the year of the AAAS symposium) astronomer Farouk El Baz writes,

> "One of the earliest controversies concerning depressions on the lunar surface has been whether these were created by meteoric impact (e.g., Gilbert 1893, 1896, Shoemaker 1962, Baldwin 1963) or volcanic processes (e.g., Dana 1846, Spurr 1944, 1945, 1948, Green 1962). In many of the classical papers written prior to the advent of spacecraft photography, evidence supporting both theories was scanty."[30]

However, because of lunar exploration certain proofs came to light. The large flat circular plains (mares) cover about 20 percent of the lunar surface. El Baz goes on, "Undisputed proof of the subsurface volcanic origin of the mare material came following examination of the first lunar samples returned by *Apollo XI*, as explained below. The *Apollo* missions also returned ample photographic evidence not only that the lunar maria were the products of lava flows, but also that these flows were emplaced at repeated times, allowing one flow to cool and solidify prior to emplacement of another."[31]

About half of the lunar surface we observe from the Earth and the maria, produce proof of a volcanic origin.

Nine years before the AAAS symposium, in the *Royal Astronomical Society Monthly Notices*, Gilbert Fielder analyzed the distribution of the lunar craters to determine whether they are volcanic or impact phenomena.

"A fresh attack on the vital problem of the craters has been made by analyzing the surface distribution of craters of a given diameter... This result is shown to apply equally to the lunarite, [bright regions] and then taken separately, to the lunarbase, [dark regions] and...argues against the theory that craters were produced exclusively by impact.

"In assessing the origin of the craters on the basis of the observed frequencies and distribution of craters alone, it is concluded that the ratio of the number of impact craters to the number of endogenic [volcanic] craters is not very large. If only one theory is allowed it must be that the craters are of internal origin."[32]

Allan Marcuse argued with this analysis a year later, but did not reject Fielder's evidence. He concluded that the chances for impact or volcanic craters are probably equal.

Also in *Aviation Week and Space Technology* we find, "There is a growing body of scientific opinion which holds that lunar craters may have a widespread volcanic origin."[33]

Based on Sagan's contention that craters are random impact events, one would expect to find fairly uniform crater features on Mercury, Mars and the Moon. This is what the present theory of planetary formation demands. On the other hand, Velikovsky's hypothesis requires that one hemisphere of Mercury, Mars and the Moon show greater volcanism—that is, more craters—than are found on the other hemisphere. In *Science News* for 1974, "The Mystery of the Hemisphere" we find,

"A major surprise in the early days of lunar exploration was the discovery that the soft maria visible from the Earth were far more rare on the Moon's farside, presumably because of some one-sided influence of the earth. Now refinements of *Mariner 9* data show one hemisphere of Mars to be far rougher [more greatly cratered] than the other, and *Mariner 10* suggests the same asymmetry for Mercury. Data files grow, observes Bruce Murray of the California Institute of Technology, yet so does the mystery of hemispherical asymmetry 'we now know,' he says, 'a little less about the Moon.'"[34]

The same year Mercury's asymmetrical cratering was reported in *Science,*

"Even more striking, Mercury, like the Moon and Mars, appears to have evolved asymmetrically. Rough and heavily cratered crust thought to be primordial surface, seems to cover half the planet while smooth plains seem to cover the rest. Why three of the five bodies studied among the inner planets have modal asymmetry is perhaps the greatest puzzle of all."[35]

Clearly the evidence seems to favor Velikovsky's conclusion. Sagan in his remarks does not discuss this evidence yet he must certainly be aware of it and knows the issue is certainly not settled. For in his co-authored book *Comet* he states this about craters, "There is still debate about whether such craters are of impact or volcanic origin."[36]

In fact, there is well observed evidence indicating that craters are of volcanic origin. The inner Galilean moon of Jupiter called Io is continually subjected to enormous tidal stresses by Jupiter which causes this moon to produce more volcanic activity than any other body in the solar system. The surface of Io is constantly changing from this ongoing volcanism and many craters are produced. Michael Zeilik in *Astronomy: The Evolving Universe,* (NY 1985) p. 185 describes these craters: "Io's volcanoes have a different shape from those found on the Earth, Venus and Mars. Few appear as cones or shields. They resemble collapsed volcanic craters." Billy P. Glass in *Introduction to Planetary Geology,* (Cambridge, England 1982), p. 363 writes that on Io,

> "There appears to be a complete absence of impact craters at least down to 5-10 km in diameter...
> "The surface is dominated by volcanic features... More than 100 caldera-like depressions up to 200 km in diameter have been observed. They are much larger than terrestrial calderas, but very few appear to be associated with significant volcanic constructs."

The craters on Io do not resemble volcanoes; they look like craters on the Moon, Mars and Mercury. Scientists who believe in the impact theory, like Sagan, can see the evidence before their eyes, but because they are imbued with a uniformitarian philosophy they refuse to believe that the evidence from Io should be applied to other bodies in the solar system. Thus they deny what appears obvious.

In the debate with Velikovsky, Sagan several times states that science had proven conclusively certain phenomena such as, "the Moon bears eloquent testimony to the fact that there have been abundant collisions during the history of the solar system." The ineloquent facts, however, indicate that the craters on the Moon, Mercury and Mars may well *not* be testimony of impact collisions but are of volcanic origin as observed on Io. This seeming bombastic tendency on Sagan's part could well color the understanding of evidence. Thus when he claims phenomena have been "established" or "proved" the reader will use, as Sagan states in *Broca's Brain,* p. 83, "a firm skeptical scrutiny" of it.

NOTES & REFERENCES

[1]*Broca's Brain*, (NY 1979) p. 82 henceforth "B.B."; *Scientists Confront Velikovsky*, (Cornell University Press), p. 43, henceforth "SCV".
[2]Ibid.
[3]SCV, Ibid.
[4]B.B., Ibid.

[5] *Velikovsky Reconsidered*, (NY 1976), p. 36.

[6] *Kronos* Vol. III, No. 2, pp. III-IV.

[7] B.B., p. 82.

[8] SCV, p. 44; B.B., p. 82.

[9] B.B., p. 126.

[10] *Kronos*, op. cit., pp. 26-27.

[11] Velikovsky, I.; *Worlds in Collision*, (NY 1950), pp. 306-307, henceforth "W in C".

[12] Velikovsky, I.; *Earth in Upheaval*, (Pocket Book Ed. 1955), p. 220.

[13] W in C; "13", p. 66.

[14] Velikovsky, I.; *Ages in Chaos*, (NY 1952), p. 32.

[15] Velikovsky, I.; *Stargazers and Gravediggers*, (NY 1983), p. 16.

[16] *Kronos* Vol. X, No. 1, p. 72.

[17] Gould, Stephan J.; *Hens Teeth and Horses Toes*, (NY 1983), pp. 105-106.

[18] Ibid., p. 105.

[19] *Stargazers*, op. cit., p. 291.

[20] *Kronos* Vol. III, No. 2, pp. 22-23.

[21] SCV, p. 45; B.B., p. 83.

[22] B.B., p. 84.

[23] Ransom, C.J.; *The Age of Velikovsky*, (Glassboro, NJ, 1976), pp. 214-215.

[24] SCV, p. 45; B.B., p. 84.

[25] B.B., p. 83.

[26] *Science*, (Jan. 16, 1987), pp. 271-273.

[27] SCV, p. 43; B.B., p. 82.

[28] SCV, pp. 45-46; B.B., p. 85.

[29] Clube, B. and Napier, V.; *The Cosmic Serpent*, (NY 1982), p. 78.

[30] *Annual Review of Astronomy and Astrophysics*, (Palo Alto CA 1974), p. 13.

[31] Ibid., p. 140.

[32] Fielder, Gilbert; *Royal Astronomical Society Monthly Notices*, 196 (1965) as cited in *Mysterious Universe* (Glenarm MD 1979), Corliss, William, ed., p. 209.

[33] Nieson, Himmel S, "Mariners Reveal Detail of Mars", *Aviation Week and Space Technology*, (Aug. 11, 1969).

[34] Cited in *Mysterious Universe*, op. cit., p. 209.

[35] Metz, William D.; "Mercury: More Surprises in the Second Assessment", *Science*, Vol. 184, (1974), p. 132.

[36] Sagan, Carl and Druyan, Ann; *Comet*, (NY 1985) p. 258.

THE HISTORICAL EVIDENCE

EXPERTS

Sagan relates the following anecdote, in *Broca's Brain*, page 86,

> "I can remember vividly discussing *Worlds in Collision* with a distinguished professor of Semitics at a leading university. He said something like this, 'The Assyrology, Egyptology, Biblical scholarship and all that Talmudic and Midrashic *pilpul* is, of course, nonsense; but I was impressed by the astronomy.' I had rather the opposite view. But let me not be swayed by the opinion of others."

I do not know why Sagan does not name his distinguished professor at a leading university. Prefacing his historical attack on Velikovsky with expert opinion that Velikovsky's historical and legendary evidence from the Middle East is "nonsense" certainly is interesting. But Sagan also claims he will not be swayed by the expert opinion of an unnamed authority. Were anyone to respond to this anecdotal criticism one could, of course, cite some anonymous distinguished professor from an anonymous leading university and claim that this anonymous distinguished professor is a world renown expert on historical, and legendary evidence of the ancient Middle East, and that he finds Velikovsky's evidence both brilliant and well documented. In fact, one might cite several anonymous distinguished professors from several anonymous leading universities and state that they are world famous figures of Assyriology, Egyptology, Biblical and the *Talmud* and *Midrash* as well as archaeology and that they rather contradict Carl Sagan's anonymous distinguished professor from an anonymous leading university. Then one might just add, "but let me not be swayed by the opinion of others."

It is, nevertheless, true that Velikovsky's thesis is not accepted by the vast majority of scholars who have not studied it, but assume it has been proven false. But again, consensus among scholars is not evidence. What is also true, is that there are highly distinguished scholars of Semitic studies who find Velikovsky's evidence "brilliant" and "well documented." Among them one scholar maintains that Velikovsky's global catastrophic hypothesis is correct. He is Claude F.A. Schaeffer whose archaeological work will be cited in later chapters. Schaeffer was a member of the institute at the College of France. He is considered one of the greatest archaeologists of our time, and he wrote in a published letter dated July 23, 1956 after having read *Worlds in Collision, Earth in Upheaval* and *Ages in Chaos*, the following,

"I hope you [Velikovsky] will go on with your research. You are working in the right direction and time will help to show the reality of global or near global catastrophes. Already continental or near continental catastrophes cannot be doubted as I showed in my stratigraphical work in the Near East. It will take time for your findings and mine to be acknowledged. This may make us sometimes impatient. But it will stir us to more work and more research." Signed "Claude F.A. Schaeffer."[1]

Professor Etienne Droiton, historian and world authority on Egyptol-ogy also wrote a letter to Velikovsky regarding his work.

"At the time, [May 29, 1952] Droiton held the position of d*irecteur general du service des antiquities* [in Egypt]. In this capacity he had under his care all the antiquities—the monuments in the field and in the museums, the famous Cairo Museum included—and every excavation made in Egypt, by whatever agency or learned society was under his supervision. [Later, after] the revolution in Egypt, [Droiton held the] post as chief curator of the Egyptian Department of the Louvre Museum in Paris."[2]

His letter in full goes on to state on the next page,

"Dear Doctor [Velikovsky],

You have so kindly sent me a copy of your fine book, *Ages in Chaos,* which I received this morning, and which I have already read almost in its entirety, so stirring and fascinating is it.

You certainly overthrow, and with what zest!, many of our historical assumptions, which we have considered established. But you do it with total absence of prejudice and with impartial and complete documen-tation, all of which is most gratifying. One might dispute point by point your conclusions: whether one admits them or not, they have posed the problems anew and made it necessary to discuss them in depth in the light of your new hypotheses. Your fine book will have been in every way a great use to science.

I thank you warmly for having sent it to me and I beg you to accept dear Doctor, the assurance of my sentiments of cordial devotion.

Etienne Droiton,
General Director Department of Antiquities"

Robert H. Pfeiffer, was Chairman of the Department of Semitic Lan-guages and History at Harvard University, and an authority on the *Bible*. Pfeiffer's published letter to Velikovsky deals with the contents of *Worlds in Collision*. He writes,

"Allow me, first of all, to congratulate you, not of course for the fact that your book has become a 'run-away best seller,' but for the magnificent qualities of content and form of your book. I read it with utter fasci-nation and absorption, being carried away by the cosmic drama which you have unfolded before me. I was amazed at the depth and vastness of your erudition, which I have not seen equaled except possibly in O. Spengler's *Decline of the West.*"[3]

It seems clear that three experts in the field of Semitic studies, all world respected figures, say that the three major works of Velikovsky are completely documented. Of course, other reputable experts disagree with Velikovsky, and certainly Velikovsky's work is not with-out errors. Nor do these citations prove Velikovsky's theory correct. This evidence was cited to answer Sagan's statement that appears to be an attempt to belittle Velikovsky's evidence before actually dealing with it.

Sagan adds, "My own position is that even if 20 percent of the legendary concordances that Velikovsky produces are real, there is something important to be explained."[4] According to Sagan, less than twenty percent of Velikovsky's historical and legendary evidence is valid. It is extremely difficult to conceive that world recognized authorities will lavish praise on an author of books in their own fields that are less than eighty percent accurate. Sagan apparently knows enough compared to experts to discredit eighty percent of Velikovsky's historical and legendary evidence in a field in which he possesses little or no training. This, on the face of it, seems quite incredible. Let us therefore proceed to Sagan's historical and legendary material.

What must be pointed out before proceeding though is that the fundamental legend that Velikovsky cited as evidence is the legend of the planet Venus as the cause of a world-wide catastrophe. Velikovsky found this legend telling of the same events among all of the major and minor ancient cultures. One would naturally expect Sagan to attack this evidence most forcefully.

DIFFUSION OR COMMON OBSERVATION

According to Sagan there are four ways in which the same (Venus) legend would be found among widely separated cultures.

1. Common Observation; all cultures witnessed a common event and interpreted it in a similar way.

2. Diffusion; the legend originated with one culture, but traveled to others with the wanderings of mankind.

3. Brain Wiring; psychologically human beings are so alike that their legends reflect the commonality of human hopes and fears.

4. Coincidence; purely by chance all cultures created the same (Venus) legend or myth.

Sagan chose diffusion and coincidence while Velikovsky, of course, chose common-observation. Sagan states,

"Velikovsky is clearly opting for the common-observation hypothesis, but he seems to dismiss the diffusion hypothesis far too casually; for example, he says (p. 303) 'How could unusual motifs of folklore reach isolated islands, where the aborigines do not have any means of crossing the sea?' I am not sure which islands and which aborigines Velikovsky refers to here, but it is apparent that the inhabitants of an island had to have gotten there somehow. I do not think that Velikovsky believes in a separate creation in the Gilbert and Ellice Islands say. For Polynesia and Melanesia there is now extensive evidence of abundant sea voyages of lengths of many thousands of kilometers within the last millennium, and probably much earlier."[5]

Velikovsky's Venus myth is found on island cultures throughout the globe. Thus when Sagan claims he is unaware, he should have read Velikovsky's book more carefully to know this basic fact. What is further apparent on inspecting page 303 of *Worlds in Collision,* is that Sagan withheld evidence it seems far too casually. The complete citation contains,

"The similarity of motifs in the folklore of various peoples on the five continents and on the islands of the oceans posed a difficult problem for the ethnologists and anthropologists. The migration of ideas may follow the migrations of peoples, but how can unusual motifs of folklore reach isolated islands where aborigines do not have any means of crossing the sea? AND *why did not technical civilization travel together with spiritual? Peoples still living in the stone age possess the same often strange motifs of cultured nations.* " [Capitalization and emphasis added]

By omission of the conjunction, "and" Sagan limited the meaning Velikovsky intended. If as Sagan maintains the Venus myth traveled to the Gilbert Islands and Ellice Islands from China or Japan why didn't the technical achievements of these superior cultures also travel with the people? That would be like the migration of the Europeans who followed Columbus to the Americas bringing their religious beliefs, but somehow forgetting carpentry, bricklaying, iron making, etc. Furthermore, Velikovsky did not dismiss the diffusionist argument at all. In *Worlds in Collision,* he did, in fact, write,

"If a phenomenon had been similarly described by many peoples, we might suspect that a tale, originating with one people, had spread around the world, and consequently there is no proof of the authenticity of the event related. But just because one and the same event [the Venus Myth] is embodied in traditions that are very different indeed, its authenticity becomes highly probable, especially if the records of history, ancient charts, sundials, and the physical evidence of natural history testify to the same effect."[6]

TEO—PLACE OR GOD

To support his diffusionist claim, Sagan asks,

> "...how, for example, would Velikovsky explain the fact that the Toltec
> name for 'god' seems to have been teo, as in the great city of
> Teotihuacan (City of the gods)...? There is no common celestial event
> that could conceivably explain this concordance...teo is a clear cognate
> of the common Indo-European root for 'god.'"[7]

This is most interesting because in their book of Nahuatl symbols C.
McGowan and P. Van Nice, *The Identification and Interpretation of Name
and Place Glyph of the Xolotl Codes,* (1984), p. 67 tell us the Nahuatl
word for "god" is "teotl". However, in the word Teotihuacan which
is derived from Aztec not Toltec[8] contains the root "teotia" which
according to J.E. Hardoy a Mexicologist, "Teotihuacan (from the
Nahuatl word) Teotia, to worship the "place of deification" or "place
of the gods."[9] There is no Teotl in the word Teotihuacan. To use the
Nahuatl word "Teotl", the city would have to be spelled
"Teotlhuacan". But even if we accept Sagan's analysis we ask of the
linguistic validity of Sagan's statement, "if we compare two languages,
each with tens of thousands of words, spoken by human beings with
identical larynxes, tongues and teeth, it should not be surprising if a
few words are coincidentally identical."[10] However, R.C. Padden, a
linguist deals with isolated words in different languages which are
identical stating, "competent linguists simply do not compare isolated
words of unrelated languages. No one has yet established a continuity
of linguistic relationships between [eastern and western] hemispheres
in the pre-Columbian period."[11] Thus Sagan's statement is based on
an incompetent analysis. However, Sagan then builds on his
incompetent analysis adding "Likewise, we should not be surprised if a
few elements of a few legends are coincidentally identical."[12] In this
case, Sagan seems to believe that all the Venus legends were created by
all the cultures of ancient man by coincidence.

Teotihuacan is not recognized by most scholars as the capital of the
Toltecs. The capital of the Toltecs is called Tollan and is located at a
place in Mexico called Tula. According to Nigel Davies', *The Aztecs,*
(University of Oklahoma Press, 1973), pp. 11-12, the main features of
the "Toltec capital Tula...are rich in representations of Quetzalcoatl,
the Plumed Serpent, the great deity of the Toltecs." Quetzalcoatl is
the planet Venus, and we are informed that Tula was his city, "his city
of Tula." Sagan perhaps assumes that the great city of the Toltecs was
Venus' city strictly by coincidence. But Sagan has also told us "There
is no common celestial event that could conceivably explain this
concordance." But Tula or Tollan is the city of a celestial body—

Venus. And we shall show further on that even. Sagan's evidence shows Quetzalcoatl or Venus was seen by the ancient Americans as a comet. Thus the evidence is entirely against Sagan's analysis.

In fact, even if we give credence to Sagan's views, *A Nahuatl-English Dictionary* by John Bierhorst Stanford Univ. Press, CA, 1985) p. 310, defines, "Teotihuacan-1. Place where one becomes a spirit i.e., the hereafter. 2. Famous archaeological site 40 km northeast of Mexico City...where the Sun and Moon were created..." There is no statement in this dictionary which equates Teo with god. The word refers to a "place" and that place is the sky "where the Sun and Moon were created." Therefore, Tollan is the city of Venus and refers to the heavens. On page 363 of the same dictionary "Tollan" is defined as "paradise, the other world." This definition also tells us Tollan is a "place" that is the sky. If Sagan wishes to argue linguistically, he should do so in a competent manner. His diffusionist argument is not supported by using an incompetent linguistic analysis. The entire argument is based on pure conjecture and that as evidence is worthless. The linguist argument fails and the coincidental argument is based on Sagan's beliefs; for he states, "I believe that *all* of the concordances Velikovsky produces can be explained away in this manner." [13] Sagan's beliefs are most certainly not evidence, though he may believe whatever he wishes. But Tollan-Tula is the city of Venus.

THE SHAPES OF COMETARY FIELDS
Turning to astronomy Sagan argues,

> "Velikovsky even goes so far as to believe that a close approach to the Earth by 'a star' he evidently identified with the planet Mars so distorted it that it took on the clear shape (page 264) of lions, jackals, dogs, pigs, fish" [and] "He [Velikovsky] points to certain concordant stories, directly or vaguely connected with celestial events, that refer to a witch, a mouse, a scorpion or a dragon... His explanation: divers comets upon close approach to the Earth, were tidally or electrically distorted and gave the form of a witch, a scorpion, or a dragon, clearly interpretable as the same animal to culturally isolated peoples of very different backgrounds. No attempt is made to show that such a clear form—for example, a woman riding a broom and topped with a pointed hat could have been produced in this way, even if we grant the hypothesis of a close approach to the Earth by a comet." [14]

It seems incredible that an astronomer appears to be ignorant of the fact that comets do indeed assume various shapes of animals with their comas and tails. In fact, in Sagan's popular book, *Comet,* he shows pictures of comets that do look like bestial apparitions. Indeed, his chapter is titled "A Cometary Bestiary". Sagan tells us,

"On these pages we have accumulated a kind of cometary bestiary, like the animal bestiaries assembled by medieval authors to amaze and delight, and even instruct. Most of the animals displayed were real; many were exotic; a few such as the unicorn were the result of errors in transmission, garbled accounts—in this case of the African rhinoceros."[15]

Sagan attacks Velikovsky's bestiary comet evidence with, "This is not very impressive reasoning. We might just as well assume that the whole menagerie was capable of independent flight in the second millennium B.C. and be done with it."[16] However, Sagan's own use of cometary shapes in his bestiary, of which he has assembled some 31 pictures, thoroughly contradicts his evidence and conclusion as well as *amaze and delight and even instruct!*

Guy Murchie in *Music of the Spheres,* Vol. 1, (NY 1967), p. 124, discusses a comet seen in the 16th century and gives a description of it as seen by eye witnesses.

"A typical description of the great comet of 1528 by an awed observer said it looked 'so horrible and produced such great terror in the common people that some died of fear and others fell sick. It appeared to be of excessive length and was the color of blood. At its summit [from its nucleus] rose the figure of a bent arm, holding in its hand a great scimitar as if about to strike... On both sides of the rays of this comet there appeared a great number of axes, knives and blood-drenched swords, among which were many hideous slowly-shifting faces with beards and bristling hair.'"

The drawing of the comet on the same page appears to be that of a witch without a pointed hat. The appearance of the comet could most certainly be taken as a phantasmagorical witch. Witches need not be the type with a pointed hat riding a broom that Sagan suggests. The description of this comet by the observer and the effect it had on the common people indicate that some fantastic apparition obviously caused some to die of fright and others to become ill. Thus, comets can appear as animals or as witches contrary to Sagan's claim. The question is why did he accuse Velikovsky of inventing cometary forms and then claim such forms are in fact quite real? As we proceed, this sort of evidence will emerge again and again. Sagan claims Velikovsky's evidence is invalid, but later he produces the same concept as valid.

READING CAREFULLY
Sagan only in his original paper and in *Scientists Confront Velikovsky* misrepresented Velikovsky with the following accusation, "Velikovsky claims a world-wide tendency in ancient cultures to believe at various times that the year has 360 days, that the month has thirty days, and

that of course, is inconsistent with the above two beliefs—the year has ten months." [17]

Sagan accused Velikovsky of being inconsistent because a year or ten months each of thirty days equals 300 days, not 360 days. But Velikovsky on pp. 344-345 actually states that, "the month was equal to...thirty-six days." Sagan's carelessness was pointed out by Lewis M. Greenberg in the journal *Kronos* and after a time while this misrepresentation circulated it was finally corrected in *Broca's Brain*. However, many readers of the first book which is still in circulation will be confronted by this disinformation. Sagan in *Broca's Brain* only, next introduces new criticism, saying that, "Velikovsky offers no justification in physics for this" [18] change in the calendar. However, a few lines down the page Sagan adds the justification in physics stating, "Velikovsky proposes that these aberrant calendrical conventions reflect real changes in the length of the day, month, and/or year—and that they are evidence of close approaches to the Earth-Moon system by comets, planets and other celestial visitors." [19]

Sagan has accused Velikovsky of bad arithmetic in his paper in *Scientists Confront Velikovsky* for stating that ten months of thirty days could not equal 360 days; when this argument failed Sagan then argued in *Broca's Brain* that these months may indeed have had thirty six days but that Velikovsky does not explain the physical reasons for changes in the calendar. But then Sagan tells us at the end of the paragraph the physical reasons Velikovsky has offered. Since Sagan is so obviously confused that he contradicts himself again and again on the same point in the same paragraph there is no need to debate the point. Earlier Sagan remarked,

> "In reading the critical literature in advance, I was surprised at how little of it there is and how rarely it approaches the central points of Velikov-sky's thesis. In fact, neither the critics nor the proponents of Velikovsky seem to have read him carefully, and I even seem to find some cases where Velikovsky has not read Velikovsky carefully. Perhaps...the present chapter...will help clarify the issues." [20]

This remark on the face is stunning, given the evidence thus far.

FRACTIONS—CALENDARS

To reinforce his evidence on the unaltered uniformitarian state of the calendar in the past, Sagan adds,

> "There is an alternative explanation, which derives from the fact that there are not a whole number of lunations in a solar year, nor a whole number of days in a lunation. These incommensurabilities will be galling to a culture that had recently invented arithmetic but had not yet gotten as far as large numbers or fractions... There is a clear whole-number

chauvinism in human affairs, most easily discerned in discussing arithmetic with four-year olds; and this seems to be a much more plausible
explanation of these irregularities, if they existed.

"Three hundred and sixty days a year provides an obvious (temporary)
convenience for a civilization with base 60 arithmetic as the Sumerian,
Akkadian, Assyrian and Babylonian cultures."[21]

When we turn to Evan Hadingham's book on ancient astronomy we
learn,

"The Babylonian reliance on numerical methods is understandable,
considering that they practiced sophisticated arithmetic as far back as
1800 B.C... At this early stage, there already existed tables for
multiplication, division, squares, square roots, cubes and *reciprocals,*
exponential functions and many other mathematical procedures."[22]
[emphasis added]

Now I have always been taught that reciprocals in mathematics are
fractions. In fact, Joseph E. Hofmann, Honorary Professor of Mathematics of the University of Tübingen, Germany in his book, *Geschichte
der Mathematik,* (History of Mathematics), translated by F. Gaynor and
H.O. Medonick for the Philosophical Library on page 6 of his
chapter, "The Babylonians", states that by use of reciprocals, "The
Babylonians had a clear understanding of the nature of common
fractions." Sagan's whole number chauvinism is clearly based on his
own chauvinism and not on the evidence.

Here is Velikovsky's response to the attack on the 360 day year.

"I am very proud of these chapters of mine toward the end of *Worlds in
Collision*...because I succeeded to quote from practically every ancient
civilization from Peru, to Mexico, to Rome, to Greece, to Babylonia, to
Assyria, to Persia, to Hindu, to China, to Japan, and to Egypt and to
Palestine, Judea and probably several more civilizations, always [a]
quotation not by myself, always by [a] specialist expressing the same
wonder that [there were] no intercalary days—the year was just this:
twelve months of thirty days—for a period of time; which was
discontinued at the beginning of the eighth century.

"Soon after that time, in all places, in all civilization[s] one or another
reform was done, and five or five and a quarter days were added by all
civilizations. The reform was [carried out] almost simultaneously—at
least during one and the same century."[23]

Sagan attacks the ancient astronomers, proclaiming, "...sloppy quan-
titative thinking appears to be the hallmark of this whole subject."[24]
Giorgio de Santillana and Hertha von Dechend in their book *Hamlet's
Mill,* having made years of extensive study and analysis of ancient
astronomy, make clear the view that speculation of the kind in which
Sagan indulges is of little value. They state:

"...it is an unsound approach to Mayan astronomy [or any ancient astronomy] to start from preconceived convictions about what the Maya's [or other ancients] could have known and what they could not have known: one should, instead, draw conclusion only from the data as given. That this had to be stressed explicitly reveals the steady decline of scientific ethics."[25]

Santillana accuses scholars like Sagan of having "cultivated a pristine ignorance of astronomical thought."[26] Sagan's method of dealing with this evidence "as given" from ancient astronomy by labeling it "sloppy" is not science; it is name calling, nothing more.

There is furthermore exact measured evidence that supports the conclusion that the motion of the Moon, from which we derive the length of the month, prior to 687 was different than it is today. In Benjamin Farrington's book *Science in Antiquity*, (London 1969), pp. 12 and 13 we find,

"A most impressive application of mathematics to astronomy is supported by a tablet found in the library of Assurbanipal at Nineveh. The library belongs to the middle of the seventh century, but the document may be itself much older or a copy of an older document. It is an attempt to tabulate the progress of the illumination of the surface of the Moon during its period of waxing. To this end the area of the Moon's face is divided into 240 parts over which the illumination is conceived as spreading first according to a geometrical, then to an arithmetic progression. This arrangement does not correspond to the [present day] facts" [of how long the Moon waxes.]

The question is, were the Babylonian astronomers careful and accurate observers? Arthur Koestler states in his book *The Sleepwalkers*, (NY 1963), pp. 20-21 that the Babylonian,

"...observations became amazingly precise: they computed the length of the year with a deviation of less than 0.001 percent from the correct value, and their figures relating to the motions of sun and moon have only three times the margin of error of nineteenth century astronomers armed with mammoth telescopes. In this respect, theirs was an Exact Science; their observations were verifiable, and enabled them to make precise predictions of astronomical events..."

Measuring and calculating how long it takes the Moon to go from new Moon to full Moon is an observation which the Babylonians made precisely using geometry and mathematics to explain the observation. But they say the Moon's period of waxing is different than that observed today. This means that the length of the month was different than that of the present time. However, because this careful measurement does not agree with the notion of uniformitarian astronomy in which no significant change is possible it is again disregarded. On the other hand, if this measurement were to agree with present theory, no doubt we would hear how well modern theory is supported by ancient observation.

The present period of the moon's synodic orbit about the Earth is about 29.5 days but the period according to B.L. Van Der Waerden who discussed this measurement in *Die Anfange Der Astronomie* (Groningen) p. 85, is precisely a full 30 days. This means that the moon had to be somewhat farther from the Earth than at present and its orbit, therefore, bigger. For the Moon, at this greater distance from the Earth, to complete one synodic orbit it would travel somewhat more slowly and its orbit would be larger and longer.

According to the highly accurate Mesopotamian astronomers, the Moon took a longer period of time to complete one revolution around the Earth. But again, since this measurement does not agree with the expectations of Sagan and his colleagues, it is ignored or cast aside. However, one can be quite sure that if this measurement did fit their expectations, it would be hailed as exact proof that there has not been a recent drastic change in the orbit of the Earth–Moon system. The evidence of the period required by the Moon's face to go from new Moon to full Moon and the measurement of the Moon's orbit around the Earth indicate just the opposite. But since this evidence cannot be faced, it is dismissed. Nevertheless, these observations support Velikovsky's hypothesis regarding the fact that the month was of a different period in ancient times. Sagan argues that,

> "A leading historian of ancient science and mathematics, Otto Neugebauer (1957) remarks that, both in Mesopotamia and in Egypt two separate and mutually exclusive calendars were maintained: a civil calendar...and a frequently updated agricultural calendar—messier to deal with, but much closer to the seasonal and astronomical realities." [27]

By the "astronomical realities" Sagan means the calendar as modern astronomers expect it to be. Sagan, along with modern astronomers, maintains that there absolutely cannot be a really different calendar in historical times. Thus, when they discover evidence contrary to that dogmatic view, it is force-fit to the model they say is an "astronomical reality".

This is made explicit by Robert R. Newton in *Medieval Chronicles and the Rotation of the Earth,* (Baltimore 1972), pp. 2–3 stating,

> "There have been many attempts to find changes in the day and month from ancient astronomical data. Since solar eclipses are striking phenomena that have been observed by many people and not just by professional astronomers, solar eclipses have played a large role in such attempts. When I looked into these attempts, I was astonished by what I found. Many of them, including all uses of solar eclipses that I have seen,

were based upon the logical fallacy of reasoning in a circle. Specifically, most reports used could not be dated on the basis of their texts or their historical contexts. The workers [scientists] thereupon assigned dates by finding which ones led to accelerations [of the celestial bodies] that agreed most closely with assumed values. [i.e., with what the astronomers expect to find.] It is not surprising that the resulting 'data' were self-consistent."

Elsewhere, Newton writes,

"*Virtually all studies of ancient [solar] eclipses that I know of* have used the following procedure in handling doubtful or ambiguous cases: The author [scientist] has assumed values of the accelerations [of the celestial bodies] in advance and has calculated the circumstances of the possible observations using them. He has then rejected as invalid all observations or interpretations thereof, that do not agree well with the assumed values. He has finally used the remaining set of observations to calculate the accelerations. He necessarily found good agreement with his initial assumptions. [emphasis added]

"This, of course is, reasoning in a circle." [R.R. Newton, *Ancient Astronomical Observations,* (Baltimore 1970), p. XIV.]

Thus, Newton confirms that this is how astronomers proceed. Data negative to Sagan's "astronomical realities" which are really "assumed realities" is simply disregarded or forced to fit what is assumed *a priori* to make the evidence self-consistent. How very convenient.

Sagan goes on to show that, "Many ancient cultures solved the two - calendar problem by simply adding a five-day holiday on at the end of the year."[28] What neither he nor Otto Neugebauer ever informs us is none of the ancient cultures cited by Sagan had two calendars or five days in their calendars before 686 B.C. What Sagan has failed to inform us is that Neugebauer refers to Babylonian astronomical texts seemed to have data, "equivalent to modern ephemerides" [Otto Neugebauer, *The Exact Sciences in Antiquity,* (NY 1969) pp. 105, 110, 129] That is, their data was almost as good as modern data.

Now, if this two calendar evidence (offered by Neugebauer) existed (before 686 B.C.), it would be greatly damaging to Velikovsky's hypothesis since it would imply that there were all kinds of changes added to ancient calendars prior to the last cosmic catastrophe that Velikovsky describes. However, we are told specifically by M.P. Nilsson in *Primitive Time Reckoning,* (Lund/London 1920), p. 367, that, "...we are met with the difficulty that an intercalary cycle [adding days or months to the calendar] was not introduced into Babylonia before the sixth century [B.C.]." A.E. Samuel, *Greek and Roman Chronology,* (Munich 1972) p. 21, says,

"...We have long lived with the cliché that the Greeks learned their astronomy from the Babylonians, but modern investigation has demonstrated that the sophisticated Babylonian systems were later than had hitherto been believed. *The irregular intercalations* [of adding days or months to the calendar] *exist down into the fifth century, showing that as late as 480 B.C. no [intercalary] cycle existed to control that calendar."* [emphasis added]

Benny Peiser's *Greek History Begins in the 6th Century,* (1989) privately published, makes the point emphatic stating, "it has become a *communis opinio that the intercalary cycle* [of adding days or months to the calendar] *cannot be detected anywhere in the ancient world before the 6th century...*" [emphasis added] Thus, ancient civilizations changed their calendars only after the last catastrophe that Velikovsky describes. Instead of disproving Velikovsky's hypothesis, Neugebauer and Sagan's evidence supports it. The common concept held by modern scientists who study ancient calendars asserts only after Velikovsky's last catastrophe did ancient nations begin to rearrange their calendars adding five days or intercalary months to make the older calendars, which had become obsolete, fit the new length of the year.

Again this evidence is ignored. Thus, when Sagan states, "I hardly think that the existence of 360-day years in the calendrical conven - tions of prescientific peoples is compelling evidence that then there really were 360 rather than 365 1/4 rotations in one revolution of the Earth about the Sun."[29] Sagan's conjecture is not based on any evidence since much of it must be either thrown out or ignored. Because only the data after 686 B.C. supports his uniformitarian view, his "thought" ("I hardly think") represents a "personal" rather than an "objective" analysis of the material evidence. Prior to 686 B.C., evidence supports Velikovsky's view. Sagan also argues,

> "An expert on early time reckoning (Leach, 1957) points out that in ancient cultures the first eight or ten months of the year are named, but the last few months, because of their economic unimportance in an agricultural society are not."[30]

Again this is not true of all ancient societies. For example, the ancient Roman calendar (which Sagan discusses in the next sentences) had only four months with names during its very early history. Velikovsky discussed this in *Worlds in Collision,* "According to many classical authors, in the days of Romulus [8th century B.C.] the year consisted of ten months and in the time of Numa, [7th century B.C.] his successor, two months were added: January and February. Ovid writes (Fasti i 27ff) 'When the founder of [Rome] was setting the calendar in order, he ordained that there should be twice five months in his year... The month of Mars [March] was the first.'" [31] And "March was considered the first month until the reign of Numa...wrote Procopius of Caesarea."[32] Some seven centuries later, Julius Caesar gave his name to the fifth month, *Quintilis,* and Augustus gave his name to the sixth month, *Sextilis.* Thus, in ancient Roman times, the calendar was March, April, May, June, Quintilis, Sextilis, September, October, November, December. There are ten months in all, but only the first four have god names, not eight or ten as Sagan's expert tells us. This is hardly evidence to support Sagan's belief. He has merely tried to make the evidence fit his view by ignoring it altogether. Further, Sagan earlier completely ignored the fact that the Roman calendar was made up of ten months just as Velikovsky claimed.

Sagan then turns to scientific evidence to support his view regarding the 360 day year, stating, "This question can, in principle, be resolved by examining coral growth rings, which are now known to show with some accuracy the number of days per month and the number of days per year, the former only for intertidal coral."[33] Thus, according to Sagan, recent coral ring dating should never reflect a 360 day year. Interestingly, Robert H. Dott, Jr. and Roger L. Batten in *Evolution of the Earth* state the following:

> "Biologists have observed that modern corals deposit a single, very thin layer of lime once a day. It is possible, with some difficulty, to count these diurnal (day-night) growth lines and to determine how old the coral is in days. More important, seasonal fluctuations will cause the growth lines to change their spacing yearly so that annual increments can also be recognized much as in growth rings of trees. Out of curiosity, and because he is a paleontologist, [John Wells of Cornell University] began looking for diurnal lines of fossil corals. He found several Devonian and Pennsylvanian corals that do show both annual and daily growth patterns. But he was astonished to find that the Pennsylvanian forms had an average of 387 daily growth lines per year-cycles, and that the Devonian corals had about 400 growth lines... By making counts between annual marks, Professor *Wells found an average of 360 lines per year on Modern* [within the last 5,000 years] *corals.*"[34] [emphasis added]

Here Sagan's colleague at Cornell University presents evidence which contradicts Sagan's conclusions regarding the coral calendar. When Sagan writes that "There appears to be no sign of major excursions in recent times,"[35] he has certainly not dealt with this evidence. In fact, Sagan admits the problem can be solved "in principle" but not "in fact" and he uses the term "some accuracy" not "excellent accuracy."

S. Warren Carey, in *Theories of the Earth and Universe,* (Stanford CA 1988), p. 196 became skeptical regarding the validity of coral counts of growth lines because,

"The growth lines vary in spacing from a few microns [a micron is a millionth part of a meter] to nearly zero, and it is often difficult to decide whether one should be counted or not. Under these circumstances it is notorious that total counts come out at what the investigator thinks they should be. The subjectiveness of such counts is highlighted by a report by R.G. Hipkin, an Edinburgh University geophysicist, that he counted 253 ridges and later 359 ridges in a repeat count of the same specimen."

This is a variation of 106 days for the same coral. We previously pointed out that archeo-astronomers make their data respecting the solar eclipses fit their conclusions by making the motions of the Earth and Moon fit retrospectively into their assumptions. Carey has told us that this is a notoriously common practice among scientists, namely, to make difficult-to-analyze data come out to fits what the investigator believes is correct. In our discussion of carbon-14 testing, below, we will show that this pernicious practice of making data fit preconceived theory is also applied to dating past events.

SYNCHRONISM

Sagan then raises this argument, "Another problem with Velikovsky's method is the suspicion that vaguely similar stories may refer to quite different periods. The question of the synchronism of legends is almost completely ignored in *Worlds in Collision*... "[36] Velikovsky synchronizes the Biblical story of Joshua with evidence in the Americas,

"It has been noted that this description of the position of the luminaries implies that the sun was in the forenoon position. The *Book of Joshua* says that the luminaries stood in the midst of the sky.

"Allowing for the difference in longitude, it must have been early morning or night in the Western Hemisphere [that is early morning in the Caribbean and night in Mexico].

"We go to the shelf where stand books with the historical traditions of the aborigines of Central America.

"The sailors of Columbus and Cortes, arriving in America, found there literate peoples who had books of their own. Most of these books were burned in the sixteenth century by the Dominican monks. Very few of the ancient manuscripts survived, and these are preserved in the libraries of Paris, the Vatican, the Prado, and Dresden; they are called *codici,* and their texts have been studied and partly read. However, among the Indians of the days of the conquest and also of the following century, there were literary men who had access to the knowledge written in pictographic script by their forefathers.

"In the Mexican *Annals of Cuauhtitlan*—the history of the empire of Culhuacan and Mexico, written in Nahua-Indian in the sixteenth century—it is related that during a cosmic catastrophe that occurred in the remote past, the night did not end for a long time.

"The biblical narrative describes the sun as remaining in the sky for an additional day ('about a whole day'). The *Midrashim*, the books of ancient traditions not embodied in the Scriptures, relate that the sun and the moon stood still for thirty-six *itim*, or eighteen hours, and thus from sunrise to sunset the day lasted about thirty hours.

"In the Mexican annals it is stated that the world was deprived of light and the sun did not appear for a fourfold night. In a prolonged day or night time could not be measured by the usual means at the disposal of the ancients.

"Shagun, the Spanish savant who came to America a generation after Columbus and gathered the traditions of the aborigines wrote [in *Historia General de las Cosas de Nueva Espana*, (1946) (3 vols.) French trans. p. 481] that at the time of one cosmic catastrophe the sun rose only a little way over the horizon and remained there without moving; the Moon also stood still." [37]

According to Zecharia Sitchen's *The Lost Realms*, (NY 1990) pp. 151 – 154, discussing "*The Day the Sun Stood Still*" from Inca legends:

"Completely ignored by scholars...has been the repeated statements in the Andean legends that there occurred a frightening darkness in the long-ago times. No one has wondered whether this was the same darkness—the non-appearance of the sun when it was due—of which the Mexican legends speak in the tale of Teotihuacan and its pyramids. For if there had indeed been such a phenomenon that the sun failed to rise and the night was endless, then it would have been observed throughout the Americas.

"The Mexican collective recollections and the Andean ones seem to corroborate each other on this point, and thus uphold the veracity of each other, as two witnesses to the same event...

"According to Montesinos and other chroniclers, the most unusual event took place in the reign of Titu Yupanqui Pachacuti II, the fifteenth monarch in Ancient Empire times. It was the third year of his reign when 'good customs were forgotten and people were given to all manner of vice,' that '*there was no dawn for twenty hours.*' [emphasis added]. In other words, the night did not end when it usually does and sunrise was delayed for twenty hours. After a great outcry, confessions of sins, sacrifices, and prayers the sun finally rose.

"This could not have been an eclipse: it was not that the shining of the sun was obscured by a shadow. Besides, no eclipse lasts so long, and the Peruvians were cognizant of such periodic events. The tale does not say that the sun disappeared'; it says that it did not rise—'there was no dawn'—for twenty hours...

"Scholars have struggled for generations with [the]...tale in Chapter 10 of the *Book of Joshua*. Some discount it as mere fiction; others see in it echoes of a myth; still others seek to explain it in terms of an unusually prolonged eclipse of the sun. But not only are such long eclipses unknown; the tale does not speak of the disappearance of the sun. On

the contrary, it relates to an event when the sun continued to be seen, to
hang on in the sky for 'about a whole day'—say twenty hours?

"The incident, whose uniqueness is recognized in the *Bible* ('There
was no day like it before or after') taking place on the opposite side of
the Earth relative to the Andes [and Mexico], thus describes a phenomenon that was the opposite of what happened in the Andes. In
Canaan the sun did not set for some twenty hours; in the Andes the sun
did not rise for the same length of time.

"*Do not the two tales, then, describe the same event, and by coming from
different sides of the Earth attest to its factuality*" [Sitchen's emphasis]

"...Whatever the precise cause of the phenomenon, what we are
concerned with here is its timing. The generally accepted date for the
Exodus has been the thirteenth century B.C. (circa 1230 B.C.), and
scholars have argued for a date earlier by some two centuries found
themselves in a minority... Subsequent to the publication of our
conclusion [related to Biblical dates] (in 1985), two eminent biblical
scholars and archeologists, John J. Bimson and David Livingston, reached
after an exhaustive study (*Biblical Archeology Review*, September/October
1987) the conclusion that the Exodus took place about 1460 B.C.

"Since the Israelites wandered in the deserts of Sinai for forty years, the
entry into Canaan took place in 1393 B.C.; the occurrence observed by
Joshua happened soon thereafter.

"The question now is: did the opposite phenomenon, the prolonged
night, occur in the Andes at the same time?

"Unfortunately, the shape in which the writings of Montesinos have
reached modern scholars leaves some gaps in the data concerning lengths
of reign of each monarch, and we will have to obtain the answer in a
roundabout way. The event, Montesinos advises, occurred in the third
year of the reign of Titu Yupanqui Pachacuti II. To pinpoint his time we
will have to calculate from both ends. We are told that the first 1,000
years from Point Zero [of the Andean calendar] were completed in the
reign of the fourth monarch, i.e., in 1900 B.C.; and that the thirty-
second king reigned 2,070 years from point zero, i.e., in 830 B.C.

"When did the fifteenth monarch reign? The available data suggests
that the nine kings that separated the fourth and fifteenth monarch
reigned a total of about 500 years, placing Titu Yupanqui Pachacuti II at
about 1400 B.C.. Calculating backwards from the thirty-second monarch
(830 B.C.), we arrive at 564 as the number of intervening years, giving
us a date of 1394 B.C. for Titu Yupanqui Pachacuti II. [The Israelites
enter Canaan in 1393 B.C.; the monarch Titu Yupanqui Pachacuti II
begins his reign in 1394 B.C.; and soon there after the Israelites experience a long day and the Andeans a long night.]

"Either way, we arrive at a date for the Andean event that coincides
with the Biblical date and the events date at Teotihuacan.

"The hard hitting conclusion is clear:

"THE DAY THE SUN STOOD STILL IN CANAAN WAS THE NIGHT WITHOUT SUNRISE IN THE AMERICAS. [Sitchen's capitals.]

"The occurrence thus verified, stands out as irrefutable proof of the veracity of the Andean recollections..."

This certainly is synchronism of events by descriptions of ancient men of the "astronomical realities" that they report. Thus, Sagan's state - ment that "This question of synchronism of legends is almost entirely ignored in *Worlds in Collision*" is belied by the facts.

THE WORLD AGES

Sagan goes on, "Velikovsky notes that the idea of four ancient ages terminated by catastrophes is common to Indian as well as to Western sacred writing."[38] Velikovsky introduced this material about "four ancient ages terminated by catastrophes" with the following statement at the very beginning of this chapter, "The World Ages",

"A concept of ages that were brought to their end by violent changes in nature is common all over the world. The number of ages differs from people to people and from tradition to tradition. The difference depends on the number of catastrophes that the particular people retained in its memory, or on the way it reckoned the end of an age."[39]

Lewis M. Greenberg pointed out that Velikovsky "acknowledged that there was a tradition of seven ages (Etruscan, Persian, sacred Hindu and Hebrew writings), ten ages (Chinese), and nine ages (Polynesian and Iceland) as well, while carefully pointing out that the number of years ascribed to various ages differed, (W in C. pp. 30–33). Nothing was hidden."[40] Sagan's remarks to the contrary that, "...in the *Bhagavad Gita* and in the *Vedas,* widely divergent numbers of such ages, including an infinity of them, are given; but, more interesting, the duration of the ages between major catastrophes is specified (see, for example, Campbell 1974) as billions of years. This does not match very well with Velikovsky's chronology, which requires hundreds or thousands of years."[41] This analysis has nothing to do with Velikov - sky's work. It is merely a smoke screen to avoid dealing with the evidence that Velikovsky has presented. For example, in *Worlds in Collision,* Velikovsky has a chapter titled "The Fifty-two Year Period" in which he deals with the period between two catastrophes and these are well documented in the cultures of several ancient peoples:

"The works of Fernando de Alva Ixlilxochitl, the early Mexican scholar (circa 1568-1648) who was able to read old Mexican texts, preserve the ancient tradition according to which the multitude of fifty-two year periods played an important role in the recurrence of world catastrophes.

He asserts also that fifty-two years elapsed between two great catastrophes, each of which terminated a world age.

"As I have already pointed out, the Israelite tradition counts forty years of wandering in the desert...and started the difficult task of the conquest, and the time of the battle at Beth-horon twelve years may well have passed...

"Now there exists a remarkable fact: the natives of pre-Columbian Mexico expected a new catastrophe at the end of [every period of] fifty-two years... They watched for the appearance of the planet Venus and when, on the feared day, no catastrophe occurred, the people of Maya rejoiced...this period of Venus, was observed by both the Maya and the Aztecs.

"The old Mexican custom of sacrificing to the Morning Star survived in human sacrifices by the Skidi Pawnee of Nebraska in years when the Morning Star 'appeared especially bright, or in years when there was a comet in the sky'... [Among the ancient Hebrews] the fiftieth year was a jubilee year... The jubilee of the Mayas must have had a genesis similar to that of the jubilee of the Israelites." [42]

It is clear that Velikovsky presents times and catastrophes that do match. But Sagan rather than deal with this, glosses over the material with broad unfounded statements. What does not match is Sagan's remarks about evidence and raises suspicion about his method of analysis. Sagan states,

"Despite copious references, there also seem to me to be a large number of critical and undemonstrated assumptions in Velikovsky's argument. Let me mention just a few of them. There is the very interesting idea that any mythological references by any people to any god that also corresponds to a celestial body represents in fact a direct observation of that celestial body." [43]

This is a complete distortion of Velikovsky's method. As we pointed out earlier with respect to Cuvier's method of employing ancient legends, Velikovsky maintained that only those myths that were widely corroborated by diverse cultures, which tell the same story of the same planet and are possible on the basis of scientific analysis should be considered. On p. 305 of *Worlds in Collision*, Velikovsky specifically wrote: "We shall follow this rule: if there exists a fantastic image that is projected against the sky and that repeats itself around the world, it is most probably an image that was seen on the screen of the sky by many peoples at the same time." Thus, when Sagan states, "I am not sure what one is to do with Jupiter appearing as a swan to Leda, and as a shower of gold to Danae," [44] one need do nothing. In this case the story is *not* told around the world by many cultures, and there is no scientific evidence to observe Jupiter as a swan, so one ignores this evidence as Sagan should.

APHRODITE, ATHENA—PLANET VENUS
Finally Sagan deals with the "Venus Myth",

> "In any case, when Hesiod and Homer refer to Athena being born full-grown from the head of Zeus, Velikovsky takes Hesiod and Homer at their word and assumes that the celestial body of Athena was ejected by the planet Jupiter. But what *is* the celestial body of Athena? Repeatedly it is identified with the planet Venus (Part 1, Chapter 9, and many other places in the text). One would scarcely guess from reading *Worlds in Collision* that the Greeks characteristically identified Aphrodite with Venus, and Athena with no celestial body whatsoever. What is more, Athena and Aphrodite were 'contemporaneous' goddesses, both being born at the time Zeus was king of the gods." [45]

Sagan did indeed point out an error respecting Velikovsky's interpretation of Aphrodite. Sagan states, "On page 247 we hear of Aphrodite, the goddess of the Moon. Who then was Artemis, the sister of Apollo the Sun...?" (BB p. 92) According to Sagan, then, the planet Venus is represented among the Greeks by the goddess Aphrodite, therefore, Athena could not be also, at the same time, identified with the planet Venus. How well does ancient historical and legendary evidence support this view? Bernard Lovell, Professor of Radio Astronomy at the University of Manchester informs us that, "When the ancient observers began to analyze the motion of these planets [Venus and Mercury] they did not realize that the same planets appeared sometimes in the morning sky and sometimes in the evening sky." [46] Peter James, a British scholar of antiquities develops this evidence stating,

> "...Both Aphrodite and Athena were Venus-deities—for Aphrodite we have the incontestable testimony of the Greeks themselves while for Athena we have the evidence by Velikovsky in *Worlds in Collision.*"
>
> "Yet the two goddesses had separate cults and entirely different attributes... Athena was a war goddess while Aphrodite was, by distinct contrast, the goddess of love. Both were Venus deities, but to simply identify the two as being one and the same goddess is impossible. Somehow two separate goddesses, each with her attributes and cult were developed by the Greeks from two separate personifications of the same planet. The key to this apparent dilemma is readily available if we remember that, to the observer, Venus appears as two planets, not one, in its aspects as morning and evening stars.
>
> "This explanation finds ample confirmation in examples drawn from comparative mythology. Istar, the Babylonian goddess of Venus had two distinct aspects and she was unique from amongst her fellow deities. Speaking of the Babylonian pantheon, one scholar (A. Leo Oppenheim in *Ancient Mesopotamia* p. 197) wrote, 'Istar alone stands out because of the dichotomy of her nature, associated with the planet Venus (as morning and evening star) and with divine qualities extremely difficult to

characterize. This complex embraces the functions of Istar as a battle loving, armed goddess, who gives victory to the king she loves, at the same time it links her as the personification of sexual power in all its aspects. In all these roles, she appears in Mesopotamian myths as well as in corresponding texts from the west, from Anatolia to Egypt, under similar or foreign names. This dichotomy of her two contrasting roles has long been understood as being connected with the planet Venus, since the Babylonians considered the morning star as male, reflecting Istar's warlike aspect, and the evening star as female, reflecting her aspect as love-goddess.'"[47]

James goes on to cite Stephen R. Langdon, one of the great author-ities on the history and religion of ancient Mesopotamia from his tome *Semitic Mythology,* Vol. 5 who found this double aspect of the goddess of the ancient East, Venus, unique. This can be said for the Sumerian Venus goddess Istar and the Canaanite Venus goddess, Anath (Athena) etc. James then states,

"We are left with the conclusion that the Greeks personified the morning and evening stars separately, (the morning star, Athena) as the goddess of war, the other (the evening star, Aphrodite) as the goddess of love, much as the Babylonians and Hittites gave their Venus-deity two distinct aspects. Yet, the Babylonians and Hittites had recognized that the goddess, despite her dual nature, was in fact, one planet. In Mesopo-tamia, the science of astronomy had already begun by the time of the Venus catastrophe in the fifteenth century B.C. Recognizing the identity of the morning and evening stars they worshipped both as Istar. The Greeks, on the other hand, comparative youngsters to the civilization of Mesopotamia, only began to develop a science of astronomy in the seventh century B.C. Unaware that the two stars were one and the same, they personified each separately, and the cults of two deities were developed. Confirmation of this suspicion may be found in several explicit statements of ancient writers."

"The Roman author Pliny wrote of Venus, 'When in advance and rising before dawn it receives the name of Lucifer, as being another Sun and bringing the dawn, whereas when it shines after sunset, it is named Vesper as prolonging the daylight, or as deputy for the Moon. This property of Venus was first discovered by Pythagoras of Samos about the 42nd Olympiad, 142 years after the founding of Rome.' (Pliny, *Natural History* 2, 36)"

"Pliny's statement is corroborated by Diogenes Laertius in his essay on Pythagoras: 'It was first declared that the evening and morning stars are the same as Parmenides maintains.' A variant tradition, also reported by Diogenes ascribes the discovery to Parmenides, whom he dated to the 69th Olympiad (504-500 B.C.). Others thought that Ibycus of Rhegium, who flourished in the 61st Olympiad (536-533 B.C.) had made the discovery.

"Whoever was the first to make the discovery, the sources agree that it was not made until the late seventh or sixth century B.C., well after the pantheon of Homer and Hesiod had been established. And if that was the case, then *the conclusion that the Venus-worship of the Greeks would be divided into two separate cults becomes inescapable.*"[48] [emphasis added]

Athena and Aphrodite were both planet Venus deities. Sagan stated, "It does not increase our confidence in the presentation of less familiar myths when the celestial identification of Athena [with the planet Venus] is glossed over so lightly" and "...it is far from prevailing wisdom either now or two thousand years ago, and it is central to Velikovsky's argument."[49] Sagan, therefore, tells us that the identification of Athena with the planet Venus is contradicted by the ancient sources. He states that on page 251 in *Worlds in Collision,* Velikovsky notes that Lucian "is unaware that Athena is the goddess of the planet Venus." Sagan adds, "Poor Lucian seems to be under the misconception that Aphrodite is the goddess of the planet Venus."[50] This is sheer ignorance. The Roman author Pliny who wrote two thousand years ago flatly contradicts Sagan's assertion. Diogenes Laertius who wrote over two thousand years ago also contradicts Sagan on this central point. The evidence of the ancient sources leads directly to the conclusion that the identification of Athena with the planet Venus by Velikovsky is correct. One of the great modern authorities of ancient Greece, Gilbert Murray, Regius Professor of Greek at Oxford University, discussed the identity of the Greek goddess Pallas Athena:

"The case [of the identity] of Pallas Athena is even simpler though it leads to a somewhat surprising result...her whole appearance in history and literature tells the same story as her name... As Pallas she seems to be the thunder-maiden... It seems clear that the old *Archaioi* [Greeks] cannot have called their warrior-maiden, daughter of Zeus, by the name Athena or Athenaia... If we try to conjecture whose place it is that Athena has taken, it is worth remarking that her regular epithet 'daughter of Zeus' belongs in Sanskrit to the Dawn-goddess, Eos."[51]

Eos is identified as the mother of the morning star, or the planet Venus.[52]

Also on this point there are other identifications of Athena with the planet Venus. Some of the ancient Semitic cultures had goddesses for Venus with clear cognate names of Athena. Athena of the Greeks is derived from and the same as the goddess "Ana-hita" of the Persians, who was the planet Venus. Athena of the Greeks is derived from and the same as the goddess "Anath" of the Canaanites who is the planet Venus. Athena of the Greeks is the same as "Anat" of the Babylonians who is the planet Venus. Linguistically and mythologically this is well known among scholars of middle-eastern, ancient mythology.

In fact, the *Encyclopedia Britinnica, Micropaedia,* Vol. I, (London, 1982), p. 336, makes it quite clear that Anahita who is Athena is also the same as Aphrodite—the Greek Venus. It states that *"In Greece Anahita was identified with Athena and Aphrodite"* [emphasis added]. We therefore, have one goddess identified as both Athena and Aphrodite which is celestially identified as the planet Venus. Thus, the ancient Greeks in complete contradiction to Sagan believed that both Athena and Aphrodite were associated with the planet Venus. The same dichotomy is also found for the planet Mercury in it two aspects as morning and evening star. Guy Murchie in *Music of the Spheres,* Vol. I (NY 1967), p. 77, tells us that "Although the Greeks named him Mercury when they saw him setting just after sunset, some of them also called him Apollo when he rose at dawn, even though the better educated among them were well aware that he was one and the same." Thus, the Greeks identified Venus and Mercury with more than one god.

Sagan adds, "There may be good justification, for all I know, in identifying Athena with Venus."[53] The justification has been in print for several years before *Broca's Brain* was published and it seems that Sagan is somehow unwilling to admit forthrightly this identification and thus, it seems that the unwary reader has been disinformed by the evidence in Sagan's book.

PALLAS—TYPHON

In pursuit of Velikovsky and the Venus Myth, Sagan claims that Velikovsky has, "given extremely inadequate justification [for] the contention... (p. 85) 'as is known, Pallas was another name for Typhon.'"[54] Part of the literature which Sagan claims to have read "carefully" prior to the AAAS symposium on Velikovsky is a major work, *The Velikovsky Affair* which incensed the public about the shabby manner in which the scientific establishment had treated Velikovsky. In it, Livio Stecchini, a historian and philosopher of science, reported on Father Franz Xavier Kugler. Kugler was an authority of ancient astronomy and mythology. In 1927 he published a small book *Sybillinischer Sternkampf und Phaethon in naturgeschichticher Beleuchtung,* which in English is, "Sybilline Battle of the Stars and Phaethon Seen as Natural History." The book deals with Venus. Kugler stated that Venus as a "sun-like meteor" approached the Earth and caused a cosmological crisis—a catastrophe. However, Stecchini pointed out that "Wilhelm Grundel, a specialist in Hellenistic astromythology, in his review of Kugler's book sharply rebuked Kugler for not mentioning that all the texts similar to those examined by Kugler ascribed the catastrophe to a comet, and specifically to the comet

Typhon."[55] Thus, both Kugler and Grundel well knew "Typhon is the planet Venus." Peter James, as we noted earlier, cited detailed, ancient evidence that the goddess Pallas Athena is the planet Venus; and Kugler and Grundel tell us Venus is Typhon. Thus, the identification of Pallas Athena with Typhon is made complete. Had Sagan truly read this material carefully he would know this. This material is stated flatly by Stecchini. "Athena who was the planet Venus"[56] and James "Athena who was the planet Venus".[57] Both knew Athena was Typhon.

J. Norman Lockyer in his book, *The Dawn of Astronomy*, (Cambridge MA, 1964), which is a copy of the 1884 first edition, on p. 361 writes, "I suppose that there is now no question among Egyptologists that the gods Set...[and]...Typhon are identical." However, Robert Graves in *The Greek Myths*, Vol. I, (Baltimore 1955), pp. 153-154 informs us that in Egypt "Anat or Anatha was confusingly identified with...Set." Thus, again we find that: Set is the same as Typhon who is the same as Pallas Athena or that Typhon is Pallas Athena.

However, there is another modern authority on Greek mythology who read *Worlds in Collision*. Moses Hadas, Jay Professor of Greek at Columbia University" [stated] "...I know that he [Velikovsky] is not dishonest. What bothered me was the violence of the attack upon him."[58] "Hadas had remarked in a published book review that 'in our time Immanuel Velikovsky...appears to be approaching vindication.'"[59] Hadas gives several examples of...misrepresentations of Velikovsky's correct quotations and writes, "It is his critic, not Velikovsky, who is uninformed and rash..."[60] Hadas *had read Velikovsky carefully* and was not so ignorant as to believe that Velikovsky falsely claimed that Pallas was another name for Typhon, especially when an entire chapter is devoted to this identification. Sagan thus ignores the evidence and this is as near as he approaches the "Venus Myth" delineated by Velikovsky.

METEORITE THUNDER

Sagan states, "the statement (p. 283) [in *Worlds in Collision*] that 'Meteorites when entering the Earth's atmosphere, make a dreadful din,' when they are generally observed to be silent."[61] Velikovsky answered this, remarking that George P. Merrill, Head Curator, Department of Geology, U.S. National Museum, part of the Smithsonian Institution, wrote,

> "a long series of reports of loud explosions accompanying the fall of meteorites. Meteorites are a subject that belongs in Sagan's own field, but he does not know that they can make noise. For example, in Emmet County Iowa, on May 10, 1879, 'The sounds produced by the

explosions incidental to its [the meteor's] breaking up were referred to as terrible and indescribable... The first explosion, for there were several, was louder than the loudest artillery.' This is only one of a number of illustrative cases described by the Smithsonian Institution."[62]

Lewis M. Greenberg informs us that,

"Opening the Nov. 1979 *Griffith Observer* [journal of astronomy] to page 9, one reads: 'A typical meteorite fall produces a brilliant fireball or meteor, leaves a smoke trail, and creates a series of sonic booms resembling the sounds of firing cannon, or of thunderclaps.' And, in July 1977, Madagascar reported a meteorite fall that was accompanies by noise 'variously described as sounding like sonic booms, artillery shots, bomb explosions or quarrying detonations' (*Science News,* Vol. 112, 8/6/77, p. 86 and 8/13/77, p. 102."[63]

This kind of astronomical evidence has been known for a long time. Even the ancients knew that meteors can make loud explosive noise. In *Planet Earth,* Jonathan Weiner informs us that, "Pliny the Elder, the Roman naturalist writing in the first century A.D., called the falling rocks 'thunderstones' because he said, they make a great roar when dropped from the sky." [64] And neither is Sagan so naive an astronomer not to know this extremely well known fact that meteors create explosive noise. In his book, *Comet,* he writes, "Meteorites... *can* be heard; they and the fireballs produce on occasion a *sonic boom or a deep rumbling roar...*"[65] [emphasis added] Sagan feels that Velikovsky has given extremely "inadequate justification" for his statements that "Meteorites when entering the Earth's atmosphere make a dreadful din." Why should Velikovsky justify what all competent astronomers know?

LIGHTNING AND MAGNETS

Next Sagan wishes Velikovsky to justify his claim that "a thunderbolt when striking a magnet, reverses the poles of a magnet."[66] What Velikovsky had postulated is that the Earth's magnetism can be reversed by immense cosmological lightning strokes. In fact, the very year that Sagan first delivered his paper on Velikovsky, 1974, Michael Purcker published a paper in *EOS* on just this question. Purcker's paper, "Lightning Strike in Sandstone", deals with magnetism in rocks. Rocks as they cool from a molten state will allow the iron particles that they contain to become aligned with the Earth's magnetic field. The magnetism is termed *remanent* magnetism. Even a sedimentary rock such as sandstone if heated and cooled slowly does this. However, according to Purcker the direction of the magnetic field will change if the rock is struck by a thunderbolt.[67] Not only that, but Purcker cites two earlier papers that document that lightning changes the magnetic direction in other types of rocks. These articles are by Cox A; "Anomalous Remanent Magnetization of Basalt," in the *U.S. Geological Survey Bulletin,* No. 1083E, (1961) and Graham K.W.T.; "Re-Magnetization of a Surface Outcrop by Lightning Curr..." *Geophysical Journal,* Vol. 6 (1961), pp. 85-102.

S.K. Runcorn of the University of Newcastle upon Tyne in *Scientific American,* Vol. 257 (Dec. 1987), p. 65 categorically states that, "...a lightning bolt can magnetize a rock outcrop." But Sagan does not seem to know this. Therefore, Velikovsky's advice that "Sagan wonders that a thunderbolt, when striking a magnet, reverses the poles of the magnet. This explains the reversals in paleomagnetism (*Worlds in Collision,* pp. 114-115). If Sagan has doubts, let him perform an experiment." [68]

HAIL OF BARAD

Sagan (B.B. p. 92) questions Velikovsky's view that "the translation (p. 51) of 'Barad' as meteorites." Velikovsky claimed that the hail stones that fell on Beth-horon and Egypt, described in the *Bible* as stones of "barad", were hot and thus were meteorites and not hail. Their fall was accompanied by loud thundering noises which is also in accord with meteorites, but not hail. Lewis M. Greenberg, *Kronos V,* 2, p. 91, explains that on page 9 of the astronomy journal *Griffith Observer,* is a "reference to a passage from the *Book of Joshua* (10:11) used to support the idea that people have been killed by falling mete - orites. Velikovsky cited the same passage in *Worlds in Collision,* (p. 42) when he discussed the fall of meteorites and introduced the word *barad* for the first time." (If Sagan's colleagues at the Griffith Observatory translates the word *barad* as meteorite why shouldn't Velikovsky do the same?

Willy Ley in *Watchers of the Sky,* (NY 1966) p. 233 reinforces Velikovsky's concept that *barad* should be translated as meteorites:

> "Moreover stones fallen from the sky were mentioned in the *Bible.* If *Joshua* 10:11 stated that '...as they fled before Israel and were in the going down to Beth-horon... The Lord cast down great stones [translated *barad*] from heaven upon them...and they died.' then stones did fall from the sky."

Furthermore J.B. Biot, a member of the French Academy described the spectacular shower of meteorites at L'Aigle in northern France in much the same manner as the Hebrews of the hail of *barad* that fell in Egypt. According to Biot "The inhabitants say that they saw them [meteorites] descend along the roofs of the houses like hail, break the branches of the trees, and rebound after they fell on the pavement." [J.B. Biot, *Philosophical Magazine* [Tilloch's] Vol. 16 (1803). "Account of a Fireball Which Fell in the Neighborhood of L'Aigle: In a letter to the French Minister of the Interior." pp. 224-228.]

SAGAN'S PRINCIPLE

Sagan claims that "On page 179 a principle [in SCV], is enunciated, [in BB] is implied that when two gods are hyphenated in a joint name, it indicates an attribute of a celestial body—as, for example, Ashteroth-Karnaim, a horned Venus." [69] In *Scientists Confront Velikovsky,* Sagan was so sure of his evidence that he claimed that the hyphenated name principle was "enunciated." Then when he rewrote his piece of *Broca's Brain,* he wasn't so sure after all; he changed the "enunciated" principle into an "implied" principle. If Sagan wishes to be more accurate, the hyphenated name principle should be an "inferred" principle and to be quite precise a "misinferred" principle; it seems that Sagan needed his hyphenated principle to ask "But what does this principle imply, for example, for the god Amon-Ra? Did the Egyptians see the sun (Ra) as a ram (Ammon)?" [70] Again, if Sagan wishes to be quite accurate, he should understand as Velikovsky pointed out long ago, that "Amon" was the planet Jupiter." [71] Sagan's principle does nothing for anything.

THE CRESCENT SHAPE OF VENUS

However, in discussing hyphenated names, Sagan states that "...Ashteroth-Karnaim, a horned Venus which Velikovsky interprets as a crescent Venus and evidence that Venus was once close enough to the Earth to have it phases discernible to the naked eye." [72] But it is also true that the crescent shape of Venus has been observed by many people especially in the low latitudes. Gary E. Hunt and Patrick Moore discuss this in their book, *The Planet Venus* wherein we find,

> "There are many cases on record of the naked eye visibility of the crescent. [of Venus] In the clear skies of South America, Lieutenant Gillis recorded it on various occasions between 1849 and 1852. Between 1929 and 1935 it was recorded unmistakable by Carl Reinhardt, D. Howell, H.W. Cornell and Dr. and Mrs. F.W. Wood, Miss M.A. Blagg well-known for her work in connection with the Moon, was unable to make out the crescent shape, but could see that Venus was definitely elongated. The Rev. T.W. Webb relates that the crescent phase was seen by a twelve year old boy, Theodore Parker, before he knew of its existence, while W.S. Franks, a winner of the Gold Medal of the Royal Astronomical Society said that his son, E.S. Franks, had frequently seen the crescent between 1890 and 1900. All these cases are well authenticated, and there seems, therefore, little doubt that the phase really is visible to people with exceptional eyesight." [73]

Therefore, if people can distinguish the crescent phases of Venus when Venus is in its present distant orbit, why shouldn't people of ancient times, when Venus' orbit, according to Velikovsky, was elliptical and hence brought it nearer to Earth and easier to distinguish, have observed clearly what is currently difficult?

Accordingly, Livio C. Stecchini in *The Velikovsky Affair*, (NY 1966) p. 88, states "Sir Walter Raleigh in his *History of the World* (1616) wondered how it could happen that the phases of Venus just discovered by Galileo seem to have been known to ancient authors." Thus, the phases of Venus are remarked and were noted by ancient writers.

In fact, Velikovsky cites ancient sources that show Venus' crescent phase was observed by ancient men as the head of a bull with horns. On pp. 166-7 of *Worlds in Collision*, he writes and cites some of these:

"Sanchoniathon says [Cf. L. Thorndike, *A History of Magic and Experimental Science*, (1923-1941) I Chap. X] that Astarte (Venus) had a bulls' head...

"Tistrya (Venus) of the Zend-Avesta (Trans. James Darmesteter, 1883), Pt. II, p. 93) the star that attacks the planets the bright and glorious Tistryra [Venus] mingles his shape with light moving in the shape of a golden-horned bull."

"The Egyptians similarly pictured the planet (Venus) and worshipped it in the effigy of a bull. (See E. Otto, *Beiträge zur Geschichte der Stierkulte in Agypten* (1938) The cult of a bull sprang up also in Mycenanean Greece. A golden cow head with a star on its brow was found in Mycenae on the Greek mainland..."

Thus, the ancient writers corroborate the evidence that they could see the phases of Venus.

THE BIBLE

Sagan writes,

"There is a contention (p. 63, [*Worlds in Collision*]) that instead of the tenth plague of the Exodus killing, the 'firstborn' of Egypt, what is intended is the killing of the 'chosen.' This is a rather serious matter and at least raises the suspicion that when the *Bible* is inconsistent with Velikovsky's hypothesis, Velikovsky retranslates the *Bible*. The forgoing... may...have simple answers, but the answers are not found easily in *Worlds in Collision*."[74]

What Sagan had difficulty finding in *Worlds in Collision* is found on page 63, the same page he cited. In fact, the answer about "first born" and "Chosen" is in the very passage Sagan cited. Here Velikovsky wrote, "In *Ages in Chaos* (my reconstruction of ancient history), I shall show that 'firstborn' (*bkhor*) in the text of the plague is a corruption of 'chosen' (*bchor*). All the flower of Egypt succumbed in the catastrophe." This is indeed serious and raises the suspicion that Sagan attempted to hide Velikovsky's evidence.

When we turn to *Ages in Chaos,* there is indeed a chapter titled "Firstborn or Chosen" on pp. 32–34. It states,

"The biblical story of the last plague has a distinctly supernatural quality in that all the firstborn and only the firstborn were killed on the night of the plagues. An earthquake that destroys only the firstborn is inconceivable, because events can never attain that degree of coincidence. No credit should be given to such a record.

"Either the story of the last plague, in its canonized form, is a fiction, or it conceals a corruption of the text. Before proclaiming the whole a strange tale interpolated later, it would be wise to inquire whether or not the incredible part alone is corrupted. It may be that the firstborn stands for some other word.

"*Isaiah* 43:16 Thus saith the Lord, which maketh a way in the sea, and a path in the mighty water;

20…I give waters in the wilderness, and rivers in the desert to give drink to and my people, my chosen.

"In the *Book of Exodus,* it is said that Moses was commanded:

"*Exodus* 4:22-23 And thou shalt say unto Pharaoh, Thus saith the Lord, Israel is my son, even my firstborn.

…and if thou refuse to let him go, behold, I will slay thy son, even thy firstborn.

"The 'chosen' are here called 'firstborn.' If Israel was the firstborn, then revenge was to be taken against Egypt by the death of its firstborn. But if Israel was the chosen, then revenge was to be taken against Egypt by the death of its chosen.

'Israel my chosen,' is *Israel bechiri or bechori.*

'Israel my firstborn,' is *Israel bekhori.*

"It is the first root which was supposed to determine the relation between God and his people. Therefore: 'at midnight the Lord smote all the firstborn in the land of Egypt' (*Exodus* 12:29) must be read 'all the select of Egypt' as one would say 'all the flower of Egypt' or 'all the strength of Egypt.' 'Israel is my chosen: I shall let fall all the chosen of Egypt.'

"Naturally death would usually choose the weak, the sick, the old. The earthquake is different; the walls fall upon the strong and the weak alike. Actually the *Midrashim* say that 'as many as nine tenths of the inhabitants had perished.'

"In *Psalms* 135 my idea is illustrated by the use of both roots where two words of the same root would have been expected.

"For the Lord hath chosen Jacob unto himself, and Israel for his peculiar treasure…who smote the firstborn of Egypt.

"In *Psalms* 78 the history of the Exodus is told once more.

> "*Psalms* 78:43 How he hath wrought his signs in Egypt.
> 51 And smote all the firstborn in Egypt...
> 52 But made his own people to go forth...
> 56 Yet they tempted and provoked the most high God...
> 31 The wrath of God came upon them, and slew the fattest of them, and smote down the chosen men of Israel...

"Were the firstborn destroyed when the wrath was turned against Egypt, and were the chosen destroyed when the wrath was turned against Israel?

> "*Amos* 4:10 I have sent among you the pestilence [plague] after the manner of Egypt: your young men [chosen] have I slain.

"In the days of *raash* (commotion) during the reign of Uzziah, the select and the flower of the Jewish people perish as perished the chosen, the strength of Egypt was the prophecy of *Amos*.

"It is possible that the king's firstborn died on the night of the upheaval. The death of the prince would have been an outward reason for changing the text. The intrinsic reason lies in the same source that interrupted the story of the Exodus at the most exciting place—after the houses of the Egyptians had crumbled—with these sentences:

> "*Exodus* 13:2 Sanctify unto me all the firstborn, whatsoever openeth the womb among the children of Israel, both of man and of beast: it is mine
> 13 ...and all the firstborn of man among thy children shalt thou redeem."

"Jeremiah testifies to the fact that burnt offerings and sacrifices were not ordered on the day Israel left Egypt.

> "*Jeremiah* 7:22 For I spake not unto your fathers, nor commanded them in the day that I brought them out of the land of Egypt, concerning burnt offerings or sacrifices.

"This is in contradiction to the text of *Exodus* 12:43 to 13:16. To free the people from this bondage is the task of Amos, Isaiah and Jeremiah.

> "*Amos* 5:22 Though ye offer me burnt offerings and your meat offerings, I will not accept them: Neither will I regard the peace offerings of your fat beasts.
> 24 But let judgment run down as waters, and righteousness as a mighty stream.
> 25 Have ye offered unto me sacrifices and offerings in the wilderness forty years, O house of Israel?"

Sagan has accused Velikovsky of changing the *Bible*, when it is in contradiction to his hypothesis. Yet here we see that Velikovsky, contrary to Sagan's assertion, goes to the *Bible* to show that there is ample evidence in it for "firstborn" to be read as "chosen". In fact, Velikovsky cites several places in the *Bible* to illustrate just this point.

Velikovsky showed his manuscript of *Ages in Chaos* to an acknowledged expert and biblical scholar. In the "Acknowledgments" p. XIII of *Ages in Chaos,* we find,

> "I am also indebted to Dr. Robert H. Pfeiffer, outstanding authority on the *Bible.* Director of the Harvard excavation at Nuzi, curator of the Semitic Museum at Harvard University, professor of ancient history at Boston University, editor of the *Journal of Biblical Literature* (1943-1947), and author of a distinguished standard work on the *Old Testament,* he is eminently qualified to pass judgment. In the summer of 1942, when the manuscript was still in its first draft, he read *Ages in Chaos...* He read later drafts, too, and showed a great interest in the progress of my work. Neither subscribing to my thesis nor rejecting it, he kept an open mind, believing that only objective and free discussion could clarify the issue."[75]

Pfeiffer wrote the following on *Ages in Chaos,* "Dr. Velikovsky discloses immense erudition and extraordinary ingenuity. He writes well and *documents all his statements with original sources.*"[76] [emphasis added] While Sagan maintains that Velikovsky's use of the *Bible* "...is a serious matter and at least raises the suspicion that when the *Bible* is inconsistent with his hypothesis, he retranslates the *Bible*"; it appears that Sagan's approach, on the other hand, is not serious and at least raises the suspicion that when his assertions are inconsistent with the documented evidence presented by Velikovsky, Sagan ignores the documented evidence.

Sagan's scholarship of the *Bible* is itself curious. In *Scientists Confront Velikovsky,* he writes, "at the moment that Moses strikes his staff upon the rock, the Red Sea parts."[77] Velikovsky had to correct Sagan's usage of the *Bible* stating "In the Biblical story, Moses did not hit the rock with his rod at the Sea of Passage; the striking of the rod against the rock is from the story of finding water in the desert."[78] Therefore, Sagan corrected it in *Broca's Brain.* Velikovsky tells us that "Biblical scholarship is not Sagan's field." This is quite certain. However, one of Velikovsky's loudest critics, Patrick Moore, the British astronomer, remarks, "All his [Velikovsky's] three books are heavily annotated and every Biblical reference is correct. In fact, Dr. Velikovsky has done his homework extremely well in this respect."[79] What a pity Sagan has not done his homework at all in this regard.

EARTH IN UPHEAVAL

Sagan then states that he finds "the situation in legend and myth...fuzzy" and that "any corroboratory evidence from other sources would be welcome by those who support Velikovsky's argument."[80] In fact, there *is* corroboratory evidence from geology, archaeology, and paleontology. It is contained in Velikovsky's book, *Earth in Upheaval*. Albert Einstein supplied Velikovsky with marginal notes and handwritten comments on chapters VIII through XII and he had read the entire text. Velikovsky in the 'Author's Note' states,

> "As early as the 1960's, I found that *Earth in Upheaval* was displacing *The Origin of Species* in the courses of a number of geophysicists—as in the case of my visit to Oberlin College in 1965. At Princeton University, *Earth in Upheaval*, from its publication and for two decades was required reading in the paleontology course of Professor Glenn Jepsen. H.H. Hess, Chairman of the Department of Geology (later Geophysics), told me that he knew *Earth in Upheaval* by heart."[81]

Here is what Norman Macbeth in *Darwin Retried* states about Velikovsky's *Earth in Upheaval*.

> "...he [Velikovsky] marshals the original field reports on a large number of phenomena that point inexorably to catastrophes...of fairly recent dates... The impact of the details and the number of phenomena (close to forty) is shattering. I hold no brief for Velikovsky's theories, but I am indebted to him for collecting material that has never been assembled in one place before. The topics in the book are discussed on the basis of reports by orthodox and reputable scientists with Velikovsky merely acting as master of ceremonies."[82]

After discussion and summary of some of the topics Velikovsky presented Macbeth adds that, "The reader should peruse Velikovsky himself so as to get the cumulative effect of his evidence... The wealth of specific cases pointing toward catastrophes make it impossible for me to accept the uniformitarian theory."[83] This seems splendid advice for Sagan who would welcome corroboratory evidence from sources other than *Worlds in Collision*.

Velikovsky presented a highly cursory description of some of the topics of *Earth in Upheaval* in his book *Stargazers and Gravediggers*:

> "When *Worlds in Collision* was published, numerous scientists repeatedly claimed that events of such magnitude and at such comparatively recent dates must have left vestiges not only in folklore, but even more so in geology and archaeology. Actually in the Epilogue to *Worlds in Collision*, I wrote 'Geological, paleontological and anthropological material related to the problem of cosmic catastrophes is vast and may give a complete picture of past events no less than historical material.' My new book, *Earth in Upheaval*, published in 1955, was a collection of this material, where I brought together evidence from geology, paleontology and archaeology. I excluded from this book every reference to ancient literature, traditions and folklore; and this I did purposely so that careless critics would not decry the entire work as 'tales and legends.'
>
> "I could show—always quoting academic sources—that the level of all oceans dropped suddenly thirty-four centuries ago; that mountains rose

in spasmodic movements in the time of advanced man, who developed advanced cultures and built cities. Abandoned cities like Tiahuanacu, and agricultural terraces, are now covered with perennial snow. The deserts of Arabia, Sahara and Gobi were covered by forests and pastures, and man's neolithic relics and rock drawing show how recently these wastes were richly watered and were inhabited. The remains of whales are found on mountains; fig trees and corals are found in polar regions and signs of ice in Equatorial Africa. Widespread extinctions in America occurred 'virtually within the last few thousand years.'

"I gave the history of the theory of catastrophism versus the theory of gradualism and evolution. The Agassiz theory of the ice ages was originally also a catastrophic theory. Agassiz spoke of the sudden arrival of the ice cover seizing the mammoths of Siberia. The north Siberian islands consist of trunks of uprooted trees and bones of mammoths, rhinoceroses, horses and buffaloes—when today lichen and moss show themselves for two months in a year—and sea is fettered in ice from September to July. In Alaska, too, gold digging machines, slicing the ground by the mile, disclosed all over the peninsula immense heaps of animals of species both extinct and extant, forms that do not belong together in a melee with millions of broken and uprooted trees.

"The fissures of rocks in Britain, France, Spain and also the Mediterranean islands are filled with bones of animals—and their state and position suggest that the land and the sea repeatedly changed places. Also on the American continent, North and South, caverns in the hills are found filled with animals of various habitats, entombed in conditions of catastrophe. Actually, Darwin could be quoted from his *Journal of the Voyage of the Beagle*. After observing the immense heaps of fossil bones in South America, he wrote: 'The greater number, if not all, of these extinct quadrupeds lived at a late period... Since they lived, no very great change in the form of the land can have taken place. What, then, has exterminated so many species and whole genera? The mind at first is irresistibly hurried into the belief of some great catastrophe; but thus to destroy animals, both large and small, in Southern Patagonia, in Brazil, on the Cordilleras of Peru, in North American up to the Behring's Straits, we must shake the entire framework of the globe' [*Voyage of the Beagle*, Charles Darwin, Appleton & Co., pp. 169-170]...

"Actually poles were displaced and the terrestrial axis did shift under violent conditions. In this connection, in Chapter IX—'Axis Shifted'— of *Earth in Upheaval* (published in November) it was possible to quote a very recent article, 'The Earth's Magnetism' by Professor S.K. Runcorn of Cambridge, which appeared in the September 1955 issue of *Scientific American*... In it he wrote that the lavas and igneous rocks in various parts of the world disclose that during the Tertiary period 'The North and South geomagnetic poles reversed places several times...' After long periods of stability 'the field would suddenly break up and reform with opposite polarity.'

"The unavoidable conclusion according to Runcorn is that, 'the earth's axis of rotation had changed also. In other words, the planet had rolled about, changing the location of its geographical poles.'"[84]

Kenneth Hsu in his book, *The Great Dying,* discusses the scientific establishment's attitude toward the theory of catastrophism. Hsu who rejects Velikovsky's hypothesis, writes, (p. 41) "To scientists, the notion of an unusual catastrophic event to explain phenomena in Earth history has become paramount to invoking the supernatural. My beloved teacher, Ed Spiecker of Ohio State [University], went so far as to exclaim that the very word revolution should be expunged from all geology textbooks..."[85] Since Sagan cannot expunge *Earth in Upheaval,* he acts as if it does not exist. As an approach to evidence, Sagan's attitude is completely unscientific!

COMETS AND SWASTIKAS

Sagan goes on,

"I am struck by the absence of any confirming evidence in art. There is a wide range of paintings, bas-reliefs, cylinder seals and other *objects d'art* produced by humanity [in SCV: going back tens of thousands of years B.C.] [in BB: going back to at least 10,000 B.C.] They represent all of the subjects...important to the cultures that created them."[86]

Velikovsky answered Sagan succinctly on this point, stating,

"Sagan moves to cave painting (where he finds only a picture of a supernova) and to ancient art generally, and asks: 'If the Velikovskian catastrophes occurred, why are there no contemporary graphic records of them?' As a novice in the field, Sagan should perceive that the great majority of ancient contemporary art is *dominated* by the theme of global catastrophes and celestial planetary deities in battle. In my lecture I referred to the Mayan, Olmec and Toltec art—and whoever visits Yucatan knows that virtually *no other theme* exists in this art. No dynastic or military exploits, but battles between planetary deities, and sacrifices to them—almost to the exclusion of other themes. The caveman pictures animals in global conflict; serpents fighting planets *are* a frequent theme in cave and mural art; and in literary art—from the *Iliad* to the Assyrian prayers, to the *Old Testament,* its prophets and psalms, to Hindi and to Icelandic epics—it (celestial catastrophe) is the all-pervading motif. So it goes in this domain which is foreign to Sagan."[87]

In this respect, we briefly examine the "Venus Myth". Sagan states in *Broca's Brain* about Velikovsky's theory that "The planet Jupiter disgorged a large comet" (p. 93); "that Venus was once close enough to Earth to have its [appearance] discernible to the naked eye" (p. 92). If this is so, then there should exist *"objects d'art* produced by humanity" to describe this. In Sagan's book *Comet,* he presents drawings of comets made by ancient man on pp. 20, 168 and 186.

However, Sagan fails to present ancient Babylonian drawings of Venus. Inanna was the Babylonian goddess who personified Venus and the Babylonians produced pictures of this goddess as she appeared in the sky. A. Falkenstein has twelve of these Venus drawings in his book, *Archaische Texte Aus Uruk,* (Ancient Texts of Uruk), Leipzig, 1936. I let the reader decide whether or not these pictures of Venus are "confirming evidence" that the Babylonians observed Venus as a comet. Lynn E. Rose has collected, organized and presented these in, "Just Plainly Wrong", *Kronos,* III:2, p. 111.

Somehow Sagan is unable to bring himself to display this evidence in his book *Comet.* Actually, Velikovsky presented a great many statements of ancient peoples from all over the world. Let us peruse some of the evidence. The footnotes that Velikovsky used will be noted with the citation of [*W in C*].

"The early traditions of the people of Mexico, written down in pre-Columbian days, relate that Venus smoked. 'The star that smoked, *la estrella que humeava,* was Sitlae choloha, which the Spaniards call Venus.' [1]

'Now I ask,' says Alexander Humboldt, 'what optical illusion could give Venus the appearance of a star throwing out smoke?' [2]

"Sahagun, the sixteenth century Spanish authority of Mexico, wrote that the Mexicans called a comet 'a star that smoked.' [3] It may thus be concluded that since the Mexicans called Venus 'a star that smoked' they considered it a comet.

"It is also said in the *Vedas* that the star Venus looks like fire with smoke [4] ...in the *Talmud,* in the Tractate Shabbat: 'Fire is hanging down from the planet Venus.' [5] This phenomenon was described by the Chaldeans. The planet Venus 'was said to have a beard.' [6] This same technical expression ('beard') is used in modern astronomy in the description of comets.

"These parallels in observations made in the valley of the Ganges on the shores of the Euphrates, and on the coast of the Mexican Gulf prove their objectivity...

"Venus, with its glowing train, was a very brilliant body; it is therefore not strange that the Chaldeans described it as a 'bright torch of heaven,'[7] also as a 'diamond that illuminates like the sun,' and compared its light with the light of the rising Sun. [8]

"At present, the light of Venus is less than one millionth of the light of the Sun. 'A stupendous prodigy in the sky,' the Chaldeans called it. [9]

"The Hebrews similarly described the planet: 'The brilliant light of Venus blazes from one end of the cosmos to the other.' [10]

"The Chinese astronomical text from Soochow refers to the past when 'Venus was visible in full daylight and, while moving across the sky, rivaled the sun in brightness.' [11]

"As late as the seventh century [B.C.], Assurbanipal wrote about Venus (Ishtar) 'who is clothed with fire and bears aloft a crown of awful splendor'[12] The Egyptians under Seti thus described Venus (Sekhmet): 'A circling star which scatters its flame in fire...a flame of fire in her tempest.'[13]...the Mexican...also called it by the name of Tzontemocque, or 'the mane.'[14] The Arabs called Ishtar (Venus) by the name Zebbaj or 'one with hair,' as did the Babylonians. [15]

"'Sometimes there are hairs attached to the planets,' wrote Pliny;[16] an old description of Venus must have served as a basis for his assertion. But hair or *coma* is a characteristic of comets, and in fact 'comet' is derived from the Greek word for 'hair.' The Peruvian name 'Chaska' (wavy haired)[17] is still the name for Venus though at present the Morning Star is definitely a planet and has no tail attached to it."88

These reports, and others, come from the entire globe. In Sagan's book *Comet,* pp. 181-187 are devoted to an analysis of the swastika and how this symbol came to be found among all peoples of ancient times. Sagan discusses the difficulties of explaining this symbol which he informs us "appears to be connected with something brilliant in the sky, and on the other hand it is clearly something separate from the Sun."89 Sagan goes on to explain,

"...these difficulties seem to be resolved if there once was a bright swastika rotating in the skies of Earth, witnessed by people all over the world. Ordinarily, the notion seems far from *astronomical reality...* [emphasis added]

"What we are imagining is something like this: It is early in the second millennium B.C... While all the people on Earth are going about their daily business, a rapidly spinning comet with four active streamers appears. When the people look up at the comet, they are looking down on the axis of rotation. The four jets, symmetrically placed around the equator on the daylight side, generate—because of the comet's rapid rotation-curved streamers, as you can easily see in the pattern formed by a rotary garden sprinkler [which give]...the usual representation of the swastika."90

To cinch this analysis, Sagan reports,

"Under these circumstances it is arresting to find, in the culture with the longest tradition of careful observation of comets, a straightforward, apparently unambiguous description of a swastika as just another comet. Such is the case of the twenty-ninth and final comet to appear in the ancient silk atlas of cometary forms that was unearthed in a Han Dynasty tomb at Mawangdui, China (Chapter 2). It dates from the third or fourth century B.C..."[91]

And, indeed, on the same page as the above comment Sagan has presented the Chinese drawing of the comet in the form of a swastika.

When we recall that we have a series of drawings of Venus in the shape of a comet, made by the Babylonian culture with a very long tradition of careful observations, giving a straight-forward, apparently unambiguous description of Venus as a comet, we see that Velikovsky's theory is underpinned by exactly the same sort of evidence that Sagan employs to explain the swastika; except Velikovsky has twelve drawings to support his view, not just one and has presented a volume of evidence, not just a few pages. Thus, it is interesting to see that Sagan does not deal with this evidence.

In this respect, it is interesting to note that Sagan describes Quetzalcoatl on page 28 of his book *Comet,* as "the great white-bearded god" and elsewhere he writes, "The Tshi people of Zaire call comets 'hair stars' and the word comet—the same in modern languages—comes from the Greek work for hair."[92] Quetzalcoatl, the god of the ancient Mexicans having a "great white beard" according to Sagan would be a comet. But nearly any encyclopedia will inform the reader that, "Quetzalcoatl [was] god of civilization [and] of the planet Venus."[93] Thus, even in Sagan's book *Comet,* he has unwittingly and indirectly given evidence to support Velikovsky's hypothesis that ancient man saw Venus as a comet. By the way, Quetzalcoatl means "feathered serpent" which to ancient people looking into the sky would be a good interpretation of a comet. As Sagan states on the same page [14] of *Comet,* "In other cultures they are 'tail stars' or 'stars with long feathers.'"

Therefore, when Sagan states, "I do not mean to suggest that all of Velikovsky's legendary concordances and ancient scholarship are... flawed but many of them seem to be, and the remainder may well have alternative, for example, diffusionist origins,"[94] it is again suggested that Sagan read his own work more carefully.

FROGS, FLIES, VERMIN
In his description of Velikovsky's hypothesis Sagan writes, "The vermin described in *Exodus* [according to Velikovsky] are produced by the comet-flies and perhaps scarabs drop out of the comet, while indigenous terrestrial frogs are induced by the heat of the comet to multiply."[95] L. M. Greenberg pointed out that,

"Sagan on December 2, 1973 before a group of scientists at a NASA Ames Research Center news conference. At the later get-together, Sagan said, 'Velikovsky explicitly [sic] predicts the presence of frogs and flies in the clouds of Jupiter...' [So first Sagan says Velikovsky has frogs on Jupiter and presumably on comet Venus, then he removes the frogs at the AAAS symposium.] ...When questioned about his remark concerning 'Velikovskian Frogs' by Thomas Ferte, in a letter dated February 1974, Sagan replied (letter dated March 6, 1974) that 'Velikovsky is equivocal about frogs, but quite explicit [sic] about flies.'... (see *CHIRON*, T. Ferte, "Velikovskian Frogs: The Unscientific Reception of *Worlds in Collision*" (1950-1970) Vol. 1, Nos. 1 & 2, Winter-Spring 1974, p. 12).

"In case anyone is confused by Sagan's critical legerdemain, let us recapitulate: 1) In December of 1973, Sagan publicly claims that Velikovsky ascribes frogs to the Jovian clouds; 2) In February of 1974, Sagan properly refers to the frogs as being 'indigenous terrestrial'; 3) In March of 1974 Sagan takes a middle ground and says that 'Velikovsky is equivocal about frogs.'[96]

Therefore, one can only agree wholeheartedly with Sagan when he states, "Scientists, like other human beings, have their hopes and fears, their passions and despondencies—and their strong emotions may sometimes interrupt the course of clear thinking."[97]

HURRICANES, EARTHQUAKES AND HOUSES

Sagan questions "Earthquakes produced by the comet level Egyptian but not Hebrew dwellings."[98] Had Sagan read page 63 of *Worlds in Collision*, he would find,

"The reason why the Israelites were more fortunate in this plague than the Egyptians probably lies in the kind of material of which their dwellings were constructed. Occupying a marshy district and working on clay, the captives must have lived in huts made of clay and reeds, which are more resilient than brick or stone... An example of the selective action of a natural agent upon various kinds of construction is narrated also in Mexican annals [by Diego de Landa's *Yucatan, Before and After the Conquest* (translated by W. Gates, 1937), p. 18]. During a catastrophe accompanied by hurricane and earthquake, only the people who lived in small log cabins remained uninjured; the larger buildings were swept away [according to de Landa]. 'They found that those who lived in small houses had escaped, as well as the newly-married couples, whose custom it was to live for a few years in front of those of their fathers-in-law.'"

Professor Greenberg states,

"There is absolutely nothing supernatural in Velikovsky's straight-forward discussion. One merely has to think of the comparison between the oak and the willow during a strong windstorm. Furthermore, there are modern parallels to the events described in *Exodus*. The leveling of

Hiroshima by an atomic bomb resulted in selective destruction to various structures; and during World War II, American-built Quonset huts of steel were torn apart by typhoon winds on Okinawa while native huts of reeds and straw remained basically unscathed."[99]

Sagan fails to deal with this, but on page 59 of *Worlds in Collision*, Velikovsky informs us that, "The rabbinical tradition, contradicting the spirit of the scriptural narrative, states that during the plague of darkness [when the earthquake occurred] the vast majority of the Israelites perished and that only a small fraction of the original Israelite population of Egypt was spared to leave Egypt." Hence, Velikovsky made it clear that, although some Israelites survived the earthquake better than their Egyptian masters because they inhabited houses better suited to withstand an earthquake, many Israelites were not so fortunate. The Egyptians who lived in stone houses, not the mud huts of the slaves, probably suffered more from an earthquake.

Sagan then delivers his *coup de grace*, "The only thing that does not seem to drop from the comet is cholesterol to harden pharaoh's heart."[100] Albert Einstein when reading coarse *ad hominem* remarks written by Velikovsky's critics wrote two words to describe what he thought about such statements. Einstein called such remarks "mean" and "miserable."[101] I can think of none better to describe Sagan's vulgar remark because nowhere in *Worlds in Collision* does Velikovsky ever mention, as Sagan claims, that "scarabs drop out of the comet..."[102] This kind of senseless and tasteless misrepresentation is the only thing that is hard-hearted!

Sagan states, "Then, when the Hebrews have successfully crossed [The Red Sea] the comet has evidently passed sufficiently further on for the parted water to flow back and drown the host of pharaoh"[103] Although many Hebrews crossed the "Sea of Passage", many were not so fortunate. Velikovsky informs us that "Although the larger part of the Israelite fugitives were already out of the reach of the falling tidal waves, a great number of them perished in this disaster, as in the previous ones."[104]

Thus ends Sagan's criticisms of the historical and legendary evidence in *Worlds in Collision* and Velikovskian Literature. Sagan stated,

> "In this chapter I have done my best to analyze critically the thesis of *Worlds in Collision*, to approach the problem both on Velikovsky's terms and on mine—that is, to keep firmly in mind the ancient writings that are the focus of his argument, but at the same time to confront his conclusions with the facts and the logic I have at my command."[105]

My opinion of Sagan's critique to this point is that it is intellectually shallow and naive regarding historical, artistic and legendary evidence. To paraphrase Isaac Asimov, "All this shows that as an art historian,

Sagan may quite possibly be an excellent astronomer."[106] One might have expected more from Sagan since he promised to confront Velikovsky's conclusions with facts and logic. In this respect he has failed; nor has he adequately dealt with the Venus evidence in depth.

Thus far it has required many pages to respond to this small part of Sagan's criticism. This is much more space than Velikovsky was to be given to respond to all four critics at the AAAS symposium. Essentially it is a debating technique. If Sagan raised more questions than Veli-kovsky could possibly answer the scientists would claim that he has not been able to answer or squarely face the questions put to him. The length of this book is what was required to respond to Sagan alone.

This procedure of raising so many points that Velikovsky could not answer in the limited space made available to him shows that the playing field which the scientific establishment created was not even, but biased against Velikovsky. When one can be accused of so much, but then so greatly restricted that one cannot answer the accusation we do not have a fair or just interaction. In a court of law such a procedure would be intolerable—except in totalitarian nations where those in power control the justice system to suppress all opposition to the views of the leadership (In this sense the AAAS symposium held on Velikovsky was no different.) When the response to an accusation is suppressed the accusation is equivalent to the verdict.

Hence the reader can see for himself that the entire approach of Sagan and the AAAS scientists was not to examine objectively the concepts that Velikovsky presented in his books but to make it impossible for him to respond appropriately and fully to what was merely a debating technique. Further, we have shown even in this small portion of our response that a great deal of Sagan's criticism is without substance and several of his statements are even contradicted by what he has written elsewhere. Criticism of this sort clearly indicates that the whole approach to Velikovsky has little if anything to do with scientific objectivity. It is simply scientific warfare which has as its goal a specific purpose—to discredit the thesis and character of the gentleman who the scientists invited to the AAAS symposium. The character of this kind of behavior is anything but gracious.

NOTES & REFERENCES

[1] *Stargazers*, op. cit., p. 318.
[2] Ibid., pp. 263-264.
[3] Ibid., p. 207.
[4] SCV, p. 48; B.B., p. 86.
[5] SCV, p. 49; B.B., p. 87.
[6] W in C, pp. 308.

[7]SCV, p. 49; B.B., p. 88.

[8]Greenberg, Lewis M.; *Kronos* III, 2 pp. 63-64 and Kubler George; *The Art and Architecture of Ancient America*, (1962), p. 327, ref. #14.

[9]Greenberg, L.M., *Kronos* III 2, pp. 63-64 and Hardoy, J.E.; *Precolumbian Cities*; (1973), p. 38.

[10]SCV, p. 49; B.B., p. 88.

[11]Padden, R.C.; *American Historical Review*, Vol. 78, No. 4, (Oct. 1973), pp. 996-997 cited by Greenberg, L.M., *Kronos* III, 2, p. 64.

[12]Ibid.

[13]B.B., p. 88.

[14]Ibid.

[15]Sagan and Druyan, *Comet*, op. cit., p. 174.

[16]SCV, p. 51; B.B., p. 88.

[17]SCV, p. 51.

[18]B.B., p. 89.

[19]Ibid.

[20]SCV, p. 45; B.B., p. 84.

[21]SCV, p. 51; B.B., p 89.

[22]Haddingham, Evans; *Early Man in the Cosmos*, (NY 1984), p. 13.

[23]Velikovsky, I.; *Kronos* XI, 3 pp. 67-68.

[24]B.B., p. 90.

[25]deSantillana, Giorgio and von Dechend, Hertha; *Hamlet's Mill*, (Boston 1969), p. 67.

[26]Tompkins, Peter; *Secrets of the Great Pyramid*, (NY 1978) paperback, p. 175.

[27]B.B., p. 90.

[28]Ibid.

[29]Ibid.

[30]Ibid.

[31]W in C, pp. 345-346.

[32]Ibid.

[33]SCV, p. 53; B.B., p. 90.

[34]Dott, R.H. Jr., Batten, R.L.; *Evolution of the Earth*, (NY 1981), p . 288.

[35]SCV, p. 53; B.B., p. 90.

[36]SCV, p. 53; B.B., p. 90-91.

[37]W in C, pp. 45-46.

[38]SCV, p. 53; B.B., p. 91.

[39]W in C, p. 29.

[40]*Kronos* III, 2, pp. 70-71.

[41]SCV, pp. 53-54; B.B., p. 91.

[42]W in C, pp. 153-156.

[43]SCV, p. 53; B.B., p. 91.

[44]Ibid.

[45]SCV, p. 54; B.B., p. 91.

[46]Lovell, Bernard, *Emerging Cosmology*, (NY 1981), p. 27.

[47]James, P. *SIS Review*, Vol. 1, No. 1, 1978, pp. 4-5.

[48]Ibid.

[49] SCV, p. 54; B.B., p. 92.
[50] SCV, p. 54; B.B., p. 91.
[51] Murray, Gilbert; *Five Stages of Greek Religion*, (Garden City, NY 1951), pp. 50- 51.
[52] Graves, Robert, *The Greek Myths*, Vol. 1, (1955), p. 150.
[53] B.B., p. 92.
[54] SCV, p. 55; B.B., p. 92.
[55] Stecchini, Livio C.; "Astronomical Theory and Historical Data", *The Velikovsky Affair* (NY 1966), p. 144.
[56] Ibid, p. 145.
[57] James P. *SIS Review*, op. cit., pp. 4-5.
[58] *The Velikovsky Affair*, op. cit., p. 64.
[59] Ibid.
[60] Ibid.
[61] SCV, p. 55; B.B., p. 92.
[62] *Kronos* III, 2, p. 26 citing Merrill, George .P. *Minerals From Earth and Sky*, (1929)
[63] *Kronos* V, 2, p. 91.
[64] Weiner, Jonathan; *Planet Earth*, (NY 1986), p. 161.
[65] Sagan and Druyan, *Comet*, op. cit., p. 225.
[66] SCV, p. 55; B.B., p. 92.
[67] Purcker, Michael; *EOS*, Vol. 55, (1974), p. 1112.
[68] *Kronos* III, 2, p. 26.
[69] SCV, p. 55; B.B., p. 92.
[70] Ibid.
[71] *Kronos* III, 2, p. 26.
[72] SCV, p. 55; B.B., p. 92.
[73] Moore, Patrick, Hunt, Gary E.; *The Planet Venus*, (London 1982), pp. 31-33.
[74] SCV, p. 55; B.B., p. 92.
[75] Velikovsky, I.; *Ages in Chaos*, op. cit., p. XIII.
[76] Ibid. Jacket blurb (back cover).
[77] SCV, p. 53.
[78] *Kronos* III, 2, p. 27.
[79] Moore, Patrick; *Can You Speak Venusian*, (NY 1972), p. 59.
[80] SCV, p. 55; B.B., pp. 92-93.
[81] Velikovsky, I.; *Earth in Upheaval*, op. cit., p. XX.
[82] Macbeth, Norman; *Darwin Retried*, (Boston 1971), p. 111.
[83] Ibid., p. 116.
[84] Velikovsky, I.; *Stargazers*, op. cit., pp. 304-306.
[85] Hsu, Kenneth; *The Great Dying*, (NY 1986), p. 41.
[86] SCV, p. 55; B.B., p. 93.
[87] *Kronos* III, 2, p. 25.
   [1] Humboldt, *Researchers*, II, 174; see E.T. Hammy, *Codex Telleriano-Remensis*, (1889).
   [2] Humboldt, *Researchers*, II, 174.

[3] Sahagun, *Historia general de las cosas de Nueva Espana*, Bk VII, chap. 4.

[4] J. Scheftelowitz, *Die Zeit als Schicksalsgottheit in der iranischen Religion*, (1929), p. 4; Venus "aussieht wie ein mit Rauch versehenes Feuer" ("looks like a fire accompanied by smoke") CF *Atharva-Veda* vi. 3, 15.

[5] *Babylonian Talmud*, Tractate Shabbat 156a.

[6] M. Jastrow, *Religious Belief in Babylonia and Assyria*, (1911), p. 221; [and] F.X. Kugler, C.F. J. Schamberger "Der Bart der Venus", *Sternkunde und Sterndienst in Babel*, (3rd supp. 1935), p. 303.

[7] "A Prayer of the Raising of the Hand to Ishtar" in *Seven Tablets of Creation*, ed. L.W. King.

[8] Schaumberger in Kugler, *Sternkunde und Sterndienst in Babel*, 3rd supp., p. 291.

[9] *Ibid.*

[10] Midrash Rabba, Numeri 21, 245a: "Noga shezivo mavhik me'sof haolam ad softo". Cf. "Mazel" and "Noga" in J. Levy, *Woerterbuch ueber die Talmudin und Midrashim*, (2nd ed., 1924).

[11] W.C. Rufas and Hsing-chih-tien, *The Soochow Astronomical Chart*, (1945).

[12] D.D. Luckenbill, *Ancient Records of Assyria*, (1926-1927), II, Sec 829.

[13] Breasted, *Records of Egypt*, III, Sec 117.

[14] Brasseur, *Sources de l'histoire primitive du Mexique*, p. 48, note.

[15] H. Winckler, *Himmels-und Weltenbild der Babylonier*, (1901), p. 43.

[16] Pliny, *Natural History*, ii. 23.

[17] "The Peruvians call the planet Venus by the name Chaska, the wavy-haired", H. Kunike, "Sternmythologie auf ethnologischer Grunlage" in *Welt und Mensch*, IX-X. E. Nordenskiold, *The Secret of Peruvian Qulpus*, (1925), pp. 533ff.

[88] *Worlds in Collision*, pp. 163-165.

[89] Sagan and Druyan, *Comet*, op. cit., p. 184ff.

[90] Ibid., pp. 184-185.

[91] Ibid., p. 186.

[92] Ibid., p. 14.

[93] *Illustrated Columbian Encyclopedia*, Vol. 17 (1969), p. 5119.

[94] B.B., p. 92.

[95] SCV, p. 54; B.B., p. 93.

[96] *Kronos* III, 2, pp. 74-75.

[97] SCV, p. 43; B.B., p. 82.

[98] SCV, p. 56; B.B., p. 93.

[99] *Kronos* III, 2, pp. 75-76.

[100] SCV, p. 56; B.B., p. 93.

[101] Velikovsky, I.; *Stargazers*, op. cit., p. 291.

[102] SCV, p. 54; B.B., p. 93.

[103] SCV, p. 57; B.B., p. 94.

[104] W in C, p. 88.

[105] SCV, p. 46; B.B., p. 84.

[106] *Kronos* III, 2, p. 74.

VELIKOVSKY'S THEORY

ORIGINALITY AND PREDICTIONS

The most powerful test of a theory is its predictive value... [Most] scientists agree that a theory that predicts something that has not yet been observed is science. Such a theory is falsifiable because it may predict something that then cannot be found, or is shown not to exist, or when found, does not accord with the prediction. On the other hand, its approximation of truth increases with each new discovery that confirms the prediction.

Kenneth J. Hsu, *The Great Dying*, (NY 1986), p. 14

It has been said that the reception of an original contribution to knowledge may be divided into three phases: during the first, it is ridiculed as not true, impossible or useless; during the second, people say there may be something in it but it would never be of any practical use; and in the third and final phase, when the discovery has received general recognition, there are usually people who say that it is not original and has been anticipated by others.

W.I.B. Beveridge, *The Art of Scientific Investigation*,
(NY 1950), pp. 151-152

Einstein stated toward the end of his life, "it has always hurt me to think that [Galileo] Galilei did not acknowledge the work of Kepler... That alas is vanity." Einstein concluded, "You find it in so many scientists."

I. Bernard Cohen, *An Interview With Einstein*, (in French 1979), p. 41

[Question] "Why do you think people [scientists] resisted [this new concept in astronomy?]"
[Answer: Gerald D. Vaucouleurs] "Number one, it did not come from a member of the establishment. As one of them told me years later, 'If it doesn't come from us, I don't believe it,' There is only one true church."
[Question] "Do you think that if [the astronomer] Oort had [offered this new concept] people [scientists] would have believed it?"
[G.D. Vaucouleurs] "Yes, of course. They would have acclaimed it as something great. The greatest discovery of a great man."

Alan Lightman, Roberta Brower,
Origins... The Lives and World of Modern Cosmologists,
(Cambridge MA 1990) p. 93.

Velikovsky's theory is based on celestial catastrophes that he claims occurred in ancient historical times. From his analysis of celestial events described in ancient legends and myths, Velikovsky drew conclusions and made advanced claims—predictions. Lionel Rubinov, professor of philosophy at Trent University in Canada explains,

> "He [Velikovsky] starts with myth and literature, developing hypotheses from these areas which he then applies to the interpretation of natural phenomena. His approach has been to speculate rather than to perform experiments. The incredible thing is that when experimental data finally is produced, it tends to confirm his hypotheses."[1]

Harry H. Hess, President of the American Geological Society, wrote,

> "Some of these predictions were said to be impossible when you [Velikovsky] made them. All of them were predicted long before proof that they were correct came to hand. Conversely, I do not know of any specific prediction you made that has since been proven to be false."[2]

One of the fundamental aspects that makes scientific theory valuable is its ability to predict correctly. Therefore, by comparing and contrast-ing Velikovsky's predictions with those of establishment science one can get an idea about these conflicting theories. The scientific estab-lishment is so passionately committed to certain theories that questions are considered heresy. When theories are held as absolute authority and questions bring forth abuse, then science as open inquiry becomes restrictive and established theory becomes established dogma. In dealing with Velikovsky's predictions, Sagan states,

> "My conclusion is that when Velikovsky is original he is very likely wrong, and that when right, the idea has been pre-empted by earlier workers. There are a large number of cases where he is neither right nor original [and] that the surface of Venus is hot, which is clearly less central to his hypothesis."[3]

Velikovsky on page 371 of *Worlds in Collision,* explains why he predicts that planet Venus must be hot:

> "Venus experienced in quick succession its birth and expulsion under violent conditions; an existence as a comet on an ellipse which approached the sun closely; two encounters with the earth accompanied by discharges of potentials between these two bodies and with a thermal effect caused by conversion of momentum into heat; a number of contacts with Mars, and probably also with Jupiter. Since all this happened between the third and first millennia before the present era, the core of the planet Venus must still be hot."

The first cause of Venus' heat stated by Velikovsky is its birth by expulsion from Jupiter under violent means. According to Isaac Asimov, the temperature of Jupiter changes with depth.

"The distance from the outer cloud layer of Jupiter to the center is 71,400 kilometers. By the time a depth of 2,900 kilometers below the cloud surface is reached (only 4 percent of the way to the center), the temperature is already 10,000 degrees C[elsius], twice as high as Earth's central point.

"At a depth of 24,000 kilometers below the cloud surface, a third of the way to Jupiter's center, the temperature is 20,000 degrees C[elsius]. At the center itself the temperature has reached a whopping 54,000 degrees C[elsius], nine times that of the surface of the Sun."[4]

Therefore, it is clear that any body ejected from the core of Jupiter will be so hot that it will be incandescent. Sagan apparently seems to be ignorant of this basic knowledge because he states, "...any event that ejected a comet or a planet from Jupiter would have brought it to a temperature of at least several thousands of degrees."[5] Sagan has failed to inform his readers that any body ejected from the core of Jupiter would be immensely hot because the core of Jupiter is so hot; and a few thousand degrees of temperature would be added to the body because it was under great stress as it left the core.

Is Sagan really ignorant of the fact that Jupiter is very hot and any body ejected from its core will also be extremely hot? No, he is not. In *Broca's Brain,* p. 180, he states, "In the case of Venus, the surface temperatures are about 480 degrees C[elsius]; for the Jovian planets, [which include Jupiter] many thousands of degrees centigrade." Thus, Sagan is once again advised to read his own work more carefully.

Velikovsky has cited several ancient cultures that describe Venus as a great flowing fiery comet. "The feather arrangement of Quetzal-cohuatl (Venus) 'represented flames of fire.'"[6] "Phaethon, (Venus) means 'the blazing star.'"[7] "On the island of Crete, Atymnios was the unlucky driver of the Sun's chariot, he was worshipped as the Evening Star which is the same as the Morning Star" (Venus).[8] Thus it is clear that Venus, if it was born from Jupiter, had to be incandescently hot just as Velikovsky claimed. However, Sagan denies that Velikovsky gave this birth and expulsion from Jupiter as the cause of Venus' heat. Sagan states, "this would appear to be a good Velikovskian argument for the high temperature of the surface of Venus, but...this is not his argument."[9] This is disingenuous, to say the least, because it is undoubtedly Velikovsky's argument. Sagan argues,

"The question of originality is important because of circumstances—for example, the high surface temperature of Venus—which are said to have been predicted by Velikovsky at a time when everyone else was imagining something very different. As we shall see, this is not quite the case."[10]

Sagan states,

"Velikovsky writes in the 1965 preface that his claim of a high surface temperature [for Venus] was 'in total disagreement with what was known in 1946.' This turns out to be not quite the case. The dominant figure of Rupert Wildt...looms over the astronomical side of Velikovsky's hypothesis. Wildt, who unlike Velikovsky, understood the nature of the problem, predicted correctly that Venus...would be 'hot.' In a 1940 paper in the *Astrophysical Journal,* Wildt argued that the surface of Venus was much hotter than conventional astronomical opinion had held because of a carbon–dioxide greenhouse effect. Carbon dioxide had recently been discovered spectroscopically in the atmosphere of Venus, and Wildt correctly pointed out that the observed large quantity of CO_2 would trap infrared radiation given off by the surface of the planet until the surface temperature rose to a higher value, so that the incoming visible sunlight just balanced the outgoing infrared planetary emission. Wildt calculated that the temperature would be almost 400K or around the normal boiling point of water (373K = 212 degrees F[ahrenheit] = 100 degrees C[elsius]). There is no doubt that this was the most careful treatment of the surface temperature of Venus prior to the 1950's and it is...odd that Velikovsky, who seems to have read all the papers on Venus, published in the *Astrophysical Journal* in the 1920's, 1930's and 1940's, somehow overlooked this historically significant work."[11]

Let us examine just how significant Wildt's paper is by seeing what the scientific community says regarding Wildt's paper on Venus. We first turn to Isaac Asimov whose figure looms over the astronomical side of Sagan's evidence. Asimov, who rejects Velikovsky's views, neverthe - less has this to say in his book, *Venus, Near Neighbor of the Sun,*

"To be sure Velikovsky made some predictions that seemed to be close to what astronomers eventually discovered to be so... For instance, Velikovsky stated that since Venus was formed from Jupiter's interior which must be very hot, Venus itself would be very hot. He said this in 1950, when astronomers believed that Venus' temperature, while warmer than Earth's might not be very much warmer."[12]

The reader will notice two things. Asimov tells us Venus is hot, in terms of Velikovsky's theory, because it came from Jupiter, contrary to Sagan's assertion, and he also informs us that scientists in 1950 "believed that Venus' temperature...might not be very much warmer" than Earth. There is clearly no mention of Rupert Wildt's historically significant work at all by Asimov. Isaac Asimov's books are usually very precise about citing contributions made by members of the scientific community; so it is indeed strange that he did not see fit to even mention Wildt with respect to Venus' temperature.

When the high surface temperature of Venus was reported in 1956, Dr. Francis D. Drake, a highly respected scientist, wrote the following in *Physics Today,* "We would have expected a temperature only slightly greater than that of Earth (for Venus), whereas the actual temperature is several hundred degrees above the boiling point of water. The finding was 'a surprise'...in a field in which the fewest surprises were expected." [13] Again it is strange that Drake somehow overlooked Rupert Wildt's historical significant work that, as Sagan informed us, "predicted correctly...that Venus would be hot." Ben Bova informs us that, "The first radio measurement of Venus' surface temperature startled astronomers so much that they refused to believe them." [14] Apparently they had somehow overlooked Rupert Wildt's historical significant work.

Let us turn to a book, *The New Solar System,* for which Carl Sagan wrote an "*Introduction*". Rupert Wildt is, in fact, mentioned twice and, indeed given credit for pioneering work with the bulk chemistry of Jupiter and Saturn on pages 169 and 171. Nowhere in this modern work is a single word written about Rupert Wildt's historical significant work looming over the astronomical side of the heat of Venus. This too seems strange. C.J. Ransom states,

> "At the AAAS meeting, Sagan claimed that the heat of Venus was not only anticipated by scientists but was well explained long before the publication of *Worlds in Collision.* He referred to the work of Rupert Wildt, who in 1940 was probably the first to suggest a greenhouse effect on Venus. (Before the AAAS meeting, Wildt's work was twice brought to the attention of scientists by Velikovsky-related publications.) Curiously enough, Wildt does not seem to be remembered when Sagan graciously accepts credit for being the originator of the ["Runaway"] greenhouse theory, and Wildt was not even referenced in one of two articles which Sagan claims as his announcement of the greenhouse effect. Could this be for either of the following reasons: First, Sagan may know Wildt's work has nothing to do with the subject; second, he may think it has something to do with the subject, but prefers credit for the idea unless his image can be enhanced by admitting that someone else first suggested the idea. Wildt died in 1976, and several science publications mentioned his major contributions to astronomy. Suggesting the greenhouse effect was not listed among them." [15]

In fact, on page 153 of *Broca's Brain,* Sagan states that "one now fashionable suggestion, I *first* proposed in 1960, is that the high temperatures on the surface of Venus are due to a runaway greenhouse effect..." [emphasis added] Here, Sagan suggests he is the first to offer a greenhouse mechanism for Venus' high surface temperatures without so much as a breath of mention that Rupert Wildt earlier offered a similar theory.

Finally, let us see what Sagan has to say about the discovery of Venus' high surface temperature in his co-authored book, *Intelligent Life in the Universe:*

"In 1956, a team of American radio astronomers at the U.S. Naval Research Laboratory, headed by Cornell H. Meyer, first turned a large radiotelescope toward Venus. The observations were made near inferior conjunction, the time when Venus is nearest the Earth, and when, also, we are looking almost exclusively at the dark hemisphere of the planet. Meyer and his colleagues were *astounded* to find that Venus radiated as if it were a *hot object* at a temperature of 300 degrees C[elsius]. Subsequent observations at a variety of wave lengths have confirmed these observations and have shown that the deduced temperature of Venus increases away from inferior conjunction—that is, as we see more of the illuminated hemisphere. The most natural explanation of the observations is that the surface of Venus is hot—*far hotter than anyone had previously imagined.*"[16] [emphasis added]

What must have been the cause of C.H. Meyer and his colleagues' astonishment? Didn't they realize that Rupert Wildt had correctly predicted Venus would be "hot"? Sagan confidently assured us that Wildt had shown "that the surface of Venus was much hotter than conventional astronomical opinion had held because of a carbon dioxide greenhouse effect." Obviously Meyer and his colleagues were as misinformed or unaware as everyone else in science was. However, the above citation is by Sagan who was, at the time of his writing this passage, strangely unaware that someone had already known Venus was "far hotter than anyone had previously imagined." Sagan at that time simply did not know that the dominant figure of Rupert Wildt looms over the heat of Venus. What could ever be the matter with Sagan and his colleague and everyone else? How did Sagan somehow overlook the historically significant 1940 paper of Rupert Wildt on Venus' heat? The suspicion grows that the only place Wildt's figure will ever loom over the heat of Venus is in critical papers on Velikovsky. In this respect, this does not seem very strange at all.

Sagan, turning to the radio emission from Jupiter predicted by Velikovsky states,

"The existence of strong radio emission from Jupiter is sometimes pointed to as the most striking example of a correct prediction by Velikovsky, but all objects give off radio waves if they are at temperatures above absolute zero. The essential character of the Jovian radio emission—that it is nonthermal, polarized, intermittent radiation connected with the vast belts of charged particles which surround Jupiter, trapped by its strong magnetic field—are nowhere predicted by Velikovsky. Indeed, his "prediction" is clearly not linked in its essentials to the fundamental Velikovskian theses."[17]

To discredit Velikovsky's prediction, Sagan claims that this is not essential to the fundamental Velikovskian theses. This is odd because it *is* essential and it is difficult to believe that Sagan does not understand this. Velikovsky maintains that not only gravity but also electromagnetism must play a major role in celestial motion. Velikovsky argued that the Sun and planets are not only gravitational bodies, but also electromagnetic bodies. In *Worlds in Collision* he wrote,

"The accepted celestial mechanics, not withstanding the many calculations that have been carried out to many decimal places, or verified by celestial motions, stands only *if* the Sun, the source of light, warmth, and other radiation produced by fusion and fission of atoms, *is as a whole an electrically neutral body,* and also if the planets, in their usual orbits, are neutral bodies." [18] [Velikovsky's emphasis]

The reason these electrical phenomena are essential to Velikovsky's theory is that he proposed that Venus' orbit was changed from an elliptical orbit to a circular one, in part by electromagnetism. There - fore, Sagan seems to be saying he knows Velikovsky's theory better than Velikovsky. Velikovsky, in October of 1953, at a lecture before the Graduate Forum of Princeton University stated, "In Jupiter and its moons we have a system not unlike the solar family. The planet is cold, yet its gases are in motion. It appears probable to me that it sends out radio noises as do the Sun and stars. I suggest this be investigated." [19] On April 6, 1955, The *New York Times* reported, "Sound on Jupiter Picked Up in U.S." The article reported,

"Radio waves from the giant planet Jupiter have been detected by astronomers at the Carnegie Institution in Washington... *No radio sounds from planets in our solar system have been reported previously*... The existence of the mysterious Jovian waves was disclosed by Dr. Bernard F. Burke and Dr. Kenneth L. Franklin. The two scientists said that they did not have an explanation for the observed radio emission." [emphasis added]

Sagan would have his readers believe that Velikovsky's prediction of radio noise from Jupiter is based on ignorance. Dr. James Warwick is a radio astronomer, and a noted authority on the radio emission from Jupiter. At a conference held at McMaster University he actually claimed that Velikovsky had *correctly* predicted non-thermal radio noise from Jupiter as "valid." Warwick then asked, "I who am a specialist in the field am moved to ask myself, 'Did this physician writing in 1954 know more about physics of radio emissions from planets than this astrophysicist 20 years later?'" [20] Sagan remarks, "Merely guessing something right does not necessarily demonstrate prior knowledge or a correct theory." [21] That is quite true, but as been shown, Velikovsky's predictions are derived from his theory and when scientists had, like Sagan, said his predictions would be wrong, the scientists were wrong.

Dr. Bruce Murray, Professor of Planetary Science at the California Institute of Technology in *National Geographic* for August 1970, p. 151

states, "We find that most of the ideas we [astronomers] had about Mars were wrong; in fact, most of the ideas we have about any celestial body prove wrong when we get real knowledge about it." The fact is that nearly every prediction the scientists have made about the planets shows that, when scientists like Sagan made predictions, they were not only likely to be wrong, but they were nearly always wrong whether their predictions were original or not. Even their most highly regarded predictions were not right. Why are the scientists who study these matters most closely nearly always wrong?

I can understand Sagan and his scientific colleagues' dismay, chagrin, and frustration as experts on these matters always making wrong predictions. What is difficult to understand is the niggardly and disin-genuous attempt on his part to withhold recognition.

However, several scientists were more honest and generous, giving Velikovsky recognition for his originality and priority of prediction. Valentin Bargmann of the Department of Physics of Princeton Uni-versity and Lloyd Motz of the Department of Astronomy of Columbia University wrote the following letter, published in the December 21, 1962 issue of *Science,* the journal of the AAAS.

"In light of recent discoveries of radio waves from Jupiter and the high surface temperature of Venus, we think it proper and just to make the following statement.

"On 14 October 1953, Immanuel Velikovsky addressing the Forum of the Graduate College of Princeton University in a lecture entitled "*Worlds in Collision* in the Light of Recent Finds in Archaeology, Geology and Astronomy: Refuted or Verified?" concluded the lecture as follows: 'The planet Jupiter is cold, yet its gases are in motion. It appears probable to me that it sends out radio noise as do the sun and stars. I suggest this be investigated.'

"Soon after that date, the text of the lecture was deposited with each of us. [It is printed as supplement to Velikovsky's *Earth in Upheaval* (Doubleday, 1955)]. Eight months later in June 1954, Velikovsky, in a letter, requested Albert Einstein to use his influence to have Jupiter surveyed for radio emission. This letter with Einstein's marginal notes commenting on this proposal is before us. Ten more months passed and on 5 April 1955, B.F. Burke and K.L. Franklin of the Carnegie Institution announced the chance detection of strong radio signals ema-nating from Jupiter. They recorded the signals for several weeks before they correctly identified the source.

"This discovery came as something of a surprise because radio astronomers had never expected a body as cold as Jupiter to emit radio waves.

"In 1960, V. Radhakrishnah of India and J.A. Roberts of Australia, working at California Institute of Technology, established the existence of a radiation belt encompassing Jupiter, 'giving 10^{14} times as much radio energy as the Van Allen belts around the Earth.'

"On 5 December 1956, through the kind services of H.H. Hess, chairman of the department of geology of Princeton University, Velikovsky submitted a memorandum to the U.S. National Committee for the (planned) IGY [International Geophysical Year] in which he suggested the terrestrial magnetosphere reached to the Moon. Receipt of the memorandum was acknowledged by E.O. Hulburt for the committee. The magnetosphere was discovered in 1958 by Van Allen.

"In the last chapter of his *Worlds in Collision*, (1950), Velikovsky stated that the surface of Venus must be very hot, even though in 1950 the temperature of the cloud surface of Venus was known to be -25 degrees Celsius on the day and night side alike.

"In 1954 N.A. Kozyrev observed an emission spectrum from the night side of Venus but ascribed it to discharges in the upper layers of its atmosphere. He calculated that the temperature of the surface of Venus must be +30 degrees Celsius; somewhat higher values were found earlier by Adel and Herzberg. As late as 1959, V.A. Firsoff arrived at a figure of +17.5 degrees Celsius for the mean surface temperature of Venus, only a little above the mean annual temperature of the Earth (+14.2 degrees Celsius).

"However, by 1961 it became known that the surface temperature of Venus is 'almost 600 degrees [K].' F.D. Drake described this discovery as 'a surprise...in a field in which the fewest surprises were expected.' 'We would have expected a temperature only slightly greater than that of the Earth... Sources of internal heating [radioactivity] will not produce an enhanced surface temperature.' Cornell H. Mayer writes, 'All the observations are consistent with a temperature of almost 600 degrees,' and admits that, 'the temperature is much higher than anyone would have predicted.'

"Although we disagree with Velikovsky's theories, we feel impelled to make this statement to establish Velikovsky's priority of prediction of these two points and to urge, in view of these prognostications that his other conclusions be objectively re-examined."

Thus write V. Bargmann of the Department of Physics of Princeton University and Lloyd Motz of the Department of Astronomy of Columbia University. Whatever could be the matter with these two respected scientists? Hadn't they realized that Carl Sagan had said that "Merely guessing something right does not necessarily demonstrate prior knowledge or a correct theory." Apparently Velikovsky somehow was able to fool these scientists, but not Sagan.

In the *New York Times* for December 22, 1979, p. 22E, Robert Jastrow of NASA's Goddard Institute for Space Studies wrote that Velikovsky was correct for these three predictions: "Venus is hot; *Jupiter emits radio noise;* and the Moon's rocks are magnetic." [emphasis added] Whatever could be the matter with Dr. Jastrow? Hadn't he realized that Sagan said, "...the vast belts of charged particles which surround Jupiter are nowhere predicted by Velikovsky?" Apparently Velikovsky was somehow able to fool Jastrow, but not Sagan.

After learning about the discovery of radio noise from Jupiter, Albert Einstein was so impressed that he asked Velikovsky how he could help further his research. Einstein said, "Which experiment would you like to have performed?"

Einstein was very emphatic in his desire to help [Velikovsky]. Velikovsky asked to have ancient relics radiocarbon dated. A few days later, however, Einstein died. Einstein's secretary though in fulfillment of his wish, a letter went from his home after his death to the Metropolitan Museum of Art with the request that Egyptian relics be submitted for radiocarbon analysis. [22]

Whatever could be the matter with Einstein? Hadn't he realized as Sagan stated that, "...all objects give off the radio waves if they are at temperatures above absolute zero." Apparently Einstein was unaware of this and Velikovsky was capable of fooling him, but not Sagan.

Dr. William T. Plummer, a member of the Department of Physics and Astronomy at the University of Massachusetts and then Senior Scientist of the Polaroid Corporation submitted a paper which was published in *Science,* titled "Venus' Clouds: Test for Hydrocarbons" stating,

> "Some of the least expected discoveries in recent years were correctly predicted by Velikovsky. He argued that *Jupiter should be a strong source of radio waves,* that the Earth should have a magnetosphere, that the surface of Venus should be hot, that Venus might exhibit an anomalous rotation, and that Venus should be surrounded by a blanket of petroleum hydrocarbons. All except the last of these predictions have been verified. Most of them by accident." [23] [emphasis added]

What is also the matter with Plummer? Hadn't he understood that, "Indeed his "prediction" [of Jupiter's radio waves] is clearly not linked in its essentials to the fundamental Velikovskian theses." Apparently Plummer was duped by Velikovsky also, but not Sagan.

Professor Harry H. Hess, Professor of Geology, Princeton University, President of the American Geological Society and Chairman of the Space Science Board of the National Academy of Science, wrote the following as quoted in *Velikovsky Reconsidered,* (NY 1966), pp. 46.

March 15, 1963

Dear Velikovsky,

We are philosophically miles apart because we do not accept each other's forms of reasoning—logic. I am of course quite convinced of your sincerity and also admire the vast fund of information which you have painstakingly acquired over the years.

I am not about to be converted to your form of reasoning though it certainly has had its successes. You have after all, predicted that *Jupiter would be a source of radio noise,* that Venus would have a high surface temperature, that the sun and bodies of the solar system would have large electrical charges and several other such predictions. Some of the predictions were said to be impossible when you made them. All of them were predicted long before proof that they were correct came to hand. Conversely I do not know of any specific prediction you made that has since been proven to be false. I suspect the merit lies in that you have a good basic background in the natural sciences and you are quite uninhibited by the prejudices and probability taboos which confine the thinking of most of us.

Whether you are right or wrong I believe you deserve a fair hearing.

> Kindest regards,
> (signed) H.H. Hess [emphasis added]

What must be wrong with Dr. Hess's understanding? Couldn't he realize as had Carl Sagan who wrote,

"My conclusion is that when Velikovsky is original he is very likely wrong, and that when he is right, the idea has been pre-empted by earlier workers. There is a large number of cases where he is neither right nor original."[24]

Apparently poor Hess was so naive as to state, "I do not know of any specific prediction you [Velikovsky] made that has since been proven to be false" and that "All of them were predicted long before proof that they were correct came to hand to hand." Apparently Hess was fooled by Velikovsky. Lastly, one is forced to wonder how Velikovsky came to fool James Warwick, who, after all, is an expert and authority on Jupiter's radio emissions and who claimed Velikovsky's originality and correctness of prediction respecting Jupiter's emissions are original and correct. Thus it seems that Velikovsky fooled Robert Jastrow, Director of NASA's Goddard Institute, V. Bargmann, a physicist at Princeton, Lloyd Motz, an astronomer at Columbia, James Warwick, an expert authority on Jupiter's radio emissions, William T. Plummer, a physicist and astronomer at the University of Massachusetts, Harry H. Hess, Chairman of the Space Science Board of the National Academy of Science, and Albert Einstein, also, but somehow with all of his efforts Velikovsky was unsuccessful at fooling Sagan.

Vine Doloria, Jr., in *God is Red,* (NY 1973), pp. 145-146, sums things up succinctly.

"As Velikovsky unveiled a concept of the solar system, respectable scholars guffawed at his apparently wild predictions and suppositions.

Practically every point he suggested was derided as being totally contrary to what science had already 'proved' to be true. Scholars in the major disciplines affected by the thesis ridiculed Velikovsky, announcing satirically that if his thesis were true, it would require certain phenomena to be present, which everyone knew was not the case. All of these wild predictions made in 1950 by Velikovsky were universally rejected.

"Then the evidence began to come in. Science had new opportunities to conduct sophisticated experiments with the beginning of the space probes. New methods of dating materials began to be developed, the International Geophysical Year of 1958 was held to determine systematically certain facts about the planets, and eventually the Mars and Venus probes by space rockets were made. Universally and without exception Velikovsky's predictions and suggestions about the planets were confirmed. No other comprehensive explanation of the solar system had returned as many different accurate results as had the theory espoused in *Worlds in Collision*.

"Naturally the scholars who had derided Velikovsky did not credit him with the results of his creative thought. They continued the curtain of silence while stealing his ideas as fast as they could read his books. Some of the more prominent scientists had made dramatic announcements that if Velikovsky were right, then Earth, the Sun, Venus, the Moon, Mars, and other heavenly bodies would have certain characteristics. When Velikovsky was proved correct they promptly hedged rhetorically and dodged their embarrassment in double-talk, too chagrined or perhaps too stupid to apologize."

Doloria adds on page 148 that,

"The most common attack now leveled against Velikovsky is that he simply made a series of lucky guesses and hit on quite a few of them. The point that this attack misses is that every prediction that he made had to fit into his general interpretation of the nature of the solar system. He was not simply spinning a tale and casually throwing off unrelated predictions. Rather, everything suggested by Velikovsky originated from the implications of his thesis. His predictions involve pulling together the meaning of numerous fields of interest to form a unified view of the universe."

Sagan is also guilty of suggesting that certain predictions of Velikovsky are not properly derived from his thesis; these will be discussed as, for example, Sagan's claim that Venus' high surface temperature is not "central to his [Velikovsky's] hypothesis" or that the magnetic fields of Jupiter are "not linked in its essentials to the fundamental Velikovskian theses." In all respects Sagan's assessment of Velikovsky's predictions is no more than a political ploy without substance or value.

One is led to ask: Why is it that so many people with modest backgrounds in science and ancient history have been influenced by Sagan's criticisms of Velikovsky's theses? I suspect that it is Sagan's reputation that has been the convincing influence. However, based strictly on the evidence this criticism clearly fails. As stated earlier when entrenched theories are held as *absolute* authority, that to question them brings forth abuse, then science as open inquiry becomes restrictive and established theory becomes established dogma. Therefore, I can only assume that Sagan is so attached to this dogma that with the uniformitarian axe he has to grind, he is determined to cut Velikovsky from his predictions to save his views of what science should be.

SAGAN AND GRAVITY

As a kind of preface to the part of his criticism of Velikovsky based on the scientific evidence, Sagan adds,

> "There is one further point about the scientific method that must be made. Not all scientific statements have equal weight. Newtonian dynamics and the laws of conservation of energy and angular momentum have extremely firm footing."[25]

Thus, one would expect Sagan to employ this "extremely firm" evidence and not ignore it in his criticism. This, however, does not seem to be the case. In his criticism of Velikovsky's evidence, Sagan did indeed state that if Venus' orbit brought it into near collision with the Earth, the probability of other near collisions, based on Sagan's understanding of Newtonian theory, would be "independent" of each other; that is, Venus would not, because of gravitational theory, return for more near collisions. In this respect Robert Jastrow, founder and director of NASA's Goddard Institute for Space Studies, wrote about this very point of independence raised by Sagan about Velikovsky in the *New York Times,*

> "...Dr. Velikovsky had his day when he spotted a major scientific boner in Professor Sagan's argument. Calculating the probability of several collisions involving Venus, Mars and Earth, Dr. Sagan estimated 1 chance in 10^{23} (10 followed by 22 zeroes) that the collisions could occur. This number was widely quoted by reporters as proof of the absurdity of Velikovsky's thesis. Professor Sagan's error lay in the assumption that the collisions were independent of one another, so that the probability of a series of collisions would be the product of separate probabilities for each collision. Dr. Velikovsky pointed out that the collisions were not independent; in fact, if two bodies orbiting the Sun under the influence of gravity collide once, that encounter enhances the chance of another, a well-known fact in celestial mechanics. *Professor Sagan's calculations, in effect, ignore the law of gravity.* Here, Dr. Velikovsky was the better astronomer."[26] [emphasis added]

But Sagan argued that Newtonian dynamics have superior weight to other forms of evidence. Why then did he misrepresent this evidence?

Here was an opportunity for Sagan to defend his argument. What Sagan did in his letter of rebuttal to the *New York Times* of Saturday, December 29, 1979 was, accuse Jastrow of scientific incompetence. Jastrow, unimpressed by Sagan's criticism, repeated his attack in *Science Digest* (Special Edition) for Sep./Oct. 1980 maintaining that Sagan ignored the laws of gravity.

The same criticism of Sagan was raised by Dr. Robert W. Bass, who is a Cambridge University trained astronomer, whose specialty is gravitational theory and its application to celestial bodies, that is "celestial mechanics". Bass is a Rhodes Scholar, who took his doctorate in 1955 under the late Aurel Wintner—then the world's leading authority on celestial mechanics. Bass did post doctoral research in non-linear mechanics at Princeton University under National Medal of Science winner, Solomon Lefschetz. In the late fifties he developed a new principle in celestial mechanics which not only gives a dynamical explanation for Bode's Law, as shown in 1972 by M. Ovendon, but predicts current planetary disturbances with an average inaccuracy of less than one percent. Bass was, at the time, a Professor of Physics and Astronomy at Brigham Young University. [27]

With these credentials it is extremely difficult to believe that Bass would be in any way scientifically incompetent of judging Sagan's use of gravitational theory. Bass stated the following,

> "At the AAAS symposium on Velikovsky, Sagan claimed that the odds against multiple planetary near collisions were 10^{23} to 1. When I asked him afterwards how he could have computed this without employing 'ergodic theory,' Sagan told me that the proof would appear as an Appendix to a forthcoming paper by him based on his AAAS presentation. He mentioned that he had followed a published method, used by such scientists as Opik and Urey, to obtain apparently reasonable statistics about meteoric collisions with the Moon, Mars and Venus; but in such calculations it is assumed (as an approximation) that the collisions were *statistically independent* events. Because the planetary motions inherently tend under their mutual gravitational attractions toward some sort of quasi periodicity, in which future near misses can be causally related to past near misses, this assumption is *absolutely identical* to the assumption that Newton's Law of Gravity may be ignored. (That is, the planets are regarded as *non-interacting billiard balls,* an approximation used in the kinetic theory of gases)."[28]

Sagan gave other replies to Bass,

> "One reply was to the effect that it was unfair for Bass to ask such complicated questions, since Bass knew more about the subject than did

Sagan. Another reply was that Bass should talk to Mulholland [another scientist at the meeting who disputed Velikovsky] since Mulholland knew more about the subject than did Sagan. The third reply was that Sagan had assumed that the events were independent. Concerning this last point, Bass remarked: 'This Sagan assumption is so disingenuous that I do not hesitate to label it as either a deliberate fraud on the public or else a manifestation of unbelievable incompetence or hastiness combined with desperation and wretchedly poor judgment.'"[29]

Sagan, as cited earlier, had stated, "Indeed the reasoned criticism of a prevailing belief is a service to the proponents of that belief; if they are incapable of defending it, they are well advised to abandon it. This self-questioning and error-correcting aspect of the scientific method is its most striking property." On this point as on others, one would suggest that Sagan take these words to heart because it is true that, "not all scientific statements have equal weight," and it seems that for Sagan, Newtonian dynamics not only do not have equal weight, but when required, do not even exist. One can only wonder what Sagan means when he states "Arguments based on Newtonian dynamics... must be given very [SCV] substantial [BB] great weight." [30]

Therefore, one must suspect that Sagan's scientific bias is so great that he does not hesitate to abuse the very laws which he so ardently affirms and thus his objectivity should be most carefully evaluated with a "firm skeptical scrutiny."

Sagan has emphasized that science is a self-correcting endeavor. However, in researching his criticisms of Velikovsky, we have dis-covered a host of problems associated with his evidence that show it is contradicted by long-known well-established scientific evidence that has never been resolved. Not only have these problems not been solved, they have not been faced squarely by the scientific community. Stanley L. Jaki, the renowned historian of science, in *The Paradox of Obler's Paradox,* (NY 1969), pp. 243-245, discusses this point. Not squarely facing up to contradictory evidence that opposes their models,

"illustrates...the paradoxically unscientific habits of often first rate scientific workers and writers. It shows their often perplexing reluctance to face grave implications of clear-cut situations. It also serves as evidence for the fact that the proverbial respect of scientists for the facts of the laboratory does not necessarily include respect for the facts of scientific history, closely related as these may be to the most avidly discussed areas of research... For those who picture science as the unmatched iconoclast of false ideas, superstitions and myths, it may come as a shock to learn that there is ample room for iconoclasm within science itself. The scientific enterprise too has its foibles, biases and myths. Far from being the always dependable ultimate arbiter of any or all issues that may arise in the context of human inquiry, science does not necessarily recognize in due time outstanding problems which are its own. Inversely, it has no built-in mechanism that would remove, and again in due time, the shackles that hinder the vision of its practitioners. In other words, science, like any other area of inquiry, needs its own school of criticism, if it is to lessen substantially its own share of myths."

The reader will discover in the following pages that in addition to all Sagan's other problems in dealing with the evidence, he too supports scientific myths based on his uniformitarian biases that often have a long standing history. This often denies the validity of Sagan's evidence and shows that what he conceives as established facts all too often are merely establishment myths.

NOTES & REFERENCES

[1] Warshofsky, Fred; *Doomsday: The Science of Catastrophe*, (NY 1977), p. 40.
[2] *Velikovsky Reconsidered*, op. cit., pp. 46.
[3] SCV, p. 58; B.B., p. 95.
[4] Asimov, Isaac, *The Collapsing Universe*, (pocket ed.), (NY 1977), pp. 58–59.
[5] SCV, p. 60; B.B., p. 97.
[6] W in C, p. 157.
[7] W in C, p. 154.
[8] W in C, p. 160.
[9] SCV, p. 61; B.B., p. 97.
[10] SCV, p. 58; B.B., p. 95.
[11] SCV, p. 81; B.B. p. 116.
[12] Asimov, Isaac; *Venus Near Neighbor of the Sun*, (NY 1981), p. 125.
[13] *Physics Today*, Vol. 14, No. 4, 1961.
[14] Bova, Ben; *The New Astronomies*, (NY 1972), pp. 163-164.
[15] Ransom, C.J.; *The Age of Velikovsky*, op. cit., p. 231.
[16] Sagan, C, Shklovski, I.; *Intelligent Life in the Universe* (Delta ed. (NY 1966), p. 319.
[17] SCV, p. 91; B.B., p. 125.
[18] W in C, p. 387.
[19] Velikovsky, I.; *Stargazers*, op. cit., p. 293.
[20] *Pensée*, Vol. 4, No. 3, p. 42.
[21] SCV, p. 91; .B.B., p. 125.
[22] Velikovsky, I.; *Stargazers*, op. cit., p. 293.
[23] Plummer, William T.; "Venus Clouds: Test for Hydrocarbons", *Science*, (March 14,. 1969), p. 1191.
[24] SCV, p. 58; B.B., p. 95.
[25] SCV, pp. 59-60; B.B., pp. 95-96.
[26] *The New York Times*, (Dec. 22, 1979), p. 22E
[27] *Kronos*, III, 2, p. 135.
[28] Ibid.
[29] Ransom, C.J.; *Age of Velikovsky*, op. cit., pp. 225-226.
[30] SCV, pp. 59-60; B.B., p. 96.

PART II

THE SCIENTIFIC
EVIDENCE

SAGAN'S FIRST PROBLEM

"THE EJECTION OF VENUS BY JUPITER"

ANCIENT OBSERVATIONS

Sagan states, "Velikovsky's hypothesis begins with an event that has never been observed by astronomers and that is inconsistent with much that we know about planetary and cometary physics, namely the ejection of an object of planetary dimensions from Jupiter." [1]

Although modern astronomers have not observed such an event ancient man reports the birth of the planet Venus. Evan Hadingham, in fact, informs us that the ancient Mexicans give the precise number of days in the past when Venus was born. [2] Velikovsky tells us,

> "Ancient Mexican records give the order of the occurrences. The Sun was attacked by Quetzal-cohuatl; after the disappearance of this serpent-shaped heavenly body, the sun refused to shine, and during four days the world was deprived of its light... Thereafter the snakelike body transformed itself into a great star. The star retained the name of Quetzal-cohuatl [Quetzal-coatl] [Brasseur in *Histoire des nations civilisees de Mexique I*, p. 181 informs]. This great and brilliant star appeared for the first time in the east. Quetzal-cohuatl is the well-known name of the planet Venus." [3]

Velikovsky then goes on to cite other ancient authorities that describe the birth of Venus and its description as a "Blazing Star and a "Comet." He also cites authorities that claim at an early time, ancient man reported a solar system of only four planets. Velikovsky states, "only four planets could have been seen, and that in astronomical charts of this early period the planet Venus cannot be found.

> "In an ancient Hindu table of planets, attributed to the year-3102 Venus among the visible planets is absent. [This according to J.B.J. Delambre, *Historie de l'astronomie ancienne*, (1817), I, p. 407: "Venus alone is not found there."] The Brahmans of the early period did not know the five-planet system. [This according to G. Thibaut, "Astronomie, Astrologie und Mathematik" in *Grundriss der indoarischen Philol und Altertumskunde*, III (1899).
>
> "Babylonian astronomy, too, had a four-planet system. In ancient prayers the planets Saturn, Jupiter, Mars and Mercury are invoked; the planet Venus is missing; and one speaks of 'the four-planet system of the ancient astronomers of Babylonia.' [This according to E.F. Weidner, *Handbuch der babylonischen Astronomie* (1915), p. 61, who writes of the star list found in Boghaz Keui in Asia Minor: 'That the planet Venus is

missing will not startle anybody who knows the eminent importance of the four-planet system in the Babylonian astronomy.' Weidner supposes that Venus is missing in the list of planets because 'she belongs to a triad with the Moon and the Sun.'] These four-planet systems and the inability of the ancient Hindus and Babylonians to see Venus in the sky, even though it is more conspicuous than the other planets, are puzzling unless Venus was not among the planets. On a later date the planet Venus receives the appellative: 'The great star that joins the great stars.' The great stars are, of course, the four planets Mercury, Mars, Jupiter and Saturn...and Venus joins them as the fifth planet. [according to E.F. Weidner ibid. p. 83]

"Apollonius Rhodius refers to a time 'when not all the orbs were yet in the heavens.'" [4]

Sagan makes a point of refuting Velikovsky's claim that Venus was a new born planet by claiming in his book *Cosmos* that, "The Adda cylinder seal, dating from the middle of the third millennium B.C., prominently displays Inanna, the goddess of Venus." We have already shown the Babylonians saw Venus as a comet. However, Sagan is somehow unable to explain why ancient man describes Venus' birth as a new star that was a comet nor why ancient civilizations had a four-planet solar system with Venus missing. I thus cannot find his view that it was "an event that has never been observed" very compelling nor his refusal to deal with ancient solar system descriptions in which Venus is missing as adequate refutation.

THE BIRTH OF VENUS

The second part of Sagan's opening remarks that the Venus' birth, "is inconsistent with much that we know about planetary physics, namely, the ejection of an object of planetary dimensions from Jupiter." Since astronomy is Sagan's field of study, let us examine whether or not the birth of Venus from Jupiter is "consistent with much that we know about planetary physics."

In 1960, the Astronomer Royal of Great Britain, W.H. McCrea made a careful astronomical calculation regarding the birth of planets between the orbit of Jupiter and the Sun based on currently accepted gravitational physics. He calculated that, based on the Nebula hypothesis, it is impossible for the planets Mars, Earth, Venus and Mercury to have formed inside the orbit of Jupiter. [5] Thus, the present theory of planetary formation that Sagan seems to believe as consistent with planetary physics is, in McCrea's calculation, inconsistent with gravitational physics. However, there is more. Velikovsky wrote,

"In my paper under review, I quote the noted British cosmologist, R.A. Lyttleton, from his *Man's View of the Universe,* to the effect that Venus (and the other terrestrial planets) must have been born from Jupiter by disruption. In the *Monthly Notices,* of the Royal Astronomical Society for 1960, Lyttleton after pointing to insurmountable physical handicaps in both the nebular and tidal theory of the origin of the solar system, demonstrated mathematically the very process that I reconstructed from the annals of the past."[6]

What was W.H. McCrea's reason for concluding that the terrestrial planets—Mercury, Venus, Earth, Moon and Mars—could not be born between the Sun and Jupiter? The answer is gravitational law, which Sagan holds in highest esteem. According to McCrea's calculations, both the nebular and tidal theories of planetary formation would not permit planets to form between the Sun and Jupiter because they would be pulled to pieces by tidal forces. Isaac Asimov in *The Collapsing Universe,* (pocket book ed.) (NY 1977), p. 173 states,

> "In 1849 the French mathematician Edouard A. Roche (1830-1883) showed that if a satellite is held together only by gravitational pull—if it is a liquid for instance—it will break up if it approaches the planet it circles by a distance less than that of 2.44 times the radius of the planet. This is called the *Roche limit.* If a satellite is held together by electromagnetic forces, as our Moon is for instance, it can come a tiny bit closer than 2.44 times the radius of the Earth before tidal stretching overwhelms and destroys it."

W.H. McCrea calculated the volume of the Sun and also of Jupiter during the period of their formation. They would have taken up a considerably larger volume at that time although their masses were roughly the same as they are today. Given their volume and mass they would still exert strong gravitational fields and would have destroyed any incipient planets forming within their Roche limits. McCrea's calculation showed that the terrestrial planets would have had to form then but they would have to have formed inside the Sun's and Jupiter's Roche limit and thus be pulled to pieces.

There are two main theories for the formation of the planets. Immanuel Velikovsky in *Worlds in Collision,* pp. 7-12 sums them up in the following.

> "All theories of the origin of the planetary system and the motive forces that sustain the motion of its members go back to the gravitational theory and the celestial mechanics of Newton. The sun attracts the planets, and if it were not for a second urge, they would fall into the sun; but each planet is impelled by its momentum to proceed in a direction away from the sun, and as a result, an orbit is formed. Similarly, a satellite or a moon is subject to an urge that drives it away from its primary, but the attraction of the primary bends the path on which the satellite would have proceeded if there had been no attraction between the bodies, and out of these urges a satellite orbit is traced. The inertia or persistence of

motion implanted in planets and satellites was postulated by Newton, but he did not explain how or when the initial pull or push occurred...

"Hundreds of millions of years ago the sun was nebulous and very large and had a form approaching that of a disc. This disc was as wide as the whole orbit of the farthest of the planets. It rotated around its center. Owing to the process of compression caused by gravitation, a globular sun shaped itself in the center of the disc. Because of the rotating motion of the whole nebula, a centrifugal force was in action; parts of matter more on the periphery resisted the retracting action directed toward the center and broke up into rings which balled into globes—these were the planets in the process of shaping. In other words, as a result of the shrinkage of the rotating sun, matter broke away and portions of this solar material developed into planets. The plane in which the planets revolve is the equatorial plane of the sun.

"This theory is now regarded as unsatisfactory. Three objections stand out above others. First, the velocity of the axial rotation of the sun at the time the planetary system was built could not have been sufficient to enable bands of matter to break away; but even if they had broken away, they would not have balled into globes. Second, the Laplace theory does not explain why the planets have larger angular velocity of daily rotation and yearly revolution than the sun could have imparted to them. Third, what made some of the satellites revolve retrogradely, or in a direction opposite to that of most of the members of the solar system?

"It appears to be clearly established that, whatever structure we assign to a primitive sun, a planetary system cannot come into being merely as a result of the sun's rotation. If a sun, rotating alone in space, is not able of itself to produce its family of planets and satellites, it becomes necessary to invoke the presence and assistance of some second body. This brings us at once to the tidal theory.

"The tidal theory, which in its earlier stage, was called the planetesimal theory, assumes that a star passed close to the sun. An immense tide of matter arose from the sun in the direction of the passing star and was torn from the body of the sun but remained in its domain, being the material out of which the planets were built. In the planetesimal theory, the mass that was torn out broke into small parts which solidified in space; some were driven out of the solar system, and some fell back into the sun, but the rest moved around it because of its gravitational pull. Sweeping in elongated orbits around the sun, they conglomerated, rounded out their orbits as a result of mutual collisions, and grew to form planets and satellites around the planets.

"The tidal theory does not allow the matter torn from the sun to disperse first and reunite later; the tide broke into a few portions that rather quickly changed from gaseous to fluid, and then to the solid state. In support of this theory it was indicated that such a tide, when broken into a number of 'drops,' would probably build the largest 'drops' out of its middle portion, and small 'drops' from its beginning (near the sun) and its end (most remote from the sun)... Actually, Mercury, nearest to the sun, is a small planet. Venus is larger; earth is a little larger than Venus; Jupiter is three hundred and twenty times as large as the earth (in mass); Saturn is somewhat smaller than Jupiter; Uranus and Neptune, though large planets, are not as large as Jupiter and Saturn. Pluto is quite as small as Mercury. The first difficulty of the tidal hypothesis lies in the very point adduced in its support, the mass of the planets. Between the earth and Jupiter there revolves a small planet, Mars, a tenth part of the earth in mass, where, according to the scheme, a planet ten to fifty times as large as the earth should be expected. Again, Neptune is larger and not smaller than Uranus. Another difficulty is the allegedly rare chance of an encounter between two stars. One of the authors of the tidal theory gave this estimate of its probability:

"'At a rough estimate we may suppose that a given star's chance of forming a planetary system is one in 5,000,000,000,000,000,000 years.' But since the life span of a star is much shorter than this figure, 'only about one star in 100,000 can have formed a planetary system in the whole of its life.' In the galactic system of one hundred million stars, planetary systems 'form at the rate of about one per five billions years. ...our own system, with an age of the order of two billion years, is probably the youngest system in the whole galactic system of stars.'

"The nebular and tidal theories alike regard the planets as derivatives of the sun, and the satellites as derivatives of the planets.

"The problem of the origin of the moon can be regarded as disturbing to the tidal theory. Being smaller than the earth, the moon completed earlier the process of cooling and shrinking, and the lunar volcanoes had already ceased to be active. It is calculated that the moon possesses a lighter specific weight than the earth. It is assumed that the moon was produced from the superficial layers of the earth's body, which are rich in light silicon, whereas the core of the earth, the main portion of its body, is made of heavy metals, particularly iron. But this assumption postulates the origin of the moon as not simultaneous with the origin of the earth; the earth, being formed out of a mass ejected from the sun, had to undergo a process of leveling, which placed the heavy metals in the core and silicon at the periphery, before the moon parted from the earth by a new tidal distortion. This would mean two consecutive tidal distortions in a system where the chance of even one is held extremely rare. If the passing of one star near another happens among one hundred million stars once in five billion years, two occurrences like this one and the same star seem quite incredible. Therefore, as no better explanation is available, the satellites are supposed to have been torn from the planets by the sun's attraction on their first perihelion passage, when sweeping along on stretched orbits, the planets came close to the sun.

"The circling of the satellites around the planets also confronts existing cosmological theories with difficulties. Laplace built his theory of the origin of the solar system on the assumption that all planets and satellites revolve in the same direction. He wrote that the axial rotations of the sun and the orbital revolutions and axial rotations of the six planets, the moon, the satellites, and the rings of Saturn, present forty-three movements, all in the same direction. 'One finds by the analysis of the probabilities that there are more than four thousand billion chances to one that this arrangement is not the result of chance; this probability is considerably higher than that of the reality of historical events with regard to which no one would venture a doubt.' He deduced that a common and primal cause directed the movements of the planets and satellites.

"Since the time of Laplace, new members of the solar system have been discovered. Now we know that though the majority of the satellites revolve in the same direction as the planets revolve and the sun rotates, the moons of Uranus revolve in a plane almost perpendicular to the orbital plane of their planet, and three of the eleven moons of Jupiter, one of the nine moons of Saturn, and the one moon of Neptune, revolve retrogradely. These facts contradict the main argument of the Laplace theory: a rotating nebula could not produce satellites revolving in two directions.

"In the tidal theory the direction of the planets' movements depended on the star that passed: it passed in the plane in which the planets now revolve and in a direction which determined their circling from west to east. But why should the satellites of Uranus revolve perpendicularly to that plane and some moons of Jupiter and Saturn in reverse directions? This, the tidal theory fails to explain.

"According to all existing theories, the angular velocity of the revolution of a satellite must be slower than the velocity of rotation of its parent. But the inner satellite of Mars revolved more rapidly than Mars rotates.

"Some of the difficulties that confront the nebular and tidal theories also confront another theory that has been proposed in recent years. According to it, the sun is supposed to have been a member of a double star system. A passing star crushed the companion of the sun, and out of its debris, planets were formed. In further development of this hypothesis, it is maintained that the larger planets were built out of the debris and the smaller ones, the so-called 'terrestrial' planets, were formed from the larger ones by a process of cleavage.

"The birth of smaller, solid planets out of the larger, gaseous ones is conjectured in order to explain the difference in the relation of weight to volume in the larger and smaller planets; but this theory is unable to explain the difference in the specific weights of the smaller planets and their satellites. By a process of cleavage, the moon was born of the earth; but since the specific weight of the moon is greater than that of the larger planets and smaller than that of the earth, it would seem to be more in accord with the theory that the earth was born of the moon, despite its smallness. This confuses the argument.

"The origin of the planets and their satellites remains unsolved. The theories not only contradict one another, but each of them bears within itself its own contradictions. 'If the sun had been unattended by planets, its origin and evolution would have presented no difficulty.'"

The *Encyclopedia Britannica, Macropedia,* (London 1982), Vol. 16, p. 1032, states explicitly, "It should be emphasized that no theory of the origin of the solar system has yet won general acceptance. All involve highly improbable assumptions. But the difficulty is in trying to find a theory with any degree of probability at all."

Thus, the theories of Sagan and his colleagues generally accept for the formation of the planets, especially the terrestrial ones are based on a process that has never been observed by astronomers and that is inconsistent with much that we know about planetary and stellar physics and gravity. In short, it is a myth.

Velikovsky in an article "Venus A Youthful Planet", published in *Yale Scientific Magazine,* for April 1967 answered Sagan's assertion in part saying that, "The origin of Venus from Jupiter is by itself no absurdity and actually is claimed by Lyttleton. Analyzing the quantitative elements of the tidal theory, he came to the conclusion that the so-called terrestrial planets, Venus included, must have erupted from the giant planets, actually from Jupiter, by cleavage." This information respecting R.A. Lyttleton's theory for the birth of the terrestrial planets from Jupiter was published seven years prior to Sagan's attack on Velikovsky's views for the birth of Venus. Shouldn't Sagan at the very least have discussed this evidence? Since he has not seen fit to deal with it, we shall.

Let us examine Lyttleton's work which is based on "fluid dynamics" that illustrates how planets born from Jupiter is "consistent with much that we know about planetary physics." Lyttleton states,

> "In explaining the origin of the solar system, there is the possibility that only four really large planets, Jupiter, Saturn, Uranus and Neptune need be regarded as primitive. If as condensation slowly formed from interplanetary material to give a large planet at somewhere near Jupiter's present distance from the Sun, the resulting body would rotate in a few hours because of the indestructible rotational momentum drawn into it. With increasing size, its power to draw in material would increase and its resulting speed of rotation would do so too, and eventually render it unstable as a single mass because of centrifugal force. It can only get out of this embarrassing condition by breaking into two very unequal pieces (mass ratio 10 to 1) with the smaller one thrown completely away from the larger portion, to be identified with the present Jupiter. At the surface of Jupiter the escape speed is now about 40 miles a second (59 km/sec) so the smaller piece would easily be thrown right out of the solar system. *The same process of breaking up would produce a string of droplets* between the two pieces as they separated, *and it is even possible that the whole of the terrestrial group of planets [Mercury, Venus, Earth, Mars, Pluto]*

and Jupiter's four great satellites were produced this way as droplets. We have seen that their combined mass is less than one percent of that of Jupiter."[7] [emphasis added]

Thus, according to Lyttleton, the birth of all planets from Jupiter is "consistent" with "planetary physics." Apparently, neither McCrea nor Lyttleton understood that Sagan had known that their analysis and calculations are "inconsistent" with the laws of gravity.

Sagan has stated that "Velikovsky's hypothesis...is inconsistent with much that we know about planetary and cometary physics, namely, the ejection of an object of planetary dimensions from Jupiter." On the other hand, W.H. McCrea states in *Nature*, Vol. 224, for 1969, p. 28, that "Littleton [Lyttleton] has suggested that the Earth, Moon and Mars may originally have formed (from) a single rotationally unstable planet... *He has shown that this is possible in accordance with the theory of rotating fluid masses and with the dynamics of the solar system.*" [emphasis added] McCrea tells us that the ejection of an object of planetary dimensions from Jupiter is completely consistent with what we know about planetary and cometary physics. Thus, it seems Sagan's opening statement is inconsistent and contradictory and requires an explanation from him about this evidence.

In fact, the fissioning concept of planet formation was also suggested by Harold C. Urey, a Nobel Prize Laureate. Patrick Moore in the *New Guide to the Moon*, (NY 1976), pp. 34–36 discusses the birth of Mars and the Moon as a fissioning process that occurred to the Earth. He states, "One variant [fissioning process] involves Mars, whose diameter is just over 4,000 miles... It has been suggested that Mars was thrown off the Earth and moved away independently, while the Moon is merely a droplet which was formed between the two bodies during the process of separation. But the main support for a fission theory has come from H.C. Urey and John O'Keefe, in America, whose ideas are based upon studies of the Moon's composition." Moore goes on to add, "There is no doubt that Urey and O'Keefe have made many interesting points. The fission theory cannot be dismissed; but it is fair to say that on the majority view, the Earth and the Moon have always been separate bodies." Thus, four well respected astronomers claim planets were born by a fissioning process, even though "it has never been seen," and is "inconsistent" with what Sagan knows!

JUPITER CATCHES COMETS

Sagan goes on to argue that comets are not born from Jupiter stating,

"From the fact that the aphelia (the greatest distances from the Sun) of the orbits of short-period comets have a statistical tendency to lie near

Jupiter. Laplace and other early astronomers hypothesized that Jupiter was the source of such comets. This is an unnecessary hypothesis because we now know that long-period comets may be transferred to a short-period trajectories by the perturbations of Jupiter…"[8]

In 1976, NASA published a two volume analysis of the most up-to-date information on comets titled, *The Study of Comets,* as part of *The Proceedings of International Astronomical Union Colloquium* No. 25, which was originally held in Greenbelt, Maryland between October 28 through November 1, 1974, the same year that Sagan delivered his analysis on Velikovsky. In Volume I, Edgar Everhart delivered a paper, "The Evolution of Comet Orbits". On page 450, Everhart states succinctly,

> *"Although it is possible for an orbit of short-period to be the result after a parabolic comet makes a single close encounter with Jupiter, this mechanism does not explain the existence of the short-period comets."* [Everhart's emphasis]
>
> "This was shown by H.A. Newton (1893). Not wanting to believe his results, and being a little dubious about Newton's procedures, I redid the problem as a numerical experiment and came to exactly the same conclusion."

Thus, since 1893, the concept that Sagan believes, that Jupiter captures long-period comets and converts them to short-period comets, has been known to be unsupported by the evidence. Yet, Sagan offers this as a solution for evolution of cometary orbits. It is quite clear that Sagan's concept is only a myth; and there are several reasons that deny this myth contains one shred of reality. Let us examine some of this evidence. A report in *Science* states,

> "The 'capture theory' held by many astronomers, supposes that these [near Jupiter] comets originally came into the solar system in parabolic orbits from vast distances. When one happened to pass close to Jupiter, that planet with its great mass pulled it out of its former orbit by gravitational attraction. After that the comet moved in an elliptical path between the Sun and the orbit of Jupiter. Dr. Vsessviatsky [sic] points out that if this were the case it would be very rarely that a comet entering the solar system would happen to pass close enough to Jupiter to be pulled into the elliptical orbit. He estimates that it would only happen one in 100,000 comets, actually he declares, that there are about sixteen of these short-period comets to a hundred parabolic ones."[9]

S.K. Vsekhsviatsky also points out that all of these Jovian comets revolve around the Sun in the same direction as Jupiter and the rest of the planets; however if they had been captured at least some would be revolving in the opposite (retrograde) direction of orbit. In an article Vsekhsviatsky published in the *Astronomical Society of the Pacific Publications,* Vol. 74 (1962) p. 106 he states specifically that,

"The absence of retrograde motions in the Jupiter family [of comets] can also not be understood from the point of view of the capture theory. It was H. Newton [in 1893] who found that 30% of all short-period comets (that is, in our case about 20 comets) should, on the capture theory, have inclinations greater than 90%. According to Scigolev, half of all short-period comets should have retrograde motion. Lately we have made a new calculation of the approaches of comets to Jupiter and have *found that no less than 10 to 15 comets should be retrograde.*" [emphasis added]

For Sagan's analysis to hold up, the number of long-period comets would have to increase in number by 99, for every short-period comet and some Jovian comets should possess retrograde orbits. Needless to say, this increased number is not known to be the case and comets of the Jovian family do not have retrograde orbits. Furthermore, M.E. Bailey confirms this in *Nature,* that the problem of short-period comets is that there are 100 times too small a number of long-period comets entering the solar system for Jupiter's gravity to affect them. [10]

According to *The Fact on the File Dictionary of Astronomy,* 2nd edition, ed. V. Illingworth, (Oxford England, 1985), p. 196, there are about 70 comets in the Jupiter family, but this authoritative dictionary also states, "No comet can remain in the Jupiter family for more than 4,000 years..." To replenish these comets, Jupiter must capture one out of 100,000 long-period comets entering the solar system continu-ally. For Jupiter to capture 70 comets over the last 4,000 years requires that 7,000,000 comets enter the solar system during this period. That makes the yearly requirement of major comets of long-period 1,750, or five comets per day. Now it takes a few years for these comets to travel into and out of the solar system. Thus, if Sagan's assertion regarding Jupiter is correct, we should be observing nightly a sky lighted by about 9,000 comets. At the very least, half of these should be extremely bright and 4,500 bright comets should be observed nightly if Sagan's assertions is valid. Where are these thousands and thousands and thousands of major comets?

According to M.E. Bailey, V. Clube and B. Napier's *The Origin of Comets,* (NY 1990) pp. 346-347,

"...the total interstellar [in] flux [of comets] with perihelia less than the radius of Jupiter's orbit is on the order of 700 comets per year...

"Such a large flux of interstellar comets, if real, would certainly have been noticed..."

The reason the comets are not noticed is that they are not there, and they cannot be seen because they simply do not exist. As the authors go on to state "This shows that interstellar comets cannot be entering the solar system n large numbers." But if Sagan's view of capture is correct these interstellar comets would be entering the solar system in large numbers so Jupiter can capture them. Where are they?

Vsekhsviatsky in the same article shows that the capture model espoused by Sagan suffers from another major defect. A comet captured by Jupiter would be placed in a fairly eccentric orbit. Thus, all the short-period Jovian comets should have relatively eccentric orbits; but this is not always the case. There are comets with almost circular orbits. Vsekhsviatsky informs us that "The observed eccentricities of the short-period comets are often smaller than the minimum values predicted by the capture theory. The past few decades have seen the discovery of comets with almost circular orbits, which cannot be explained by capture."

For comets captured by Jupiter to change from long-period to short-period ones, their elongated long-period orbits must be reduced in size. According to Edgar Everhart above,

> "*There is no evolutionary path for long-period comets of small perihelia* [close approach to the Sun] *to evolve onto orbits of 5 to 13 year periods typical of short-period comets.* [Everhart's emphasis]
>
> "...Comets that begin on parabolic orbits of small perihelia reach shorter periods very slowly. They cut across Jupiter's orbit at a large angle, the interaction is brief, and the energy perturbations are small. Those that survive the attrition of removal [that is, those that are not ejected from the solar system] on hyperbolic orbits would not also survive the solar dissipation [destruction of the comet by heat during close passages to the Sun] of hundreds of thousands of returns at small perihelia" [small distance to the Sun].

Thus, there is no explanation for long-period comets being converted to short-period comets. They would lose all their material long before their orbits were reduced in size and circularized based on the capture model.

If comets come from the Oort cloud, there is a very basic problem respecting their natal heat. A small comet would conduct nearly all its natal heat to the surface where it would radiate away into space. Interplanetary space is exceedingly cold, only a few degrees Kelvin above absolute zero or 273.15 degrees Celsius below the freezing point of water. Thus, comets should not have heat at great distance from the Sun. However, there is very well observed evidence that comets brighten by heating of their volatiles at distances from the Sun where solar radiation would be an inadequate source of energy to heat the comet. According to R.A. Lyttleton's "Does a Continuous Solid Nucleus Exist in Comets?" in *Astrophysics and Space Science,* Vol. 15, (1972), p. 175, "Numerous comets are known that have perihelion distances in excess of 2 AU [twice the Earth's distance to the Sun] and would therefore always remain at such distances that solar heating could be expected to produce little or no effect. Comet Humanson, (1962 VIII), for instance, remained far outside the orbit of Mars, and yet, showed a fine tail. The great comet of 1729 had perihelion distance over 4 AU, but was visible to the naked eye! Can it really be seriously maintained that this was no more than a dusty snowball a few kilometers in size warmed by solar heat at nearly the distance of Jupiter?"

For example, Comet P./Schwassmann-Wachmann 1 has a highly circular orbit with a perihelion distance of 5.4 AU and an aphelion distance of 6.7 AU. There is no way that solar radiation can supply adequate heat to this body to make it hot or warm. Yet, Anita L. Cochran, et. al., "Spectrophotometric Observations of P./Schwass-mann-Wachmann I During Outburst", in *Astronomical Journal*, Vol. 85, (1980), p. 85, state that the entire orbit of this comet "is beyond the distance where formation of a coma normally occurs and the comet should be inactive. However, P/SW1 displays outbursts which are much larger than those seen in other comets... This comet has been observed to outburst repeatedly over several months and then remain quiet for extended periods of time." These outbursts give off volatiles and dust.

This behavior of comets implies that they must contain more heat than the Sun can supply. There seems to be no other source known to account for this additional heat except that these small bodies were born recently within the last few thousand years and still have not conducted all their natal heat away into space. Such a phenomenon strongly implies these comets were born from a hot body inside the solar system only a few thousand years ago. This evidence cannot be explained by the Oort cloud model of comets, but this evidence fully supports Velikovsky's thesis of the recent birth of comets.

Furthermore, not only can a planet help capture a long-period comet and convert it to a short-period comet, but the opposite will also occur. In *The Cosmic Serpent*, V. Clube and B. Napier p. 131 state, "Short-period comets have characteristic lifetimes of between a few hundred and a few thousand years. Not only do they break up, they also get driven away by planetary encounters. There are at present, approximately one hundred times too many short-period comets relative to the rate at which long-period comets are captured by Jupiter..." Based on this evidence, a great many of the short-period comets should have been ejected from their orbits by close approach to the planets in the inner part of the solar system to become long-period comets or never return to the Sun.

Let us examine this more closely. Jupiter's gravitational pull would only have captured 1/100th of the present number of short-period comets and interaction with the planets would have ejected a large percent of the captured comets from the inner solar system. Based on these gravitational analyses, there are perhaps 1,000 times too many short-period comets in the solar system. Edgar Everhart "Evolution of Long and Short Period Orbits? in *Comets,* ed. L.L. Wilkening (Tucson AZ 1983) p. 663 realizes that the solar system is in a non-steady state because it loses comets. It has not gained any. Everhart states "we should note that all channels to interstellar space are one way [away from the solar system]. The idea that if comets exit to interstellar space then steady-state equilibrium requires that just as many must reenter on the same channel...not one comet has yet been observed whose orbit when it first approached the solar system was hyperbolic, although many comets leave our system on hyperbolic orbits... A truly hyperbolic comet may be discovered but this will not change the conclusion that the gain-loss of comets from and to interstellar space is far from equilibrium." This simply means that at the present rate of comet loss to interstellar space the solar system will lose all its short-period and long-period comets over time. The short-period comets will burn out or be ejected from the solar system in less than ten thousand years. The long period comets will be ejected over millions of years. The solar system is only observed to be losing comets; it is not known to have gained even one since comets have been studied scientifically. Therefore, Sagan's view of how the short-period comets develop is without support. In spite of all these unsolved problems with the capture model, Sagan clings to it.

THE OORT CLOUD

The view that Sagan seems to espouse is that comets come from outside the solar system apparently from the Oort cloud. In his book *Comet,* Sagan actually produces a diagram of the Oort cloud which is supposed to contain many millions of comets and is also supposed to reach "halfway to the nearest stars."

However, this distant Oort cloud filled with innumerable comets has never been observed by astronomers. Therefore, the short-period comets that Sagan claims are changed long-period comets came from a cloud that no astronomer has observed. If it really does exist and reaches "halfway to the nearest stars" then comets entering the solar system from this interstellar cloud should reflect this great distance by following orbits that are termed "hyperbolic". A comet with a hyperbolic orbit must have traveled to the solar system from the interstellar Oort cloud. Nevertheless, J.C. Brandt, an astronomer and expert on comets, informs us that, "very careful examination of the original orbits (that is, the comets trajectories prior to entering the inner solar system) discloses none that are hyperbolic—there are no initially interstellar comets."[11]

George W. Harper in *Closeup: New Worlds,* (NY 1977) p. 192 states "If comets were coming in from outside the [solar] system, a clear majority would have hyperbolic orbits and most would be wide hyperbolas, not just marginally so..." What Harper suggests is that a great many long period comets should show that they came from interstellar distances by their orbits. This is not the case.

Furthermore, it *is* known that the short-period comets could not have originated from the Oort cloud at all. Arman H. Delsemme's article, "Whence Come Comets?" in *Sky and Telescope,* (March 1989), pp. 260–264, discusses and explains this problem. He notes that,

"Like new comets [from the Oort cloud] the long-period [comets]...seem to come more or less from all points on the celestial sphere. Some move in *direct* orbits around the Sun (that is, in the same sense as the planets) while others follow so-called *retrograde* paths [that is, in the opposite direction as the planets]. Clearly the distribution of these comets has not been changed as their periods [of their orbits] have shrunk."

The problem, according to Delsemme is that,

"the short-period comets that orbit the Sun in 200 years or less...are easily distinguished from their long-period cousins by a different symmetry in their orbital distribution. Most of them move around the Sun on direct [prograde] paths close to the plane of the ecliptic. An inclination of only 15 degrees to 20 degrees is typical. At first glance, it might seem that the short-period orbits are just the end result of the process...

"However, Martin Duncan, Thomas Quinn and Scott Tremain have shown that common sense is wrong in this case... The distribution of the inclination of cometary orbits remains rather well preserved during their diffusion inward and eventual capture by the giant planets. Hence, the [short-period] comets cannot arise from a parent population with an all-sky distribution."

If short-period comets were captured from their long-period cousins, they too would exhibit an all-sky distribution instead of orbiting the Sun near the planetary plane of the ecliptic.

There is a further problem with capturing comets from interstellar space where the Oort Cloud is believed to lie. According to Donald K. Yeomans in *Comets,* (NY 1991), p. 338,

"The problem with all interstellar [comet] origin theories is the very low likelihood of their being captured by the solar system. In 1982 Mauri J.

Voltonen and Kimmo A. Innanen [in "The Capture of Interstellar Comets," *Astrophysical Journal,* Vol. 255 (1982) pp. 307-315] computed that the capture of interstellar comets would only be possible if comets had a relative velocity less than 0.4 kilometers-per-second, with respect to the solar system. These low velocities could only be achieved for interstellar clouds that move in very nearly the same direction and velocity as the Sun and this is exceedingly unlikely. The Sun's velocity relative to neighboring stars is approximately 17 kilometers-per-second so that the interstellar [capture] hypothesis is difficult to defend."

Since the Oort cloud of comets in interstellar space will have velocities somewhere between that of the Sun and its neighboring stars, they will be traveling much too fast or too slowly to be captured. Not only that but the Sun not only travels around the galaxy along its orbit but also rides up and down in a merry-go-round fashion as it revolves on its orbit. The clouds of comets would also have to behave exactly as does the Sun. In essence the capture of a single comet is extraordinarily small given these circumstances. And then the capture of 70 comets by Jupiter in 4,000 years appears to be basically impossible.

Since the short-period comets cannot be derived from the population of long-period comets from where do they supposedly come? According to the astronomers, short-period comets come from another cloud, different than the Oort cloud. This new cloud is called the Kuiper-belt and is found in deep space near the planetary plane. Thus, we now have two unseen clouds from which comets arrive. But the problem remains that there are still too many short-period comets. To get these Kuiper-belt comets to enter the solar system in greater numbers, the astronomers suggest that along the plane of the galaxy there is a great deal of unseen matter called dark matter. Delsemme suggests that the behavior of the short-period comets, "implies the existence of a large amount of invisible [dark] matter in the galactic disk."

Thus, we have the unseen Kuiper-belt greatly affected by invisible dark matter to perturb (gravitationally move) its comets into the solar system.

Thus, there is no orbital evidence that the Oort cloud exists; it has not been observed by anyone. The number of comets that supposedly enter the solar system from it is a hundred times too small in number to produce the number of short-period comets currently observed. If comets are captured by Jupiter they should have highly elliptical orbits while some have just the opposite—highly circular orbits and none could survive long enough to reduce the sizes of their orbits. And a small number of Jupiter family comets should have retrograde orbits, but none do. And there must be an unseen Kuiper-belt filled with comets that are influenced greatly by invisible matter in the galactic disk. And this Sagan apparently believes is consistent with much that we know about planetary and cometary physics regarding comets. S.K. Vsekhsviatsky in *Comet International Colloquium Liege* ed. P. Swings, p. 500 states,

> "It should be taken into consideration that these hypotheses [Oort cloud] explain absolutely nothing, they only remove the comet problem into indefinite past time and rather remote regions of the solar system. One can but wonder how such hypotheses possessing no intrinsic logic and no required efficiency, nevertheless, could satisfy investigators while at the same time numerous arguments are available that the processes of creation as well as development of small bodies, comets among them, occurred quite in another way."

In fact, Sagan is clearly aware of this dilemma regarding the origin of comets. In his book *Comet,* he writes,

> "There are only two possibilities: Either comets are being made today somewhere in the solar system or there is a vast repository of hidden comets that supplies a steady trickle of samples. All suggestions about how comets might be manufactured lately, in anything like sufficient numbers, have failed." [12]

Sagan also tells us, "Many scientific papers are written each year about the Oort cloud, its properties, its origin, its evolution. Yet there is not yet a shred of direct observational evidence for its existence." [13] Thus, when Sagan disputes Velikovsky's hypothesis with the remark "Velikovsky's hypothesis begins with an event that has never been observed" by astronomers he should add that the present theory of the origin of comets is also based on the hypothesis of a cloud "that has never been observed by astronomers" and that the measured orbits of comets to prove this hypothesis shows that it is "inconsistent with much that we know about planetary and cometary physics" namely the conversion of sufficient numbers of long–period comets into short–period ones." H. Alfven and A. Mendis in *Nature,* Vol. 246 (1973), p. 410, state that articles about the origin of comets emphasize predominantly the Oort cloud theory while never discussing or mentioning any of the alternative concepts and then "even sweeping under the rug those observational facts which are adverse to the dominant view." While Dr. Paul Weissman is cited in *The Solar System: Great Mysteries Opposing Viewpoints,* (St. Paul, MN 1988), pp. 97-98 stating, "To astronomers, comets are sort of misfits of the solar system. Every good idea about their origin has some major drawback."

This raises an important distinction. Based on gravitational theory, there should be far fewer short-period comets in the solar system. According to V. Clube and B. Napier in *The Cosmic Serpent,* p. 131,

> "There are at present approximately one hundred times too many short-period comets relative to the rate at which long-period comets are captured by Jupiter and fed into the observed stock of Apollo asteroids. The present number is probably due to a burst of new comets formed several thousand years ago."

Clube and Napier assume that "a single large comet fragmented during Jovian capture or perihelion passage." But the problems attendant to that capture model still remain. However, Velikovsky's theory would also fit this evidence quite well in that of the great number of comets born a few thousand years ago, some have still not dissipated all their materials and this is consistent with all the cometary evidence.

EXPLOSION

Earlier Sagan stated, "In reading the critical literature in advance, I was surprised at how little of it there was." In a prior debate Velikovsky had with Professor Motz, the point was raised that a volcanic explosion on Jupiter could not have ejected a body the size of Venus. Velikovsky's reply published in the *Yale Scientific Magazine,* stated that,

> "The basic erroneous assumption by Motz is in ascribing to me the concept of a 'volcanic eruption' [explosion] of Venus from Jupiter…that I claim Venus erupted from Jupiter in a volcanic process is wrong—and it is decisive for the argument. Not only do I not claim this…I stress that a cometary body could not have such an origin. Thus, the entirety of Motz's argument on this score, and only with corrected figures, could apply to the early version of Professor Vsekhsviatsky's theory of the origin of comets by volcanic eruption from Jupiter, but not to my concept… All calculations by Motz on a volcanic eruption and the necessary thermal state of Jupiter are not applicable…"[14]

As we learned earlier, Velikovsky's thesis is that proposed by the highly respected British scientist, R.A. Lyttleton. Therefore, since Sagan has claimed to have read this material carefully, to ascribe some kind of explosion on Jupiter for the birth of Venus, would be incorrect. This is, nevertheless, just what Sagan does. He states respecting this point that, "the comminution physics is well-known."[15] The word *comminution* means to pulverize something into a fine powder; in other words, Sagan says that Venus was born of an explosion on Jupiter that pulverized it. Sagan claims Venus, according to Velikovsky, was born by an explosion of Jupiter. Moses Hadas, the eminent scholar of Greek, had stated, "One after another of the reviews misquote him [Velikovsky] and then attacked the misquotation."[16]

Sagan has misrepresented Velikovsky's thesis for the birth of Venus from Jupiter and attacks the misrepresentation with the following.

"To escape from Jupiter, such a comet must have a kinetic energy of $1/2$ mve^2, where m is the cometary mass and ve is the escape velocity from Jupiter which is about [SCV] 70: [BB] 60 km/sec. Whatever the ejection event—volcanoes or collisions—some significant fraction, at least 10 percent, of this kinetic energy will go into heating the comet... Thus, any event that ejects a comet or a planet from Jupiter would have brought it to a temperature of at least several thousands of degrees, and whether composed of rocks, ices or organic compounds, would have completely melted it. It is even possible that it would have been entirely reduced to a rain of self-gravitating small dust particles and atoms, which does not describe the planet Venus particularly well."[17]

Apparently R.A. Lyttleton, one of England's foremost scientists, was such an ignoramus that when he postulated that all the inner planets as well as the large moons of Jupiter were produced by a fissioning of Jupiter that he could not understand as Sagan does that the material would be pulverized into a "rain of self-gravitating small dust particles and atoms which does not describe the inner planets particularly well." Sagan argues that if,

"Velikovsky has stones falling from the skies in the wake of his hypothesized planetary encounters, and imagines Venus and Mars trailing swarms of boulders...[we should be] bombarded by objects that can make craters a mile or so across, should be happening every second Tuesday."[18]

Although large meteorites are not falling in the numbers that Sagan expects, they seem to be falling as mini-sized comets. In *Sky and Telescope,* Vol. 72, for Sept. 1986, pp. 234-235, is an article titled, "Holy Moses, It's Hailing Comets!", it states,

"The Earth is under continual bombardment of mini comets... So relentless is this rain of icy space debris that it could have significantly influenced the evolution of our planet's atmosphere and oceans.

"This surprising and highly controversial claim by Louis Frank, John Sigwarth and John Craven is their interpretation of satellite observations of Earth's upper atmosphere. They calculate that, on the average, 20 icy objects, each some 40 feet across and weighing 100 tons or so, slam into the atmosphere every minute...

"Evidence for this bombardment is buried in thousands of ultraviolet images of Earth's airglow emission taken by the Dynamics Explorer 1 satellite."

The reason these objects do not strike the Earth, according to Frank, is that since they are made of frozen volatiles, they are tidally destroyed near the Earth. According to the article, "The kinetic energy of such a projectile is equivalent of that of 5,000 tons of TNT." Large ones would, of course, have much more kinetic energy. Thus, one 5 times larger than the average would have enough kinetic energy if it struck the Earth, to excavate a crater of sizable proportions. The conclusions by Louis Frank and his colleagues are supported by some important space scientists. James Van Allen, the discoverer of the Van Allen Belts states, "I am quite persuaded that the cometary hypothesis holds." Contrary to Sagan's assertion, it seems the Earth is under heavy bombardment by relatively large bodies, not every second Tuesday, but every hour, every day. And the question arises, why haven't these tiny comets dissipated all their materials unless they were born recently as Velikovsky claims?

ENERGY

Sagan tells us that,

> "The total kinetic energy required to propel Venus to Jovian escape velocity is then easily calculated to be on the order of 10^{41} ergs which is equivalent to all the energy radiated by the Sun to space in an entire year, and one hundred million times more powerful than the largest solar flare ever observed. We are asked to believe, without further evidence or discussion, an ejection event vastly more powerful than anything on the Sun, which is a far more energetic object than Jupiter."[19]

Poor Lyttleton apparently was so ignorant of the energy requirements of his fissioning process that he didn't realize it was impossible as well as ridiculous, because Sagan said that to eject a body the size of Venus requires more energy than the Sun produces in an entire year. Yet, Lyttleton's hypothesis has all the inner planets born from Jupiter at the same time as well as throwing off an object thirty-five times more massive than all these bodies in the process. Apparently Lyttleton was so naive and ignorant as not to know that Sagan's calculation would require more energy to accomplish this than the Sun would produce in thirty-five years. Yet, for some strange reason, Lyttleton produces a calculation that permits Jupiter to fission, yet does so without exploding the planet that formed into a fine powder of dust and atoms.

BIG COMETS

One point raised by Sagan in his book *Cosmos,* is that Venus is, "some 30 million times more massive than the most massive comet known."[20] Based on the capture theory that Sagan suggests the long-period comets would have to be considerably larger to survive hundreds of thousands of close passages to the Sun and become short-period comets. The question is, how large would these comets have to be not to dissipate all their material? R.A. Lyttleton in *Mysteries of the Solar System,* (Oxford 1968), p. 147 made this calculation and states,

"In the whole age of the solar system, a comet with an average period of 100,000 years would make 4.5×10^4 returns to the Sun, and if at each of these it lost only 1/1000th of its mass, through tail formation and meteor stream production, the initial mass would have been more than 10^{19} times as great as the present mass—which at a minimum means several times the mass of the Sun!"

Thus, for Jupiter to capture comets and convert them from long-period to short-period ones, the comets would have to make hundreds of thousands of close passages to the Sun, and, to survive these numerous passages, these comets would have to be immense. And this Sagan suggests is congruent with much that "we" (meaning scientists) know is consistent with planetary and cometary physics.

However, William K. Hartmann in *Moon and Planets,* 2 ed. (Belmont, CA 1983), p. 238, discusses the concept of "Giant Comets?" stating,

"Among the larger estimate of comet diameters are values such as 50 km. But if comets really originated by the ejection of planetesimals from the outer solar system, the Oort cloud would be likely to preserve that original size distribution. Collisions in the sparsely populated Oort cloud would be rare and unlikely to break up all the largest bodies.

"The sizes of the largest planetesimals are unknown, but calculations and observations of planetary obliquities...suggest that planetesimals grew to diameters as large as 0.1 or more of the diameter of the planet in their zone before being expelled by the planet or colliding with it. That could mean that among the millions of comets in the Oort cloud, many are hundreds of kilometers across and some perhaps 1,000 km or more across! Perhaps Chiron (a small planet-like body located between Saturn and Uranus) is such a body. This suggests that the comets we have seen are just the tip of the cometary iceberg. Perhaps occasionally during planetary history the inner solar system is visited by a world-scale body that blazes to life with a brilliant immense coma and tail before vanishing again into its long night in the Oort cloud."

Jupiter's diameter is 143,800 km. According to Hartmann, a planetesimal in its region could be one-tenth that diameter—about 14,000 kilometers in diameter or around 8,000 miles—the approximate size of Venus. Hence the theory most astronomers accept respecting the origin of the solar system and comets allows for gigantic comets, contrary to Sagan's analysis.

According to George W. Harper in *Closeup New Worlds,* (NY 1977) p. 192, the region occupied by comets contains, "...literally hundreds of thousands, or even millions of minor asteroids and planetoids possessing radii up to 150 to 1,500 kilometers [90 to 900 miles] with a few having radii up to roughly 3,000 kilometers and perhaps five or six with radii upward of 10,000 kilometers."

Is Sagan really ignorant of the theory for comet formation that allows for very large bodies? No, he is not! In his book *Comet,* (NY 1985) p. 217, Sagan states "It is entirely plausible that much bigger comets than those several kilometers across were ejected into the Oort Cloud. But there are far fewer of these, and much more rarely will we see one redirected into our small but well lit volume of space."

The question thus arises: has a large world-sized comet ever been observed? The answer is yes. It is a well-known fact that comets often break up and their constituent pieces then follow a common orbit. The pieces of one such comet were analyzed by N.T. Bobrovnikoff in an article titled "Comets" in *Astrophysics,* ed. Hynek (NY 1951). The comets he concluded were once "one single body" and "If put together, all these comets would make something like the mass of the Moon." Hence, the astronomers postulate huge comets and apparently these have been observed, contrary to Sagan's analysis.

Furthermore, Lyttleton has not only Venus, but Mercury, Mars and Earth, born from Jupiter as comets and taken together they are over 100 million times more massive than the most massive comet known. And W.H. McCrea, one of England's foremost astronomers, must have been an ignoramus to write that Lyttleton's work accords with the theory of rotating fluid masses and the dynamics of the solar system. It is extremely interesting and quite intriguing that Sagan never mentions R.A. Lyttleton's fissioning process or W.H. McCrea's evaluation of it, which Velikovsky claims gave rise to Venus. Without further evidence or discussion of Lyttleton's fission theory of the ejection event, Sagan seems to ask us to believe that Lyttleton did not know what he was scientifically talking about. This one may find extremely difficult to believe.

ESCAPE VELOCITY

Sagan continues,

> "Another problem is the escape velocity from the Sun's gravity at the distance of Jupiter is about 20 km/sec. The ejection mechanism from Jupiter, of course, does not know this. Thus, if the comet leaves Jupiter at velocities of less than [SCV] 70, [BB] 60 km/sec, the comet will fall back to Jupiter, if greater than about $[(20)^2 + (60)^2]^{1/2} = 63$ km/sec, it will escape from the solar system. There is only a narrow and therefore unlikely range of velocities which is consistent with Velikovsky's hypothesis."[21]

This analysis by Sagan, an astronomer, is embarrassingly quite in error. James Oberg, a mission flight controller for the McDonnell-Douglas

Aerospace Corporation at the NASA Johnson Space Center, who has a combined background of applied mathematics, computer science and astrodynamics, did the more realistic calculation respecting the velocity of Venus needed to keep it in the solar system. C. Leroy Ellenberger summarized the gist of Oberg's calculation, thus,

> "In a simple ballistic escape it can be envisioned that, for Venus to be captured by the Sun, she would have to be propelled to a distance at which the Sun's gravitational attraction at least equals Jupiter's. This would be the radius of the Tisserand sphere of action, which for Jupiter, equals 48.1 million km. The initial velocity required to lift a body just to that distance is 0.9993 of escape velocity. Therefore, for proto-Venus to escape Jupiter and be captured by the Sun implies an initial velocity very close to escape velocity."[22]

According to Oberg's calculation to escape Jupiter and be captured by the Sun, the escape velocity from Jupiter will be sufficient for Venus. Therefore, Velikovsky's analysis not is contradicted by "Newtonian Dynamics" which according to Sagan carries greater weight than other forms of evidence.

However, Velikovsky's concept is the same as that of Lyttleton. *The droplets created by a fissioning event would have formed far from Jupiter and would not be traveling at solar system escape velocity.* Both McCrea and Lyttleton understood this and accounted for this question of velocity. Sagan's whole argument begs the question. It has nothing to do with Lyttleton's concept. It is a straight-forward distortion of the concept Velikovsky suggested for the birth of Venus.

HOW LONG AGO?

Sagan previously stated, "It is about the time scale... In the 4.6 billion-year history of the solar system, many collisions must have occurred. But have there been major collisions in the last 3,500 years..."[23] Sagan has also told us that, "All suggestions about how comets might be manufactured lately in anything like sufficient numbers, have failed."[24] Velikovsky remarked,

> "I have expressed my opinion that many comets are of recent origin, and I have supported this view by reference to the frequency and luminosity of comets in the days of imperial Rome in comparison to the number of comets visible to the unaided eye in the last centuries.
>
> "This notion received vigorous confirmation in the extensive work on comets done in Soviet Russia by a leading authority on the subject, Professor S.K. Vsekhsviatsky. His research reveals that periodic comets, as observed during recent decades, are losing their luminosity and their matter at a rate so rapid that fifty or sixty revolutions suffice to disintegrate a comet completely. Thus, the Halley comet can hardly go back beyond 3,500 years, or the year 1500 before the present era. In the

last century several comets with short period have failed to return, having lost all their matter, and a few others actually fell apart before the eyes of observers."[25]

Napier and Clube, two British astronomers have come to this same conclusion in *The Cosmic Serpent*. Even Sagan is aware of this. In his book *Comet,* he writes,

"Dusty comets have been observed to pour tons of fine particles into interplanetary space every second, and for most comets several times more water is lost than solids...for most comets, layer after layer of ice is lost in successive perihelion passages. Since the material in the coma and tail is never recaptured by the comet, it gradually dwindles, successive layers are peeled off and lost to space, and its interior parts are exposed to view. One way or another, every comet we see is dying."[26]

The basis of Velikovsky's thesis is that when Jupiter fissioned, it gave rise to Venus, but also to thousands of smaller bodies namely the comets. What do other scientists say with respect to the comets and what of Venus tells us that it was a comet?

If Venus was indeed born from the planet Jupiter, there should remain some trace of the atmosphere of Jupiter in the atmosphere of Venus. Sagan stated on Dec. 2, 1973 to a group of scientists that, "...Jupiter is a kind of remnant of the chemistry which was around in the early history of the solar system..."[27] If this is the case as Sagan informs us, then this remnant of chemistry which was around in the early history of the solar system should be present on Venus. Lewis M. Greenberg reports,

"Argon-36 or 'primordial argon' as it has been termed has been found to be either 100 or 500 times as abundant on Venus as on the Earth... *New Scientist,* (80: 916, 1976) may have come closer to the truth when it reported on Venus' argon-36: 'The significance of argon-36 is that it is supposed to be primordial argon; that is an argon isotope formed when the solar system was created. Since argon-36 is radioactive, most of the originally created supply should have disintegrated and disappeared over the four-billion year history of the solar system. Indeed, the atmospheres of earth and Mars have much, much smaller quantities of argon-36 than Venus. Venus, therefore, may have an origin different from those of earth and Mars—either *a much more recent birth,* (so that the argon-36 has not disintegrated) or an altogether different kind of origin in which more argon-36 was created than for earth and Mars.'"[28] [emphasis added]

"On hearing the news about argon-36 an Edison, NJ man cabled a simple message to [Dr. John] Hoffman [Head of the mass spectrometer team for *Pioneer Venus 2*] [that] 'Emanuel [sic] Velikovsky was right about Venus.'"[29] Hence, it does seem that Venus either has "a much more recent birth or an altogether different kind of origin" and it unquestionably possesses as Sagan stated of Jupiter, "the chemistry which was around in the early history of planets." This indicates that the possibility of Venus' recent birth is not as improbable as Sagan would have us believe. There will be much more evidence regarding this concept later on.

Furthermore, *Comets,* published in 1980 [not Sagan's book] states that,

> "While Venus itself is in many respects most unlike a comet, it is very comet-like in its interaction with the solar wind... Both comets and Venus have magnetic tails which are not intrinsic... The Venus tail appears to be either striated or very dynamic and thus quite similar to a cometary tail. Plasma clouds are seen above the Venus ionosphere which may be the Venus analog of cometary tail rays."[30]

Apparently Venus' interaction with the solar magnetosphere is quite like a comet. Also, Velikovsky points out that "The zodiacal light, or the glow seen in the evening sky after sunset, streaking in the path of the sun and other planets (ecliptic), the mysterious origin of which has for a long time occupied the minds of astronomers, has been explained in recent years as the reflection of the solar light from two rings of dust particles, one following the orbit of Venus."[31] As is well-known comets also have a dust tail as well as a magneto tail. Thus, it can be seen that in some ways, Venus' recent cometary origin is confirmed by findings of modern astronomy. Therefore, let us return to how recent the event may be based on other astronomers. "[Fred] Whipple [the Harvard astronomer] upon calculating the orbits of the asteroids, came to the conclusion (1950) that two collisions occurred between these bodies and a comet, once 4,700 years ago and the second time 1,500 years ago, or within historical times."[32] Sagan questions the possibility of cometary collisions in historical times. In fact, B. Napier and V. Clube in their book, *The Cosmic Serpent,* come to the conclusion that the Earth was struck by a comet at about the same time as Velikovsky's comet. Napier and Clube's comet is reported to have struck the Earth around 1369 B.C. while Velikovsky's had a near collision around 1447 B.C. Whatever could be the matter with Napier and Clube? They not only claim that the Earth was struck by a comet called "Venus", but they even cite legendary and historical evidence, some of it from Velikovsky's *Worlds in Collision,* to prove their case.

OTHER ASTRONOMERS AND COMETS

What do other astronomers postulate as the origin of comets? The late M.W. Ovendon, a highly respected astronomer, published a paper and a discussion of it in *Nature,* in which he discusses a planet never known to exist named Aztex. It was supposedly located between the orbits of Jupiter and Mars. Ovendon concluded that for some unknown reason, Aztex exploded but then he had to explain what had happened to 99.9 percent of the debris of Aztex. Another highly respected astronomer, Thomas Van Flandern of the U.S. Naval Observatory also dealt with the question regarding Aztex's debris. Van Flandern wrote in *Science Digest* that except for small asteroids, "the only remaining debris that we have any chance of seeing today would be objects hurled to great distance from the Sun but eventually pulled back by gravity" and adds, "Only one kind of celestial object matches that description perfectly—the comet."[33] Duardu Cardona discusses this material by Ovendon and Van Flandern adding, "Thus, regardless of what else has been written about these intriguing sky wanderers, Van Flandern believes that the comets were born when Ovendon's planet exploded."[34] Not only did Ovendon suggest comets are derived from an exploding planet between Jupiter and Mars, but so did Oort, the originator of the Oort cloud theory. According to A. Mendis and H. Alfven, "On the Origin of Comets", *The Study of Comets,* part 2, (NASA SP 393, 1976), p. 642, "Oort originally made the highly unlikely supposition that these comets together with the minor planets resulted from the break up of a planet inside Jupiter's orbit." How interesting! No one saw fit to attack Oort or Van Flandern who described Ovendon's conclusion stating, "What a sight it must have been for early man: the sky ablaze with meteors, night and day for months, and comets streaming among the stars."[35] No one has seen fit to ridicule Ovendon for suggesting that a never-before-observed planet exploded. No one has seen fit to ridicule Van Flandern that all of the bodies of this exploding planet would have been thrown out of the solar system or become a rain of fine particles. And this event is clearly described as an explosion. Nor does anyone argue that the energy necessary to accomplish the explosion would be more energy than the Sun produced in hundreds of thousands of years. Sagan attacks Velikovsky in a manner which implied that such ideas are clearly preposterous. However, whether or not one agrees with Ovendon and Van Flandern, their ideas, though different, are not so terribly different than those Sagan ascribes to Velikovsky. Cardona states, "Was Velikovsky not derided when he expressed his concurrence with ancient opinion that comets were born from the planets?"[36] Again and again the same concepts proposed by Velikovsky in the 1950's were later proposed by respected and competent scientists in respected journals. In these cases these same ideas were treated, if not with acceptance, at least with respect. *There is clearly a double standard at work with this kind of criticism!*

Sagan claims that gravitational law dominates other evidence. Then perhaps he should explain Van Flandern's evidence for his conclusion that the comets originated in the solar system, in the vicinity of Jupiter. Van Flandern writes, "A great deal of other evidence also supports the 'recent' explosion hypothesis. By applying the laws of gravitation to comets, we can trace their orbits back in time. When we do, we find that comets all seem to have originated from a common point between Mars and Jupiter..." Although Velikovsky's time scale is far shorter for the fissioning of Jupiter than Ovendon and Van Flandern's explosion theory, we also have Napier and Clube's conclusion that the Earth was, indeed, involved in a catastrophe with a comet about 3,369 years ago. Sagan has raised his argument about the time scale, but here too his colleagues' (Napier and Clube) time scale is very close to Velikovsky's.

Sagan has attacked Velikovsky for concluding that a planet, namely Venus, was born from the planet Jupiter. However, R.A. Lyttleton proposed that all the small planets and the four large moons of Jupiter were born from Jupiter. Furthermore, Harold Urey, a Nobel prize winner, offered the concept of fissioning of the Earth to produce the planet Mars and also the Moon.

We do not hear anyone in this case attack Lyttleton or Urey in a manner which implies such ideas are preposterous. Again, whether or not one agrees with Velikovsky's idea which was taken directly from Lyttleton as a possible explanation for Venus' birth, it is based on sound scientific principles. Why is Velikovsky vilified but not Urey? Or Lyttleton? Or Ovendon? Or Van Flandern? Or Napier and Clube?

IT HAS NEVER BEEN OBSERVED

Sagan, himself, considers the idea that an unknown and unobserved companion star of the Sun periodically circles the Sun and produces great showers of comets. In *Comet* he writes,

> "During its plunge through the inner Oort cloud, the companion star would spray a billion comets into the inner solar system. But because they would be carried on slightly different trajectories, the comets would not all arrive at the same instance. Rather, they would be spread out over a million years or more."[37]

Sagan discussed a hypothetical star that has never been observed. He conjectures that it would gravitationally cause a billion comets to enter the inner solar system from an Oort cloud that has never been proven to exist, let alone been observed, while some astronomers suggest that an unseen planet called Planet X is responsible for putting comets into the solar system. Yet, Sagan's discussion is based on pure supposition while he implies that Velikovsky's views are impossible because it has never been seen. It appears that when Sagan or members of the scientific establishment build hypotheses based on conjecture that stars, exploding planets, Planet X, the Oort clouds, the Kuiper-belts and invisible matter that have never been observed, cause comets to invade the solar system, they are practicing science; however, when Velikovsky builds a case based on similar, but anciently observed evidence, he is not practicing science. Sagan's view of the evidence is riddled with contradiction and authoritarian conceit. Sir Fred Hoyle in his book, *From Stonehenge to Modern Cosmology,* (San Francisco 1972), p. 62, states: such "argument amounts to nothing more than the convenient supposition that something which has not been observed does not exist. It predicts that we know everything." Such a view, it must be pointed out, is totally unscientific. But if Sagan wishes to argue that that which has never been observed cannot exist, then the Oort cloud does not exist nor do any of the other celestial phantoms of astronomical thought that have never been observed.

Jonathan Eberhart in a Commentary in *Science News* for Jan. 30, 1989, p. 72 explains what is essentially wrong with Sagan's approach against Velikovsky's concept regarding the birth of planets and comets by a fissioning of Jupiter. What Sagan has offered is not scientific fact, but assumption masquerading as such. Eberhart states,

> "The conduct of science often leads not so much from answer to specific answer as it does along converging successions of questions—a matter of the often subtle but critical distinction between fact and hypothesis. Yet despite all the constraints of the scientific method, that fragile awareness sometimes has a way of evolving into what has been called 'canonical wisdom'—assumptions, in other words that are occasionally found masquerading as established fact." [As, for example, the concept that Jupiter catches long period comets and converts them to ones of short period.]

Eberhart continues,

> "*It is not that such conclusions were necessarily wrong, more to the point is that one could not know them to be correct.* [emphasis added] The problem arises that taking them for granted results in overlooking a different line of inquiry that might have led in a more meaningful direction. If you are determined enough to 'confirm' the existence of a 'fact' such as the presence of Martian canals, who can say what interpretations you might overlook-or reject out of hand-when phenomena seemingly inconsistent with that 'reality' present themselves to view."

Unfortunately, that is precisely the method employed by Sagan to argue with Velikovsky's concepts. It is not so much known that what he offers as evidence contrary to Velikovsky's thesis is *incorrect* but that what he offers throughout his criticism of that thesis is simply not known to be established as fact. When assumptions are put forth as arguments we do not have honest debate. Rather, what we have is rhetoric and polemics substituting for honest debate. What Velikovsky has offered is clearly hypothesis but the evidence offered by Sagan is also hypothesis yet throughout his criticism are phrases such as,

> "Velikovsky's hypothesis begins with an event that has never been observed by astronomers and that is inconsistent with much that *we know* [meaning scientists] about planetary and cometary physics, namely the ejection of an object of planetary dimensions from Jupiter." [emphasis added].

However, some very noted and highly respected planetary scientists have put forth just this concept not as fact but as rational scientific hypothesis and Velikovsky has offered their concept as a possible mechanism for his theory. Richard S. Lewis in *From Vinland To Mars,* (NY 1976) p. 306 points out that,

> "One of the most curious consequences of lunar exploration so far is that in spite of the mass of physical evidence accumulated about the Moon since 1964, there is not enough to exclude any of the major theories of its Origin ...the comment of Don L. Anderson, of the California Institute of Technology, summarizes this circumstance nicely: 'All the classical theories of lunar origin are still with us—capture, *fission,* and dual planet accretion,' he said." [emphasis added].

To suggest, as Sagan does, that "we know," when in fact one cannot definitely know this is inconsistent with how planets and comets are born is merely rhetoric masquerading as scientific fact.

In the introductory remarks to his criticism Sagan stated, "Appeals to authority are impermissible." But one can only wonder what Sagan means by the phrase "we know." One must assume that Sagan means "scientists know." This is clearly an appeal to the authority of *some* scientists but not to *all* scientists. Neither R.A. Lyttleton nor Harold Urey knew this to be engraved in stone, and put forth just the opposite of what Sagan suggests is indisputably disproven. And Patrick Moore, another respected astronomer who, like Sagan, disagrees with the fissioning concept nevertheless says, "the fission theory cannot be dismissed" although most scientists disagree with the idea. If the fission theory cannot be dismissed why does Sagan claim it is? Sagan has argued that assumptions are facts in order to dismiss a different line of inquiry. Although Velikovsky's thesis may or may not prove valid, Sagan's approach is unquestionably biased and irrelevant.

NOTES & REFERENCES

[1]SCV, p. 60; B.B., pp. 96–97.
[2]Haddingham, Evan; *Early Man in the Cosmos*, op. cit., p. 225.

[3]W in C, p. 157.

[4]Ibid., pp. 160-161.

[5]McCrea, W.H.; *Proceeding Royal Astronomical Society Series*, AV 256, (May 31, 1960).

[6]Velikovsky, I., "A Rejoinder to Motz", *Yale Scientific Magazine*, (April 1967).

[7]Lyttleton, R.A.; *Man's View of the Universe*, (Boston 1961), p. 36.

[8]SCV, p. 60; B.B., p. 97.

[9]Corliss, W.R.; *Mysterious Universe*, op. cit., p. 533; See also *Science*, 76 Supp. (July 1, 1932).

[10]Bailey, M.E.; *Nature*, Vol. 324, (1986), p. 350.

[11]Beatty, J.K., et. al., eds. *The New Solar System*, 2 ed. (Cambridge MA 1982), p. 182.

[12]Sagan and Druyan, *Comet*, op. cit., p. 194.

[13]Ibid., p. 201.

[14]Velikovsky, I.; *Yale Scientific Magazine*, op. cit.

[15]SCV, p. 61; B.B., p. 98.

[16]*The Velikovsky Affair*, op. cit., p. 64.

[17]SCV, pp. 60-61; B.B., p. 97.

[18]SCV, p. 62; B.B., p. 98.

[19]SCV, p. 61.

[20]Sagan, C.; *Cosmos*, (NY 1980), p. 91.

[21]SCV, p. 61; B.B. pp. 97-98.

[22]*Kronos* V, 1, p. 93.

[23]SCV, p. 48; B.B., p. 86.

[24]Sagan and Druyan, *Comet*, op. cit., p. 194.

[25]Velikovsky, I.; *Earth in Upheaval*, op. cit., p. 261.

[26]Sagan and Druyan, *Comet*, op. cit., p. 137.

[27]*Pensée*, Vol. 4, No. 1, p. 57.

[28]*Kronos* IV, 4, pp. 1-3.

[29]Ibid., p. 2.

[30]Wilkning, L.L. ed,. *Comets* (Tuson AZ 1980), Russell C.T. et. al., "Solar Wind Interactions With Comets: Lessons From Venus:, p. 561.

[31]*Earth in Upheaval*, op. cit., pp. 259-260.

[32]Ibid., p. 260.

[33]*Science Digest*, (April 1982), p. 94.

[34]*Science Digest*, (April 1982), p. 81.

[35]*Kronos* XI, 2, p. 27ff.

[36]*Kronos* XI, 2, p. 27ff.

[37]Sagan and Druyan, *Comet*, op. cit., p. 301.

SAGAN'S SECOND PROBLEM

REPEATED COLLISIONS AMONG
THE EARTH, VENUS AND MARS

30 THOUSAND, 10 MILLION, 30 MILLION

"Most short period comets may have achieved their orbits by multiple gravitational encounters with Jupiter, or even by multiple encounters with more distant planets and, eventually, with Jupiter itself."
Carl Sagan, Ann Druyan, *Comet,* (NY 1986) pp. 95-96.

Before dealing with this problem in its printed form in *Scientists Confront Velikovsky* and *Broca's Brain,* it seems proper to relate the process by which Sagan derived the figures of his proof. Ralph. E. Juergens informs us,

"In arguing at San Francisco that probabilities make Velikovsky's near-collision hypothesis untenable, Sagan neither stated his assumptions nor provided copies of his calculations for evaluation. Obviously, however, whatever calculations he had performed up to that time had given diverse (different probability) results, as indicated by handwritten changes in the transcript of his symposium presentation. His initial estimate apparently was that Venus, as a comet with aphelion near the orbit of Jupiter and perihelion, inside the present orbit of Venus, 'will take an average of some thirty thousand years before it comes to an impact [sic] with the Earth.' By hand he revised the figure to 'ten million years' before releasing the pages for duplication. Now in the 1976 version of the paper there is another increase to 'thirty million years.'"[1]

Thus, it seems when the figures do not give the results that Sagan wishes, he changes the figures and get the numbers he needs. It is suggested that he raise the 30 million years to 30 billion, or 30 trillion, or 30 quadrillion if that will make him feel more secure.

GRAVITY AND COLLISIONS

Thus, Sagan states, "let us take the number 30 million years to give the maximum quantitative bias in favor of Velikovsky."[2] But if Sagan really wanted "to give the maximum quantitative bias" to favor Velikovsky, he would have used his earlier small number of 30 thousand years or at least "10 million years." Sagan has chosen the largest number he had thus far presented; how this gives maximum quantitative bias to Velikovsky is difficult to understand. After stating,

"But Velikovsky has (see e.g., page 388) not one but five or six near-collisions among Venus, Mars and the Earth—all of which seem to be statistically independent events," he adds, "...For six encounters in the same millennium the odds rise to...about a trillion quadrillion to 1."[3]

Shulamith Kogan points out that, "Sagan's calculations *only apply to actual impact collisions.* Yet, in the above quoted text, while Sagan admits that his odds apply to *impact,* he also writes that Velikovsky gives, 'five or six *near collisions.* '"[4] Therefore, Sagan's probability results only apply if the comet—Venus—smashed into the Earth, then went around its orbit and smashed into the Earth again, then went around its orbit and smashed into the Earth again and again and did this smashing process five or six times. This is smashing of a very high order and one will find Sagan's logic quite smashing as well.

Velikovsky actually states, "Each collision between two planets caused a series of subsequent collisions"[5] in *Worlds in Collision.* In response to Sagan he said, "In his statistical approach, Sagan considers all events described in *Worlds in Collision* as independent of one another, whereas they are clearly interdependent."[6] To repeat, Robert Jastrow of NASA's Goddard Institute for Space Studies, stated that, "Dr. Velikovsky had his day when he spotted a major scientific boner in Professor Sagan's argument... Dr. Velikovsky pointed out that if two bodies orbiting the Sun under the influence of gravity collide once, that encounter enhances the chance of another, a well-known fact in celestial mechanics. Professor Sagan's calculations, in effect, ignore the law of gravity. Here, Dr. Velikovsky was the better astronomer."[7] And Robert Bass was so bold as to state that Sagan's, "assumption [regarding independent probabilities] is absolutely identical to the assumption that Newton's Law of Gravity may be ignored,"[8] and also, "This Sagan assumption is so disingenuous that I do not hesitate to label it as either a deliberate fraud on the public or else a manifestation of unbelievable incompetence or hastiness combined with desperation and wretchedly poor judgment."[9] Thus, Sagan's probability odds only work if Sagan somehow manages to repeal the laws of gravity.

On page 85 of *Worlds in Collision,* Velikovsky wrote "The head of the comet did not crash into the Earth but exchanged major electrical discharges with it." On page 372 Velikovsky wrote that planets "cushioned in the magnetic fields about them, a spark will fly from one planet to another, and thus an actual crushing collision of the lithospheres will be avoided..." These statements come directly from the book Sagan is analyzing, therefore, there was no basis for Sagan to suggest that the planets scrape each other.

Does Sagan really believe that Velikovsky had the comet Venus smash into the earth, five or six times? This crashing–smashing event had to occur five or six times for Sagan's probability analysis to give a trillion quadrillion to one odds. S. Kogan writes,

> "Incredibly, it can be shown from Sagan's own text that he actually knows that Velikovsky never thought in terms of a grazing or impact collision. For in another connection (purporting to prove Velikovsky wrong on terrestrial-tidal problems) Sagan unwittingly writes (SCV p. 67 BB p. 103)
>
>> "Velikovsky believes that the close passage of Venus (or Mars) to the Earth would have produced tides at least miles high (page 70-71 [W in C]); *in fact if these planets were even tens of thousands of kilometers away,* AS HE SEEMS TO THINK the tides" etc. [emphasis and capitals added][10]

Hence, it can easily be seen that Sagan's odds which are only valid by repealing the laws of gravity are also not valid because Sagan has let the cat out of the bag, that the planets did not smash into each other five or six times. If they did not smash into each other, then they missed each other. But if the collisions are near collisions and not smashing collisions the probabilities change.

Let us, therefore, see what happens to the odds as the distance between the two celestial bodies grows larger. Shulamith Kogan informs us,

> "Actually, the closest approach described in *Worlds in Collision* could conceivably be several lunar distances... Velikovsky also believed that the planets, because of their magnetic fields, would avoid 'an actual crushing collision of lithospheres' during a close encounter... But already for N = 100 (less than twice the distance to the Moon), we find according to Sagan's statistical methods, a probability greater than 1.0 which, of course, is absurd."[11]

The absurdity of Sagan's analysis lies in the following: What is the probability of pulling an ace of diamonds from a deck of cards? There is a mathematical calculation that can derive the odds for this. But suppose one were to then ask someone who drew from the deck of cards, "What are the odds of pulling more than one ace of diamonds?" This, of course, is absurd. However, Shulamith Kogan using Sagan's probability analysis at N = 100, has the comet during one near collision with the Earth almost hitting it more than one time.

If one were to lower the distance below N = 100, there is a point of distance, using Sagan's probability analysis, where the odds are one chance in one time, that is, the comet would always return to the Earth for a near collision at every passage. By employing Sagan's intriguing analysis for this distance, his trillion quadrillion to one odds must be reduced by one less than a trillion quadrillion.

There is an additional reason that argues against Sagan's conclusion that the Earth and proto-planet Venus scrape each other five or six times. That reason once again is the phenomenon known as gravity which for Sagan constitutes the strongest kind of evidence. For Sagan to derive his probabilities, the Earth and Venus must move inside each other's Roche limit. We have pointed out, citing Isaac Asimov, to show that if a large body moves within 2.44 Earth radii, about 9,760 miles from Earth it would be pulled to pieces by tidal gravitational forces. For Sagan's analysis to work, Venus must defy the laws of gravity not once or twice, but five or six times. It must approach the Earth and be pulled to pieces, but somehow maintain its integrity then do this same impossible gravitational trick four or five more times so that Sagan can claim the odds for such a collision are a trillion quadrillion to one. Actually, the odds against Sagan's analysis based on gravitational theory, is roughly (infinity to one). The only way Sagan's analysis can survive is again to repeal the laws of gravity.

SAGAN'S MATHEMATICS
But there is more. Raymond C. Vaughan examined Sagan's probability analysis and had this to say, "Various examples of Sagan's slapdash scholarship were discussed...by S. Kogan... The following sentence from Sagan's,

> "'An Analysis of *Worlds in Collision*' provides yet another example: [page 97, Sagan writes] 'Thus, the chance of Velikovsky's comet making a single full or grazing collision with the Earth *within the last few thousand years* is
>
> $$(3 \times 10^4) \,/\, (3 \times 10^7) = \; 10^{-3} \text{ or one chance in a thousand.}'$$
>
> "What do Sagan's numbers represent? Thirty million evidently represents the number of years before an impact occurs, or the odds per year against an impact, but there's no clue to the meaning of 3×10^4. If it's supposed to represent *'the last few thousand years'* then it's wrong: it should be 3×10^3 and the quotient should be 10^{-4}. If it's supposed to represent the number of millennia before an impact...then the number seems correct, but its use in this calculation is not what Sagan suggests; the quotient has nothing to do with odds or probability but is simply a mathematical truism that expresses the number of millennia per year [or 0.001 millennia]. Once again the numbers Sagan rattles off don't stand up to close scrutiny." [12]

GRAVITATIONAL EVIDENCE
Is there evidence, not probability mathematics but actual evidence that argues that Venus must have had a near collision with the Earth?

Gravitational theory holds that when celestial bodies come close and interact, as Velikovsky states about the Earth and Venus, then there should remain some lingering remnant in some part of the orbital pattern of both bodies. Velikovsky reports that,

"...a discovery was announced by P. Goldreich and S.J. Peale of the University of California, Los Angeles, and reported at the annual meeting of the American Geophysical Union on April 23, 1966. The surprising discovery dealt with the axial rotation of Venus, already known to be slow and retrograde. Every time Venus passes between the sun and the Earth, it turns the same face to the earth. Gravitationally, this phenomenon cannot be explained even if Venus were lopsided, as some science writers have offered as the explanation, it would have been locked with the very same face toward the sun, whose gravitational pull on Venus is so much stronger than that of the earth; this 'resonance' as the discoverers of the phenomenon termed it, if confirmed, is a sure piece of evidence of close contact in the past between Earth and Venus, evidence not erased by the passage of time, in this case time measured in a mere few thousand years."[13]

An article titled "Venus and Earth: Engaged or Divorced?" in *Astronomy* (Vol. 7 for Oct. 1979), p. 58, discussed I.I. Shapiro and his colleague's analysis of the Venus-Earth resonance. They note radar observations gathered over a 14 year period of time has permitted them to nail down Venus' rotation period with high precision.

"They find it to be 243.01 ±0.03 days. The 3 1/2 hour difference between this value and the resonance period of exactly 243.16 days; while very small, is statistically significant. On the other hand, the researchers point out that the probability of Venus' rotation period falling *by chance alone* within one-fifth of a day of a resonance period is under 1%. Therefore, they suggest that Venus could either now be evolving toward such a resonance, or was once in resonance in the recent past."

William R. Corliss who reported this article in *The Moon and the Planets*, (1985), p. 304, adds this remark,

"The possibility of a recent or imminent resonance is redolent of a recent solar system instability. It would be interesting if 'recent' means 'within the time of man' to that there would after all be astronomical explanations of many legends of celestial turmoil."

Zdenek Kopal in *The Realm of the Terrestrial Planets*, (NY 1979) p. 180 informs us that:

"The remarkable resonance...between the synodic orbit of Venus and its axial rotation with respect to the Earth *is certainly not accidental*. It strongly suggests the existence of tidal coupling [Kopal's emphasis] between the two neighboring planets, but the specific mechanism which could lead to its establishment is largely obscure...a...coupling between Venus and the

Earth—a body much less massive [than the Sun]—constitutes a real challenge to our understanding."

James Oberg further explains how difficult it is for scientists to account for this phenomenon,

"The best explanation for this close resonance (and for the fact that the Venusian year is within a few hours of being exactly 8/13 of Earth's year), to appeal to coincidence—an unsatisfactory solution at best. Nagging doubts insist that something vital is missing from the logic involved. The best current theory [for Venus' retrograde rotation] calls for a large off-centered asteroid impact late in Venus' formation phase. This presents difficulties. Such an accident could reverse the spin but could not account for the spin axis being at near right angles to the plane of the orbit (an extremely unlikely result in a freak collision). If the spin reversing collision could set up nearly any new axis, but this axis would eventually wander back to its old position because of the planet's oblateness. Such oblateness could have disappeared over millions of years that passed while the new slow rotation rate no longer provided sufficient centrifugal force. If this explanation sounds like magic its the best there is. Astronomers remain completely baffled." [Oberg, "Venus" *Astronomy* (August 1976), p. 16].

Zdenek Kopal, above page 191, puts the problem this way,

"The first problem concerns the rotation of the planets. What made Venus rotate so slowly, and what tilted its axis of rotation almost upside down to give rise to its retrograde rotation. The only probable mechanism would be a very close encounter with another celestial body whose gravitational attraction played havoc with Venus and altered some of its kinematics [motions] and at the same time cause it to lock onto the Earth gravitationally?"

The answer is an interaction with the Earth that Velikovsky suggests. Here is what appears to be clear evidence based not on the probability theory, but on "gravitational theory" that Carl Sagan stated must be given greatest weight. It indicates that Venus' axial rotation is locked onto the Earth and not onto the Sun. Hoimar Von Ditfurth in *Children of the Universe,* (NY 1976), p. 115 remarks that, "the Earth must once have exerted a braking or decelerating effect on Venus until the two planets mutual gravitational attraction brought about the 'coupling' we observe today." To do so, the Earth and Venus had to be quite close to each other for their gravitational fields to be effective in creating this couple effect. If the Earth and Venus never had a near encounter then any gravitational anomaly on Venus would cause it to lock onto the Sun. The Earth's gravitational field is far too small compared with that of the Sun to nudge Venus into such a resonance. This evidence Sagan also has not explained.

STATISTICAL INEVITABILITY

It is extremely difficult to believe that Sagan believes that a comet will not reapproach a body once it has had a near meeting along its orbit. In fact, he states that the orbit of a comet, "which intersects those of Jupiter and Earth *implies a HIGH PROBABILITY of a close reapproach to Jupiter...*"[14] [emphasis and capitalization added]. Why would a comet in orbit between the Earth and Jupiter have a high probability of close approach to Jupiter? The reason is that the laws of gravity do exist and the comet would reapproach Jupiter closely precisely as a body that nearly collided with the Earth would have a *high probability* of close reapproach to the Earth. That reason is "gravitation".

Furthermore, in dealing with statistical probability of a comet or asteroid striking the Earth, Sagan states in his book *Comet,*

"Many asteroids that cross the Earth's orbit are probably extinct comets. Because they tend to be small and dim, they are hard to find. There are fewer than a hundred known objects that cross the orbit of Mars or Earth, and this is after a decade of intense searching. These Earth-approaching asteroids are of particular interest to us, because they represent something with serious and calculable consequences: the high-speed collision of an object kilometers across with the Earth would rep-resent a major catastrophe, of a sort that must have happened from time to time during the history of the Earth. *It is a statistical inevitability.*"[15] [emphasis added]

Hannes Alfven and Gustaf Arrhenius in their book *Evolution of the Solar System,* (NASA, Wash. D.C. 1976), p. 460 discuss the probability of capture or collision of a body that approaches the Earth. They state,

"...we learn from Kepler that if a body leaves the neighborhood of the planet after an encounter, it will move in an ellipse which will bring it back to the vicinity of the orbit of the planet, once or twice for every revolution. If the body is not in resonance, it will have innumerable new opportunities to encounter the planet... Hence, even if at any specific encounter capture is 'horrendously improbable' as Kaula puts it, subsequent encounters will occur a 'horrendously' large number of times, so that the probability of a final captures becomes quite large, and may approach unity. [The event must happen]

"In fact, we can state as a general theorem (with specific exceptions) that *if two bodies move in crossing orbits and they are not in resonance, the eventual result will be either a collision or a capture.*" [their emphasis].

Thus, two distinguished scientists say that if two bodies are in overlapping orbits in the same plane, they will collide. They say that whatever the probability is against such an event occurring, is inconsequential and they state their conclusion as a theorem. A theorem is a proposition demonstrably true or acknowledged as true.

Therefore, no matter what probability or improbability Sagan offers against the Earth and proto-planet Venus colliding, becomes mute. According to Alfven, a Nobel prize laureate, and Arrhenius, they can only collide. This event then becomes a statistical inevitability.

Therefore, when Sagan cited his odds of billions and billions and billions to one that the Earth would encounter a comet with an Earth-intersecting orbit, he was, according to his own statement, arguing that a "statistical inevitability" was almost impossible because as he calculated it, the odds of collision were a "trillion-quadrillion to one." If Sagan then wishes to raise the odds against a "statistical inevitability" by all means let him do so. Francis Hitching tells us, "scientists generally rule out of consideration any event that has less than one chance in 10^{50} of occurring" because it is considered an absolute impossibility.[16] Let Sagan raise the odds against a near collision to 10^{50}; then he will have a real case against Velikovsky's several near collisions. He will then have shown that an absolutely impossible event is also "statistically inevitable".

Paolo Maffei in *Monsters in the Sky,* (Cambridge, MA 1980), p. 24 discusses comet Whipple and its interaction with Jupiter. Jupiter is a giant among the planets, being 318 times more massive than the Earth. Nevertheless, gravitational interactions between Whipple's comet and Jupiter would not be very different than those between the Earth and proto-planet Venus because both the Earth and Venus are relatively large bodies. Maffei states that the comet Whipple orbited near Saturn and Jupiter from 1660 to 1770 and adds, "Following *several close approaches to Jupiter,* the comet was captured by the planet." [emphasis added] Why did this comet have several close approaches to Jupiter? The answer is that one approach gravitationally leads to others. The close approaches are not independent probabilities as Sagan claims, but interdependent as Velikovsky claimed. On page 25 Maffei shows that, between 1770 and the year 2046, comet Whipple will have had six close approaches to Jupiter in which in each case, its orbit will change. Now these close approaches are what Sagan claims are immensely improbable. In fact, Maffei also shows (pp. 26-27) that comet Brooks will have had *seven* close approaches to Jupiter between the years 1773 and 2018 in which its orbit changes each time. While Maffei shows on page 21 that comet Wolf between 1750 and 1960 it will have had four close approaches to Jupiter in which its orbit will change. Between the years 1750 and 2046 three comets will defy the probabilities and have several near collisions with a planet. This reinforces this concept of interdependence. The entire basis of Sagan's argument of independent probability as Bass stated earlier on, seems to be a simple fraud.

In discussing the probability of the Sun colliding with another star, R.A. Lyttleton in *Mysteries of the Solar System,* (Oxford 1968), pp. 20–23 states:

"The average separations of adjacent individual stars in the galaxy are so many millions of times greater than their linear dimensions that the probability of a star colliding with another, even in a whole eon of 10^9 years, is minutely small. This consideration has been claimed again and again by many science [writers] to rule out any encounter hypothesis. But in fact, the concept of 'probability' applies only to future and not to past events. Scatter a handful of grains of sand on the floor, and the 'probability' that they come to rest where they do when each might have fallen elsewhere is infinitesimally small. But once something has happened, *probability* no longer comes into it. All that can be inferred from a consideration of this kind is that if the probability of stars colliding in the future is small, then it is unlikely that many solar systems will have been produced by such a mechanism...

"...so long as a set of events can happen it is legitimate to consider them as a possible explanation. There have been many instances of golf balls colliding in mid-air, and there is even on record an occasion when after colliding, both balls fell into the hole. But once such a thing has *happened,* it is no use offering disbelief on the ground that it is too 'improbable.'"

Lyttleton then goes on to examine the mindset of scientists or others who find a collision event among stars unpalatable and concludes,

"The idea that the Earth's existence might depend on mere chance greatly disturbs the holders of such views, with the result that they will snatch at any argument valid or not to try to rid science, as they conceive it, of an emotionally distasteful hypothesis."

Carl Sagan's argument based on the probabilities of the collision hypothesis offered by Velikovsky is of the very same nature as those who reject the concept that the Sun may have had a collision with another star. His odds against the collision hypothesis are merely an index of his emotional distaste of the concept that the Earth has had a recent collision and has nothing to do with the scientific considera-tions that must ensue from such an event.

It is quite ironic that Sagan should employ probability theory to try to disprove Velikovsky's concept that Venus had five or six near collisions with the Earth. At the start of this chapter we cited Sagan who claimed most of the short-period comets had to have multiple encounters with Jupiter and even more comets had to have multiple encounters with Jupiter, than the outer planets, and again with Jupiter to be captured. In fact, astronomers do not accept as valid that Jupiter can capture most of its comets by one encounter with the planet.

R.D. Chapman and J.C. Brandt in *The Comet Book,* (Boston 1984), p. 88 explain:

> "Originally, it was thought that a single close encounter with Jupiter would suffice [to capture a comet]. In fact, single encounters can capture comets into short-period orbits. However, such encounters are very rare and cannot account for the observed number of short-period comets. Instead, it now seems that the capture of long-period comets into the inner solar system results from accumulated perturbations of hundreds of not-so-close interactions with Jupiter and to some extent, with the other giant planets."

Based on Sagan's probability concept, hundreds of encounters with Jupiter are more than highly improbable—if we use the same concept of independent encounters, then the odds rise well above ten to the fiftieth power which is considered absolutely impossible. Sagan's argument against Velikovsky is irrational since he accepts as highly probable hundreds of encounters by comets with Jupiter and the outer planets but then rejects as too improbable the five or six encounters of Venus with the Earth. Why are hundreds of encounters by comets with planets more probable than five or six Venus–Earth encounters?

NOTES & REFERENCES

[1] *Kronos* III; 2, p. 85.
[2] SCV, p. 62; B.B., p. 99.
[3] SCV, p. 63; B.B., p. 99
[4] *Kronos* VI, 3, p. 35.
[5] W in C, p. 373.
[6] *Kronos*, III, 2, p. 34.
[7] *The New York Times*, (Dec. 22, 1979), p. 22E.
[8] *Kronos* III, 2, p. 135.
[9] *Age of Velikovsky*, op. cit., pp. 225–226.
[10] *Kronos* VI, 2, p. 36.
[11] Ibid., p. 40 Note 7.
[12] *Kronos* VI, 4, p. 93.
[13] *Kronos* IV, 3, p. 67.
[14] SCV, p. 98; B.B., p. 324.
[15] Sagan and Druyan, *Comet*, op. cit., p. 266.
[16] Hitching, Francis; *Neck of the Giraffe*, (NY 1982), p. 67.

SAGAN'S THIRD PROBLEM

THE EARTH'S ROTATION

BOILING OCEANS

In this section, Sagan questions whether the Earth ever slowed to a near halt in its rotation, and if this could possibly occur, how the Earth speeds up again. He states,

> "...the energy required to brake the Earth is not enough to melt it, although it would result in a noticeable increase in the temperature: the oceans could have been raised to the boiling point of water; an event which seems to have been overlooked by Velikovsky's ancient sources."[1]

Sagan implies that all the oceans would boil if the Earth gradually slowed its rotation over a period of a day and this boiling water would have killed all life in the seas. Since life still exists in the oceans, Sagan implies the event simply could not have occurred. The problem is, did all the oceans boil from top to bottom or did only particular areas of the world's oceans boil? If the Earth's rotation slows almost to a halt over a period of a day, the heat produced by changing rotational (motion energy) into other forms of energy would need to be calculated. However, it must be pointed out that Velikovsky places these events at the termination of the ice age which he maintains ended about 3,500 years ago. This would also have to be introduced into the calculation. The amount of ice as icebergs in the oceans would have been immense. Furthermore, the gravitational attraction between the Earth and proto-planet Venus would have caused the Earth to move (accelerate) toward Venus slightly. In so doing much of the energy would have been converted into energy of revolution. This conversion of energy to revolution would greatly reduce the amount of energy that would be converted into heat. This motion would make the Sun appear to stand still.

Sagan, the foremost advocate of nuclear winter, must certainly realize that if the oceans were heated, immense clouds of water vapor would have risen quickly. These immense clouds would have blocked much sunlight from the atmosphere and the water vapor would cool to the dew point (which would be considerable below the boiling point) and cool rain would have fallen. Kenneth Hsu in his book, *The Great Dying*, discusses what would happen if, The amount of water vapor thrown into the air would supersaturate the stratosphere... The approach of the large comet postulated by Velikovsky also brought innumerable meteors along with it. Many of these would have fallen into the oceans and thrown enormous amounts of water vapor into the atmosphere. The water vapor would rapidly recondense rain and snow out of the atmosphere. "Croft estimated that most of the vapor would return to the Earth's surface in a few months. Total precipi-tation would amount to a thousand meters (3,000 feet) or so coming down at an average rate of 5 to 10 meters or 200 to 400 inches per day"[2] or 17 to 35 feet or water per day. This would have begun long before the oceans became very warm. Velikovsky's catastrophe was smaller than that described by Hsu. Tilting the Earth's axis would also make the Sun seem to stand still.

The ancient sources however, contradict Sagan regarding the heating of some bodies of water to the boiling point which he claimed Velikovsky overlooked. In *Worlds in Collision,* Velikovsky devotes an entire chapter to this topic entitled, "Boiling Earth and Sea" wherein he states, "'The rivers steamed and even here and there... The sea boiled, all the shores of the ocean, all the middle of it boiled,'" states the *Zend-Avesta.* The star of Tistrya (Venus) made the sea boil."[3] Velikovsky also presents a chapter on the topic of rain in *Worlds in Collision* entitled, "The Hurricane" in which he discusses that the hurricane, "swept every corner of the world"[4] and, of course, snow, sleet and hail would have fallen in certain northern latitudes. Thus, Velikovsky neither overlooked the fact that the seas boiled "here and there" as he reported from his ancient sources, nor did he overlook the evidence that life in the seas would have survived, especially in the seas of the higher latitudes and polar regions where there was an ice cap. The melting of the ice cap and melting of the numerous icebergs would also have to be taken into account.

SOLAR ACTIVITY AND ROTATION
Ralph E. Juergens dealt with the gravitational aspect of Sagan's argu-ment stating,

> "Sagan stated that the most serious objection to Velikovsky's conclusion that the Earth's rotation was slowed...during the catastrophes of the second and first millennia has to do with correcting the effect: 'How does the Earth get started up again, rotating at approximately the same rate of spin? The Earth cannot do it by itself, because of the law of con-servation of angular momentum.'
>
> "At that time, [1974] I suggested this: If the deceleration were due to a transitory increase in the electrical charge of the Earth, then the resump-tion of a 'normal' rate of spin might well be laid to a subsequent loss, or drainoff, of the excess charge. The phenomenon would be entirely in

keeping with, and indeed attributable to, constraints imposed by the law of conservation of angular momentum.

"One of Velikovsky's earliest and most strident critics raising this same issue emphasized the close analogy between the rotation of the Earth and a massive flywheel. It is interesting, therefore, to note that Booker [H. Booker, *An Approach to Electric Science,* (NY 1959), p. 621] in a straight-forward work on the fundamentals of electrical science stresses the fact that electric charge placed on a rotating flywheel increases its polar moment in inertia. The presumption, of course, is that the charge is con-vected as the flywheel rotates and thus constitutes an electric current. It follows that electric charge added to the Earth and convected by its rotation must increase the planet's polar moment of inertia.

"Now should this happen—should appreciable, unaccustomed charge be emplaced on the Earth, other conditions aside—the law of conserva-tion of angular momentum must come into play. The increase in polar moment of inertia must be accompanied by a decrease in angular velocity of rotation, such that the product of the two remains constant; this is what the law says. Increase the Earth's electric charge, and the length of its day must also increase...

"The second of our premised assumptions helps to explain the restora-tion of spin following a hypothetical charging episode. If the Earth's customary burden of charge...acquired during an extraordinary event will presumably be dissipated into the environment rather quickly in the aftermath of that event. The Earth effectively 'grounded' to the inter-planetary medium must almost immediately begin to shed its excess charge, and its spin rate must increase accordingly.

"When the process is completed, assuming that the environment itself has not been significantly altered by the passing event, we may expect to find that the length of the day is just about what it was before."[5]

Juergens then proceeds to give evidence that this phenomenon has been observed and reported by reputable scientists. He adds,

"In 1960 Danjon reported a sudden deceleration of the Earth's rotation following a solar flare of record intensity. According to his observations, the length of the day increased by 0.85 milliseconds and thereafter began to decrease at a rate of 3.7 microseconds per day. Eventually the rate of spin stabilized near its pre-flare value.

"This announcement raised quite a few eyebrows. Quite impossible, said the experts. One skeptic pointed out that the phenomenon implied an increase in the Earth's polar moment of inertia of such magnitude as might be produced, for example, by instantly lifting the entire Himalayan massif to a considerable height. Danjon, anticipating such objections argued that, 'it is very likely that electromagnetism alone that will furnish the explanation for these variations'... But his claim was generally disregarded."[6]

In *Our Amazing World of Nature: Its Marvels and Mysteries,* Herbert
Friedman describes the event of that electrical storm as follows:

> "At 2:37 p.m. on November 12, 1960, astronomers detected a brilliant
> explosion on the face of the Sun. Six hours later, a gigantic cloud of solar
> hydrogen gas, ten million miles across and still trailing halfway back to
> the Sun, 93 million miles away, collided with the Earth at a speed of four
> thousand miles a second.
>
> "Though inaudible and invisible, the collision started a violent chain of
> disturbances on and around the Earth, an electric and magnetic storm of
> mammoth proportions. Compass needles wavered erratically. For hours,
> all long-distance radio communications were blacked out. Teletypes
> printed gibberish. Overhead, sheets of flaming-red northern lights flashed
> in the night sky, bright enough to be seen through overcast and clouds.
> Electric lights flickered in farmhouses as if a thunderstorm raged, yet the
> air and sky were clear and silent.
>
> "For more than a week, such chaotic conditions continued. They
> were clearly the results of our Sun on the rampage. Yet, let me assure
> you that such a storm amounts to no more than a tiny ripple in the usual
> steady flow of solar energy."[7]

The effect of this solar storm had clearly carried electromagnetic
energy to the Earth. The storm's effects were felt on the Earth for a
week. When Andre Danjon presented this evidence that Earth's rotation was affected by the electrical energy of this gigantic storm, the
scientific community, steeped in Newtonian dynamics, treated the
evidence as though it did not exist because it violated cherished scientific models of celestial motion. But Juergens goes on to show,

> "Then in 1972 it happened again, even more impressively, Danjon was
> gone (deceased 1967), but Plagmann and Gribbin were on the watch.
> They found that on August 7-8, following a week of frenzied solar
> activity, the length of the day suddenly increased by more than 10
> milliseconds (more than a hundredth of a second). And again, there was
> gradual return to normal...
>
> "Again, disbelief on the part of the establishment led to excesses.
> Neglecting the important facts that, following Danjon, Plagmann and
> Gribbin had predicted the phenomenon, had been on the alert for
> unusual flare activity, and had been rewarded for their vigilance...
>
> "The increase in the length of the day reported by Danjon is very
> nearly 10^{-8} day...such an effect must involve a temporary increase in the
> Earth's polar moment of inertia of approximately the same proportion.
> Since the 'normal' value is about 10^{30} km-m^2...
>
> "It is interesting to note that a mere 10,000 fold multiplication of the
> suggested cause of the Danjon effect...might produce the Velikovsky
> effect..."[8]

Thus, two scientists on two occasions give evidence that the Earth's rotation not only slowed, but in complete contradiction to Sagan's argument, "How does the Earth get started up again" they give evidence that the Earth's rotation started to increase over a period of days to its pre-flare rotational velocity. There is no gravitational explanation for this behavior and that is probably why Sagan rejects it. To admit that this evidence is valid would require that its cause be sought. The only other known force that could do this would be electromagnetism. Therefore, rather than deal with such evidence, Sagan chooses to disbelieve it. By disbelieving the evidence, it may go away. This is a unique scientific attitude on Sagan's part.

SOLAR MAGNETISM
Velikovsky informs us that,

> "In 1948–1949 Donald Menzel produced motion pictures of prominences [flares] or explosions of matter on the sun; they were made at the Solar Observatory in Climax, Colorado. The exploded matter rose at a very great speed to immense heights, all the time gaining in velocity, and then descended to the sun, not on a curved path, as a missiles would do, but by retreating on the path it had covered, comparable to a missile reversing its direction and returning to its point of departure. Moreover, the velocity of its descent was without the acceleration expected in a fall, and this too was in violation of gravitational mechanics.
>
> "It was observed that when the protuberances or surges of exploded matter on the sun run into one another, both of them recoil violently; such observation was made by McMath and Sawyer, and on other occasions by Lyot. The conclusion drawn by E. Petit of Mount Wilson Observatory (1951) is that solar protuberances are electrically charged."[9]

Here we observe that the rotation of the Earth seems to be in accord with Velikovsky's hypothesis. Electrified solar flares of great intensity add electric charge to the solar magnetosphere which travels to the Earth and adds electric charge to the Earth. This changes the Earth's polar moment of inertia, which slows the spin of the Earth; and as the charge on the Earth is shed back into the solar magnetosphere in which the Earth is grounded, its spin returns to normal.

GEOMAGNETISM
Sagan argues that if a magnetic storm of such intensity affected the Earth, it would have left a strong signature in the form of remanent magnetism in the rocks. He states, "there is no sign of rock magnetization of terrestrial rocks ever having been subjected to such strong field strengths…"[10]

This is incorrect because Velikovsky in his "Forum Address" states of the remanent magnetism found in rocks:

"The newly developed science of paleomagnetism brought, and daily continues to bring, confirmation of the fact that lavas and igneous rocks in all parts of the world are reversely magnetized. But what is more startling is to find that the reversely magnetized rocks are a hundred times [or a thousand times] more strongly magnetized than the earth's magnetic field could have caused them to be. H. Manley writes in his review, writes It may seem strange that a rock which is made magnetic by the earth's 'field' should become so strongly magnetized [Manley states] 'compared with the generating force. This is one of the most astonishing problems of paleomagnetism.'"[11]

Apparently Sagan when he stated that "there is no sign of rock magnetism of terrestrial rocks ever having been subjected to such strong field strengths" hasn't any problem with reporting the evidence as compared with H. Manley whose article "Paleomagnetism" in *Science News* for July 1949, finds that the rock magnetism of terrestrial rock has been subjected to such strong field strengths. Apparently Manley is unaware that the evidence he reports from his own field cannot be correct because Sagan tells us that it isn't. One may suspect that Manley, who was astonished by the Earth's high remanent magnetism would be struck dumb by Sagan's assertion. But since the rocks do contain high remanent magnetism it had to originate somewhere. Velikovsky suggests some came from the strong electromagnetic field of Venus when it was a comet. We have shown that lightning strikes in rock changes the direction of the magnetism in the rock. It also leaves strong remanent magnetism as well.

COMETARY MAGNETISM

Sagan states, "no cometary influence can make the Earth stand still"[12] and "Nor is there anything that braking the Earth to 'halt' by cometary collision is any less likely than any other resulting spin."[13] Thus, the question is: "do comets carry electric charge?" Velikovsky also addresses this point.

"In the configuration of the cometary nuclei and tails there was found 'good evidence that all particles in the comet influence the motion of each other,' and the configuration of the streamers in the tails of many comets 'strongly indicate a mutual repulsion.' Thus, writes Professor N.T. Bobrovnikoff, Director of the Perkins Observatory [in *Astrophysics* ed. J.A. Hynek (1951) pp. 327-328]. It was also calculated that the repulsion of the tails of the comets by the sun is twenty thousand times stronger than the gravitational attraction, and the implication is that it cannot be caused by the pressure of light as previously thought, and that *electrical* repulsion must be in action. From spectral analysis it is known that the cometary tails do not shine merely by reflected light, and that their light is not caused by combustion either, but most probably is an *electrical effect,* comparable to the effect of the Geissler tube."[14] [emphasis added]

Furthermore, it was reported in *Science,* for December 1973 that radio noise was detected from the comet Kohoutek as it approached perihelion. Radio noise that is derived from a body which is not a thermal source tells us that the body carries an electromagnetic field.

The next question is, "how strong is a small comet's magnetic field?" On September 15, 1982, observations of a radio star were made through the magnetic tail of comet Austin by Imke de Pater and Wing-H. If the magnetic field was weak the direction and polarization of the radio star's emissions would not change direction. But, if the electromagnetic field of the comet distorted both the direction and polarization of the radio star's emission, then the electromagnetic field of the tail would be strong. According to an article in *Sky and Telescope,* Vol. 67, (1984), p. 415, titled "Comet Austin's Puzzling Tail" we learn that the "observations were made at a wavelength of 20 centimeters (1,465 megahertz) with the Very Large Array in New Mexico. When Austin's plasma tail came within some 220,000 kilometers of the line of sight to [radio star] 1242 + 41 there was a pronounced change in the position of the radio source." To do this the magnetic fields in the comets tail had to be hundreds or thousands of times greater than the present concept of cometary magnetic fields.

So it seems comets do possess strong magnetic fields and an immense comet such as Venus, coming from the core of Jupiter, would carry a part of Jupiter's large electromagnetic energy. Thus, when Venus approached the Earth, it would have discharged great planetary thun – derbolts. Velikovsky devotes a chapter in *Worlds in Collision* titled "The Spark" in which ancient people describe immense lightning strokes loosed from the comet to the Earth. These celestial lightning strokes would have changed the Earth's polar moment of inertia, just as electrical flares from the Sun do. There is not one single point of evidence respecting the slowing and restoration of the spin of the Earth that is not supported by scientific evidence.

The journal *Science* contains an article on the interaction of magnetic fields between an asteroid and Venus which states,

"Examination of *Pioneer Venus* observations of *the interplanetary magnetic field increases to a peak* and then decreases almost symmetrically. *The largest of these events lasted almost 10 hours and rose to a peak field strength of over 140 percent of that of the surrounding magnetic field.* It was suggested that this enhancement was associated with the passage of a comet (Oljato) by Venus."[15] [emphasis added]

If as small a body as Oljato can raise the peak magnetic field strength between itself and Venus by 140 percent above the normal surround –

ing magnetic field, it seems probable an immense comet with a huge electromagnetic field would raise the field strength between itself and the Earth by a thousand percent or more.

Sagan stated earlier that Newtonian dynamics have a very firm footing; well, so does electromagnetism. Sagan believes these forces play no role in celestial mechanics. John Gribbin, in *The Death of the Sun,* presents a graph on page 131 wherein he correlates solar activity with the length of the day: the rotation of the Earth. His chart begins around 1820 and extends to about 1980. It shows that when the Sun was more active—produced more electromagnetic radiation—the length of the day increased because the rotational velocity of the Earth slowed. When the Sun was less active, the opposite occurred.

In 1981, D. Djurovic's, "Solar Activity and Earth's Rotation", in *Astron Astrophys,* Vol. 100, pp. 156-158, showed a correlation of the Earth's rotation with well-known solar Sun spot variation periods of 0.5, 3.3, 6.6, and 11 years.

In 1982, F. Carter, et. al, in "A Comparative Spectral Analysis of the Earth's Rotation and Solar Activity", in *Astron Astrophys,* Vol. 114, pp. 388-393, measured solar activity and emphasized a possible relationship with the Earth's seasonal rotation.

In 1983, D. Djurovic's "Short-period Geomagnetic Atmospheric and Earth-rotation Variations", in *Astron Astrophys,* Vol. 118, No. 1, pp. 26-28, studied 3 and 4 month rotational periods of the Earth and solar activity and showed that there is a correlation of georotation with solar behavior for short periods.

It is well known that the Sun oscillates up and down some ten kilometers periodically every 160.01 minutes. G.P. Pil'nik's "Multiple Waves in the Earth's Diurnal Rotation", in *Soviet Astronomy,* Vol. 28, No. 1 pp. 112-114, showed that this solar oscillation of 160.01 mins. was reflected in changes in the rotation of the Earth of 159.56 mins.

Each of these studies resoundingly illustrates that long and short term variations in the Sun's activity are reflected in corresponding correlated changes in the Earth's rotation.

Sagan may refuse to acknowledge this evidence, however, the concept that Velikovsky's suggestion that electromagnetism does play a role in terrestrial rotation is gradually being recognized in astronomy and is increasingly difficult to ignore. In fact, James S. Trefil in his book, *Space Time Infinity,* admits "To further complicate the problem (of the length of the day) recent research suggests that the solar wind [the Sun's electromagnetic field] can alter the length of the day." [16]

LINEAMENTS

If the Earth's rotation halted slightly and the globe actually rolled about as S.K. Runcorn has stated, then the Earth, like Mars and the Moon, should also show a global system of lineaments and this it does. G.H. Katterfield and G.V. Charushin in *Modern Geology*, Vol. 4, (1973), p. 253 discuss this fact and conclude that,

> "A comparison of Earth with Mars, Mercury and the Moon shows that global systems of lineaments determined for the Earth are common to the terrestrial planets. The origin of such regularly oriented general planetary grids is attributable to planetary causes, i.e., deformation of the planet due to changes in its size, internal constitution and *rotation rate.*" [emphasis added]

Thus, the geological evidence corroborates the remanent magnetism and the astronomical evidence to show that the Earth has indeed changed its rotational velocity. The changing of the position of the polar axis suggested by Runcorn would certainly make the sun appear to stand still during the catastrophe that Velikovsky describes. Therefore, the Earth did not have to come to a complete stop as Sagan seems to suggest and the energy of rotation did not have to all be converted into heat to kill all life in the oceans. In our discussion of Sagan's Appendix II we will show that the concept of a tectonic polar shift was well thought of by Albert Einstein, who was quite willing to give his support to just such a concept. It is only by avoiding and ignoring this concept that Sagan can raise the issue of boiling oceans.

NOTES & REFERENCES

[1] SCV, p. 64; B.B., p. 100.

[2] Hsu, Kenneth; *The Great Dying*, (NY 1986), pp. 190–191.

[3] W in C, pp. 91–93.

[4] Ibid., p. 69.

[5] *Kronos*, II, 3, pp. 12–14.

[6] Ibid., pp. 14–15.

[7] *Our Amazing World of Nature: Its Marvels and Mysteries*, (Reader's Digest Book), (Pleasantville, NY), p. 261.

[8] *Kronos*, II, 2, pp. 15–16.

[9] *Earth in Upheaval*, op. cit., p. 263.

[10] SCV, pp. 65–66.

[11] *Earth in Upheaval*, op. cit., pp. 254–255.

[12] SCV, pp. 65–68.

[13] *Earth in Upheaval*, op. cit., pp. 263–264.

[14] *Earth in Upheaval*, op. cit., p. 293.

[15] Russell, C.T., et. al., "Interplanetary Magnetic Field Enhancement and Their Association With Asteroid 2201 Oljato", *Science*, (Oct. 5, 1984), p. 43.

[16] Trefil, James S.; *Space Time Infinity*, (NY 1985), p. 159.

SAGAN'S FOURTH PROBLEM

TERRESTRIAL GEOLOGY AND LUNAR CRATERS

EARTH IN UPHEAVAL

Sagan deals here with the question of whether geology, archaeology, and craters observed on the Moon can substantiate Velikovsky's hypothesis. As shown earlier, Velikovsky wrote a book on the these aspects of his theory. Macbeth tells us,

> "Velikovsky's opponents pointed out that he was talking about events that qualified as catastrophes, transcending anything that is now going on in scale and violence. They declared that this put Velikovsky out of court because the uniformitarian doctrine provided no room for such events. Velikovsky, who...was rather innocent as to Anglo Saxon geological theory, was surprised at this reaction and at the violent feelings he had aroused. His response was admirable; without extensive public recrimination, he disappeared into the library for several years and complied a book called *Earth in Upheaval.*"[1]

This book of evidence has been available since 1955, or 19 years prior to the time Sagan first presented his attack on Velikovsky. Therefore, it is interesting to read Sagan's questioning:

> "To the best of my knowledge, there is no geological evidence for a global inundation of all parts of the world at any time between the sixth and fifteenth centuries B.C. If such floods has occurred, even if they were brief, they should have left some trace in the geological record. And what of the archaeological and paleontological evidence? Where are the extensive faunal extinctions of the correct date as a result of such floods? And where is the evidence of extensive melting in these centuries, near where the tidal distortion is greatest?"[2]

A.A. Meyerhoff and H.A. Meyerhoff in discussing geologists, who advocate continental drift theory when they encounter evidence that contradicts their uniformitarian theory state,

> "There is a failure of almost all geologists and geophysicists to face [contrary evidence]...is fatal [and] is incredible in a profession as responsible as ours.
>
> "Possibly the four greatest failings...are [1] subjectivity in the selection of data (those data which conflict are omitted or are not discussed... [2] general failure to recognize geological facts...[3] continuous *ad hoc* modifications of the hypothesis... [4] an apparent unwillingness...to discuss even among (themselves) the many contradictions..."[3]

Sagan is apparently so determined to avoid Velikovsky's geological evidence that he succumbed to the failings of other scientists. This denial that such evidence exists "to the best of my knowledge" is highly unscientific. Macbeth states "Catastrophes have been taboo for a century among the orthodox." [4]

Apparently, orthodox uniformitarians such as Sagan seems to be, do not want to be contaminated with evidence that is taboo to their views of science.

THE END OF THE ICE AGE

Sagan requests geological evidence for the "correct date" of the events that Velikovsky claims for a world-wide catastrophe. Velikovsky's date, is that 3,500 years ago, is the end of the Ice Age, the rising of mountains to their present heights and a period when the Earth was inundated by the oceans plus major extinctions of fauna and changes in climate. Thus, in terms of this probable dating, evidence for this date should exist. Based on radiocarbon dating, Velikovsky stated,

> "It was shown that ice, instead of retreating 30,000 years ago, was still advancing 10,000 or 11,000 years ago. This conflicts strongly with the figures arrived at by the varve method concerning the final phase of the Ice Age in North America.
>
> "Even this great reduction of the date of the end of the Ice Age is not final. Radiocarbon analysis according to Professor Frederick Johnson, chairman of the committee for selection of samples for analysis [*The Committee on Carbon 14 of the American Anthropological Association and the Geological Society of America*], revealed 'puzzling exceptions.' In numerous cases the shortening of the time schedule was so great that, as the only recourse, Libby assumed a 'contamination' by radiocarbon. But in many other cases 'the reason for the discrepancies cannot be explained.' Altogether the method indicates that 'geological developments were speedier than formerly supposed.'
>
> "H.E. Suess of the United States Geological Survey reported recently that wood found at the base of interbedded blue till, peat and outwash of drift, and ascribed by its finder to the Late Wisconsin (last) glaciation is, according to radiocarbon analysis, but 3,300 years old (with a margin of error up to two hundred years both ways), or of the middle of the second millennium before the present era [3,500 years ago]" [5]

According to C.E.P. Brooks in *Climate Through the Ages,* (NY 1949), p. 296, between 8,000 and 4,000 years ago, the temperature of the Earth was 5 degrees Fahrenheit warmer than now. This is termed the "hipsithermal" and there is no acceptable explanation for this higher temperature.

According to Charles Hapgood, *The Path of the Pole,* (Philadelphia 1977), pp. 106–107,

"During the Byrd Expedition of 1947-1948, Dr. Jack Hough, then of the University of Illinois, took three cores from the bottom of the ocean off the Ross Sea, and these were dated by the ionium method of radioactive dating at the Carnegie Institution on Washington by Dr. W.D. Urry who had been one of those to develop the method. [What the cores showed were] layers of fine sediment typical of temperate climates. It was the sort of sediment that is carried down by rivers from *ice-free continents*. Here was a first surprise, then. Temperate conditions had evidently prevailed in Antarctica in the not distant past... [emphasis added]

"Then when this material was dated by Urry, it was revealed that the most recent temperate period had been very recent indeed. In fact it ended only about 6,000 years ago." [See Hough, Jack; "Pleistocene Lithology of Antarctic Ocean Bottom Sediments", *Journal of Geology,* Vol. 58, pp. 257-259]

This is corroborated by Reginald Daly in *Earth's Most Challenging Mysteries,* (Nutley, NJ, 1975), pp. 227 and 264, who states that,

"Carbon 14 dating has shown that Antarctica's ice is less than 6,000 years old. [Arthur] Holmes writes, 'Algal remains, dated at 6000 B.P. [before present] have been found on the latest terminal moraines.' (*Principles of Physical Geology,* p. 718) This shows that Antarctica must have been sufficiently free from ice for green algae to grow 6,000 years ago."

But how long ago was Antarctica last glaciated? Velikovsky's theory implies that it was rapidly reglaciated in historical times, but how to prove such a concept. Charles Hapgood in *The Path of the Pole,* p. 111, gives evidence that Antarctica had no continental glacier until only a few thousand years ago.

"It is rare that geological investigations receive important confirmation from archaeology; yet, in this case, it seems that this matter of deglaciation of the Ross Sea can be confirmed by an old map that has somehow survived for many thousands of years. A group of ancient maps, including this one, was the subject of an entire book which I published in 1966 under the title *Maps of the Ancient Sea Kings.*

"In some way or other, which is still not, and may never be entirely clear, this extraordinary deglaciated map of Antarctica has come down to us. Apparently originated by some ancient people, preserved perhaps by Minoans, Phoenicians, and Greeks, it was discovered and published in 1531 by the French geographer Oronce Fine, and is part of his Map of the World... 'Impossible!' That would be the opinion of practical people of intelligence and learning. 'Utterly impossible!' But sometimes truth is really strange. It has been possible to establish the authenticity of this ancient map. In several years of research, the projection of this ancient map was worked out. It was found to have been drawn on a sophisticated map projection, with the use of spherical trigonometry, and to be so scientific that over fifty locations on the Antarctic continent have been found to be located on it with an accuracy that was not attained by modern cartographic science until the nineteenth century. And of course, when this map was first published in 1531, nothing at all was known of Antarctica. The continent was not discovered in modern times until about 1818 and was not fully mapped until after 1920."

Thus, the evidence of Brooks regarding the hipsithermal period when the Earth was 5 degrees Fahrenheit warmer 4,000 years ago is corroborated by Jack Hough's finding of sediments off the Ross Sea which are typical of temperate climates and are dated to 6,000 years ago. This evidence is further corroborated by Arthur Holmes' finding algae remains in the latest terminal moraines of Antarctica that are dated to be 6,000 years old. And all of this is corroborated by Charles Hapgood who proved ancient man had been to, and carefully mapped the continent of Antarctica. What is clearly shown by this evidence is that 4,000 years ago something stupendous happened to change that climate. All this is fully consistent with Velikovsky's theory that the climate changed about 3,500 years ago.

Evidence also comes from Arlington H. Mallery's *The Rediscovery of Lost America,* (NY 1951), pp. 194-199. In his chapter "The Zeno Map of the North", Mallery describes a map published in 1558 of Greenland. He states on page 196,

"The Greenland depicted [on the map] differs radically from the Greenland known to the modern world. The land surface is shown free of ice, almost covered by mountains, crossed by open rivers, and divided into *three* islands! A fjord marked *Ollum Lengri* on a version of the map and a flat surface, which I concluded was a strait, extending westward between the mountains, divide the land which, now hidden by ice, we know as Greenland, a single island.

"My belief that the map might be authentic was not shaken by the fact that it shows such an unusual topography for Greenland. For, as I have explained, the topography of both Greenland and Iceland has been drastically altered since the Middle Ages by erupting volcanoes, earthquakes, and ice caps. Consequently, an accurate ancient map would show both countries with contours unlike their modern contours." [A world-wide catastrophe such as that described by Velikovsky would also have changed the topography.]

Mallery continues,

"So assuming that the map was an ancient Norse sailing chart, I platted it on a polar projection. It then became obvious that the original map had been platted on a portolano projection. Drawing grids on a copy of the Greenland section of the map, I found even more than I had dared to hope I would find: all the points were platted with remarkable accuracy when compared to the same points on the latest U.S. Army maps! I was then certain that the rejected Zeno map was an accurate map of

Greenland before it was covered by its present sheet of ice more than a mile thick."

Since the ability to plot longitude was unknown even to the Norse and ancients of Greece, Egypt and Babylon Mallery concluded, page 197, "Of one thing I was certain: this accurate map of the North Atlantic was the product of a very remote, very advanced civilization." Mallery continues, p. 198, discussing the southern area of Greenland:

"Sooner than I would have thought possible, confirmation of my analysis of the map (at least the portion showing Greenland) came from an authority in the science of determining subglacial topography by seismic soundings. The authority was Paul-Emile Victor, whose French Polar Expeditions explored Greenland from 1948 to 1951. An Associated Press news dispatch announced on October 26, 1951, months after I had published my analysis of the map, the following discovery of the Victor expedition: 'A French expedition reported this week that Greenland is really three islands bridged by an ocean.' In a letter to me, dated October 22, 1953, Victor said: 'The analysis of our soundings confirms the preliminary announcement that Greenland is really three islands...'

"Victor's soundings revealed a passage westward across Greenland corresponding to the flat area between the mountains which I had pointed out as a strait dividing the land. They also showed the presence of a large fjord in a location corresponding to the location on one version of the Zeno map of a fjord marked *Ollum Lengri-Lengri,* longest of all...

"Coming almost as an anticlimax to Victor's confirmation of the accuracy of the Zeno map, a later development strengthened this confirmation. When I asked Victor to explain the presence of a single mountain in the flat area crossing Greenland, he showed me his soundings, which proved the 'mountain' to be an island!"

Thus, we have ancient maps of both Greenland and Antarctica relatively free of ice while Greenland is completely ice free. If one assumes that the Ice Age ended as the uniformitarian establishment geologists claim 10 to 12 thousand years ago, then one must draw the preposterous conclusion that civilizations capable of charting the Earth existed many tens of thousands of years prior to this time. Velikovsky claimed that there was a polar shift which redistributed the climate and moved the polar regions to their present boundaries. These ancient maps confirm that ancient man lived in a world in which the climate was different, and that 8,000 to 4,000 years ago was a period of warm climate that allowed man to visit and map Antarctica and Greenland. Civilization can be no older than 6,000 years. Therefore, the ice covering the Antarctic continent and the great island of Greenland had to have come after the hipsithermal period when it is well known the

Arctic Ocean was ice free. This corresponds fairly well with the change Velikovsky posited for this event 3,500 years ago.

If as Velikovsky suggests the ice caps of Greenland and Antarctica were built up rapidly at the time of his 3.500 year old catastrophe then one would expect that the dust in the atmosphere rained out in the tropic and temperate zones but fell as snow over Greenland and Antarctica. Velikovsky suggests the dust in the ice caps of Greenland and Antarctica was produced by immense volcanism on Earth and dust left from the raging proto-planet Venus. Therefore, we would expect the ice caps to reflect this by having an inordinate amount of dust below a certain level. Above that level the ice caps would have dust that is currently falling at the present average rate. During the ice ages it is well known that there was a great deal more rain. This is stated by Robert Silverberg whom we will cite below that during glacial epochs there was greatly increased rainfall. A rainy climate does not allow dust to remain in the atmosphere. Based on the uniformitarian concept that the ice during the ice ages was built up gradually during a period of much greater rainfall than we experience at present, then, the ice caps of Greenland and Antarctica should possess much less dust in them during the Ice Age than currently. But if Velikovsky is correct the Ice Age ice should contain much more dust than ice of the present period. This is exactly the opposite of what the Ice Age theory demands. The findings support Velikovsky's hypothesis and categorically deny the uniformitarian hypothesis. According to Hammer, Clausen, Dansgaard, Neftel, Kristimlotter and Reeh in the *American Geophysical Union Monograph,* Vol. 33 (1985), p. 90 the dust particles in the Greenland glacier "was up to 100 times as great in the last ice age as at present." With respect to Antarctica they say the dust compared to Greenland is "an order of magnitude higher." or 1000 times as great as at present. It is impossible for so much dust to accumulate during time periods when there was less dust in the atmosphere.

Sagan, in *Broca's Brain,* "The Climates of Planets", pp. 189-190, writes,

> "There are many indications of past climatic changes. Some methods reach far into the past, others have only limited applicability. The reliability of the methods also differ. One approach, which may be valid for a million years back in time, is based on the ratio of the isotopes oxygen 18 to oxygen 16 in the carbonates of shells of fossil foraminifera. These shells, belonging to species very similar to some that can be studied today, vary the oxygen 16/oxygen 18 ratio according to the temperature of the water in which they grew."

Let us examine this evidence to see what it says. The reader will recall the prior discussion respecting boiling oceans. If the Ice Age ended, as

Velikovsky claims 3,500 years ago, then there should be evidence of a quite sudden change in the temperature of the oceans. Cesare Emiliani of the University of Chicago is an Earth scientist interested in the question of the ocean's temperature. He did discover a new dating process that Sagan mentioned above developed from the field of atomic physics. His method measures the amount of an isotope of oxygen in the shells of oceanic organisms and in particular, oyster shells. We are informed by Macbeth that,

> "Cores of undisturbed sediment were brought up from the ocean floor. Careful study showed changes in temperature from layer to layer in the cores, enabling Dr. Emiliani to chart the climates of the past and trace the ups and downs of the Ice Age. The results were disturbing because they cut the duration to a third or less of what had been assumed to be the proper figure.
>
> "I do not assert that Dr. Emiliani was correct. I only want to show that geological and biological projections into the past (like most extrapolations) have a precarious base and that a shift in chronology would shake the foundations on which Lyell and Darwin constructed their theories."[6]

Thus, it is seen that the absolute dating of past events can give diverse results.

CLIMATE EVIDENCE

Another example of climate changes that appears to be in accord with Velikovsky's claim that the climate changed twice—once drastically 3,500 years ago and a second less drastic change 2,800 years ago—is seen in the following. According to Jonathan Weiner in *Planet Earth,* p. 99, Reid Bryson, a climatologist at the University of Wisconsin, has concluded that the Harrappan empire ended because of a major climatic change that occurred 3,600 years ago. From analysis of pollen from varves (mud layers) in salt marshes around Harrappa in India he claims,

> "Between 10,000 and 3,600 years ago...rainfall was at least three times what it is now. Then, about 3,600 years ago, again judging by the pollen record, plants began changing from those characteristics of fresh water to those that grow around salt water. And then there's no lake at all—the lake dried up and there's barren sand.""
>
> 'That period lasted about 700 years and ended some 2,900 years ago,' says Bryson. 'That is a very long drought.' All the vegetation in the region disappeared. Rainfall afterward increased but never returned to the level it had reached in the heyday of the Harrappan's...they were so totally dependent upon the monsoon rains. When, about 3,600 years ago...drought began, their agriculture, the economic base of their culture, disappeared."

This evidence appears to support Velikovsky. There is also evidence that during the period of the Ice Age, there was much greater rainfall than during the interglacial period such as the one we are currently enjoying. Robert Silverberg in *Clocks of Ages,* (NY 1971), p. 94, writes, "During the glacial epochs, such regions as Africa, South America, central Asia, and the southern United States experienced 'pluvial' periods [of greatly increased rainfall]... During these pro-longed rainy spells, lakes and rivers grew, basins now dry filled with water." Thus, there also appears to be a close correlation between the fall of the Harrappan civilization and the end of the Ice Age.

VOLCANOES

Sagan states that if, as Velikovsky claims,

> "'that in the days of the Exodus, when the world was shaken and rocked...*all* volcanoes vomited lava and *all* continents quaked'... Volcanic lavas are easily dated, and what Velikovsky should produce is a histogram of the number of lava flows on Earth as a function of time."[7]

This might take 20 or 30 years to do and Velikovsky would naturally never be able to finish his work in several other areas. But since Sagan seems so intent on disproving Velikovsky's thesis, perhaps he ought to undertake a 20 or 30 year study to prove his case against Velikovsky by making a histogram of all the lava flows on Earth. Velikovsky has a chapter in *Earth in Upheaval* p. 136, that shows there was greater volcanism in the past than at present:

> "All in all only about four or five hundred volcanoes on Earth are considered active or dormant, against a multiplicity of extinct cones. Yet only five or six hundred years ago many of the presently inactive volca-noes were still active. This points to a very great activity at a time only a few thousand years ago. At the rate of extinction witnessed by modern man, the greater part of the still active volcanoes will become extinct in a matter of several centuries."

Francois Derrey in *Our Unknown Earth,* (NY 1967), p. 186 confirms this:

> "K. Sapper, a geologist tried to add up the volcanic activity of the last five centuries. Retaining only very moderate figures, and supposing that this activity was always kept at its present rate—we have seen that it was certainly much more considerable in the past."

Charles Hapgood concluded that the last Earth-wide upheaval occurred 3,400 years ago, around the same time as the catastrophe described by Velikovsky and shows that there was a volcanic maximum that coincides with the end of the Ice Age. He writes,

> "There have been times in the past when the quantity of volcanic action has been extraordinary. As an example of this, there appears to be

evidence that in a small area of only 300 square miles in Scandinavia, during Tertiary times, there may have been as many as 70 active volcanoes at about the same time. Berquist, who cites the evidence remarks, 'Volcanic activity on this scale erupting through 70 channels and concentrated in a relatively short period must have been very impressive.'"[8]

Hapgood goes on to show that the period that ended the Ice Age was one of volcanic maxima but he does not date every lava flow. He claims that based on his list,

"it is, of course impossible to come to any conclusion as to the amount of volcanism throughout the world at any one time during the glacial period; perhaps 5,000 radiocarbon dates would be necessary for this."[9]

But, just as Velikovsky, he argues that in the recent past there was much more active volcanism than now.

Frank C. Hibben in *The Lost Americans,* pp. 176-177 tell us, "There is no doubt that coincidental with the end of the Pleistocene [Ice Age] at least in Alaska, there were volcanic eruptions of tremendous proportions." And, J.P. Kennett and R.C. Thunell in *Science,* Vol. 187, pp. 497-502 state that, "Volcanicity in Iceland during the last 10,000 years has been three to five times more vigorous than during previous Late Cenozoic intervals." Their paper argues that the Ice Age period of volcanism was a maximal volcanic era. Francois Derrey in *Our Unknown Earth,* (NY 1967) p. 184 states,

"...it is probable that they were infinitely more frequent and above all more violent in the past. In our time, the craters of active volcanoes hardly exceed six-tenths of a mile in diameter, while the mouths of former volcanoes measure tens of miles."

However, one researcher undertook to make a histogram for Iceland only. Dr. Euan W. Mackie of the University of Glasgow and assistant keeper of the Hunterian Museum states,

"peaks in the histogram of dates of volcanic eruptions must coincide with peaks in that of sea level changes if [global catastrophic theory] is correct. If such correlations do appear, it is difficult to see how a Uniformitarian interpretation could explain them away as a coincidence."[10]

From the evidence of his histogram, Mackie concludes "some of these eruptions appear to have occurred at about the same time as the sea level changes."[11] Thus, there is some evidence albeit not full nor conclusive for Velikovsky's contention.

How accurate is the dating method that Sagan wishes Velikovsky employ on the lava flows of all the worlds volcanoes? According to Funkhouser, Barnes and Haughton in an article titled, "The Problem of Dating Volcanic Rocks by the Potassium-Argon Method", which appeared in the *Bulletin of Volcanoligique* 29, for 1966, some lava flows known to be in place for only 200 years were potassium–argon dated to be hundreds of millions of years old. There are also volcanoes whose lavas flow into the sea. These too would presumably have to be dated. According to C.S. Noble and J.J. Haughton in a paper titled, "Deep-Ocean Basalts: Inert Gas Content and Uncertainties in Age Dating" published in *Science,* Vol. 162 for 1968, pp. 265-26, "The radiogenic argon and helium content of three basalts erupted into the deep ocean from an active volcano (Kilauea) have been measured. Ages calculated from these measurements increase with sample depth up to 22 million years for lavas deduced to be recent." Thus, the dating methods Sagan asked Velikovsky to employ on all volcanic lava flows is flawed. Sagan, however, does not inform the reader of this. Why not?

Therefore, Velikovsky writes regarding radiocarbon dating,

> "Already there is an accumulation of similar results that do not fit into the accepted [dating] scheme, even if the Ice Age is brought as close to our time as 10,000 years. Professor Johnson says: 'There is no way at the moment to prove whether the valid dates, the 'invalid ones,' or the 'present ideas' are in error.' He says also, 'Until the number of measure-ments can be increased to a point permitting some explanation of contradictions with other apparently trustworthy data, it is necessary to continue to form judgments concerning validity by a combination of all available information.'"[12]

And to the best of his ability, this is what Velikovsky has done. Using several evidential routes, he unwinds the clock that brings us back to dating the events he has enumerated in *Worlds in Collision*. It is this evidence that Macbeth tells us is "shattering".

GEOMAGNETIC REVERSALS

Sagan states,

> "Velikovsky believes (p. 115) that reversals of the geomagnetic field are produced by cometary close approaches. Yet, the record is clear—such reversals occur about every million years...[though] not in the last few thousand; and they occur more or less like clock work."[13]

If this were true, it would certainly be a surprise to those working in the field. Velikovsky's chapter, "Magnetic Poles Reversed" in *Earth in Upheaval,* asks,

> "When was the terrestrial magnetic field reversed for the last time? Most interesting is the discovery that the last time that the reversal of the magnetic field took place was in the eighth century before the present era, or twenty-seven centuries ago. The observation was made on clay fired in kilns by the Etruscans and Greeks. The position of the ancient

vases is known. They were fired in a standing position, as the flow of the glaze testifies. The magnetic inclination or the magnetic dip of the iron particles in the fired clay indicates which was the nearest magnetic pole, the north or the south.

"In 1896, Giuseppe Folgheraiter began his careful studies of Attic (Greek) and Etruscan vases of various centuries, starting with the eighth century before the present era. His conclusion was that in the eighth century the earth's magnetic field was inverted in Italy and Greece. Italy and Greece were closer to the south than the north magnetic pole.

"P.L. Mercanton of Geneva, studying the pots of the Hallstatt age from Bavaria (about the year 1000) and from the Bronze Age caves in the neighborhood of Lake Neuchatel, came to the conclusion that about the tenth century before the present era the direction of the magnetic field differed only a little from its direction today, and yet his material was of an earlier date than the Greek and Etruscan vases examined by Folgheraiter. By checking on the method and results of Folgheraiter, Mercanton found them perfect...

"These researches continued and described in a series of papers by Professor Mercanton, presently with the Service Meteorologique Universitaire in Lausanne, show that the magnetic field of the earth, not very different from what it is today, was disturbed sometime during or immediately following the eighth century to the extent of a complete reversal." [14]

Manley speaks of, "the possibility of its (Earth's magnetic field) reversal in historic times, 2.500 years ago to be cleared up by more research."

However, since its publication in 1896 by Folgheraiter and 1907 by Mercanton, this evidence has stood firm and was reported in *Nature,* Vol. 242 for 1973, p. 518, a year prior to Sagan's presentation of his paper. It may be "clear" to Sagan that such reversals have not occurred in the last few thousand years, but this can only be clear by blinding oneself.

Furthermore, there is evidence of a magnetic reversal for 3,500 years ago. Thomas McCreedy reported that in Turner and Thompson's *Earth and Planetary Science Letters,* for 1979, (number 42, pp. 412–426), that,

"Other recent papers support the viewpoint. Turner and Thompson (1978), examining sediments from Loch Lomond, Scotland, reported a large magnetic declination swing in the middle of the first half of the first millennium B.C. This is in very good agreement with similar findings at Lake Windermere, England (1971)." [15]

In *Science News,* Vol. 125 No. 24 (June 16, 1984, pp. 374ff) is evidence of geomagnetic excursions on Crete.

"Researchers (Downey and Tarling) from the University of Newcastle upon Tyne in England, report that the destruction of late Minoan sites on Crete occurred in two events separated by as much as 30 years... The researchers found that the magnetic signatures at archaeological sites on eastern Crete were identical to one another. They differed though from sites on central Crete where the ages were consistent... They also were surprised to observe that the direction and intensity of magnetic signals for central Crete matched those of the heavy ashfall deposits on Santorini, whereas the directions from sites in eastern Crete matched those of the deposits that followed the volcano's paroxysmal eruption. During this second phase, the volcano collapsed forming the huge caldera."

This evidence corroborates the evidence from Loch Lomond and Lake Windermere of geomagnetic changes around 3,500 years ago. In fact, two West German scientists, Hans Pilcher of the University of Tübingen and Wolfgang Schiering of Mannheim University, claim that the separation in time between the Theran eruption and the earthquake that followed it is 50 years. In *New Scientist,* for July 7, 1977, p. 17, we learn that these scientists point out that the difficulty

"stems from disparities in the dating of the devastation on Santorini which undisputedly followed Theran's eruption, and that of the Cretan ruins (60 miles to the south). The latter appear to have occurred some 50 years after the former...

"The two German researchers claim that no modification of the ceramic chronology, on which the dating is based, is possible and that the calamity which struck Crete...cannot, therefore, have been connected with the Theran explosion."

The direction of the magnetic signals indicates different alignments of the magnetic field and the period between the two alignments is in complete agreement with Velikovsky's 50 year period that separates the two catastrophes.

Thus, here also Sagan's assertion is in contradiction of the evidence. There are magnetic reversals at the time of Velikovsky's catastrophes. Susan Schlee, in her book, *The Edge of an Unfamiliar World, a History of Oceanography,* tells us "North and South poles may even switch places *several times* in a million years."[16] [emphasis added]

J.P. Kennett and N.D. Watkins in *Nature,* Vol. 227 (1970) pp. 930–934 state that, "Geomagnetic field reversals occurred more than twenty times during the past 4 M.Y. [million years] and probably more than one hundred times during the Tertiary." Apparently Sagan's million year clock is in need of repair.

MOUNTAIN BUILDING
Sagan states, "Velikovsky's contention that mountain building occurred a few thousand years ago is belied by all the geological

evidence which puts those times at tens of millions of years ago and earlier." [17]

With respect to how long ago the major mountain ranges of the Earth rose, Sagan claims they rose millions of years ago, not during historical times. Velikovsky, however, in *Earth in Upheaval,* reports,

"In the Alps, caverns with human artifacts of stone and bone dating from the Pleistocene (Ice Age) have been found at remarkably high altitudes. During the Ice Age the slopes and valleys of the Alps, more than other parts of the continent, must have been covered by glaciers; today in central Europe, there are great glaciers only in the Alps. The presence of men at high altitudes during the Pleistocene or Paleolithic (rude stone) Age seems baffling.

"The cavern of Wildkirchli, near the top of Ebanalp, is 4,900 feet above sea level. It was occupied by man sometime during the Pleistocene. [According to G.G. MacCurdy, *Human Origins,* (1924), I, p. 77], 'Even more remarkable, in respect to altitude, is the cavern of Drachenloch at a height of 2,445 meters [8,028 feet],' near the top of Drachenberg, south of Ragaz. This is a steep, snow covered massif. [MacCurdy adds], 'Both of these stations are in the very heart of the Alpine field of glaciation.'

"A continental ice sheet thousands of feet thick filled the entire valley between the Alps and the Jura, where now Lake Geneva lies, to the height of the erratic boulders torn from the Alps and placed on the Jura Mountains. In the same geological epoch, between two advances of the ice cover, during an interglacial intermission, human beings must have occupied caverns 8,000 feet above sea level. No satisfactory explanation for such location of Stone Age man has ever been offered.

"Could it be that the mountains rose as late as in the age of man and carried up with them the caverns of early man? In recent years evidence has grown rapidly to show, in contrast to previous opinion, that the Alps and other mountains rose and attained their present heights...

"[According to R.F. Flint, *Glacial Geology and the Pleistocene Epoch,* pp. 9-10], 'Mountain uplifts amounting to many thousands of feet have occurred within the Pleistocene epoch [Ice Age] itself.' This occurred [according to Flint] with 'the Cordilleran mountain system in both North and South America, the Alps-Caucasus-Central Asian system, and many others...'

"The fact of the late upthrust of the major ridges of the world created, when recognized, great perplexity among geologists who, under the weight of much evidence, were forced to this view. The revision of the concepts is not always radical enough. Not only in the age of man, but in the age of *historical* man, mountains were thrust up, valleys were torn out, lakes were dragged uphill and emptied...

"[According to B. Willis, *Research in Asia,* II, p. 24], 'The great mountain chains challenge credulity by their extreme youth,'...[discussing the] Asian mountains.

"The Himalayas, highest mountains in the world, rise like a thousand mile-long wall north of India. This mountain wall stretches from Kashmir in the west to and beyond. Bhutan in the east, with many of its peaks towering over 20,000 feet, and Mount Everest reaching 29,000 feet, or over five miles. The summits of the lofty massifs are capped by eternal snow in those regions of the heavens where eagles do not fly nor any bird of the sky.

"Scientists of the nineteenth century were dismayed to find that, as high as they climbed, the rocks of the massifs yielded skeletons of marine animals, fish that swim in the ocean, and shells of mollusks. This was evidence that the Himalayas had risen from beneath the sea. At some time in the past, azure waters of the ocean streamed over Mount Everest, carrying fish, crabs and mollusks, and marine animals looked down to where now we look up and where man, after many unsuccessful efforts, has until now succeeded only once in putting his feet. Until recently it was assumed that the Himalayas rose from the bottom of the sea to their present height tens or perhaps hundreds of millions of years ago. Such a long period of time, so long ago, was enough even for the Himalayas to have risen to their present height. Do we not, when we tell young listeners a story about giants and monsters, begin with: 'Once upon a time, long ago...?' And the giants are no longer threatening and the monsters are no longer real.

"According to the general geological scheme, five hundred million years ago the first forms of life appeared on earth; two hundred million years ago life developed into reptilian forms that dominated the scene, achieving gigantic size. The huge reptiles died out seventy million years ago, and mammals occupied the earth—they belong to the Tertiary. According to this scheme, the last mountain uplifts took place at the end of the Tertiary, during the Pliocene; this period lasted until a million years ago, when the Quaternary period, the age of man began. The Quaternary is also the time of the Ice Age or the Pleistocene—the Paleolithic or Old Stone Age; and the very end of the Quaternary, since the end of the Ice Age, is called Recent time: the Neolithic (Late, or polished, Stone), Bronze, and Iron Cultures. Since the appearance of man on earth, or since the beginning of the Ice Age, there have been no uplifts on any substantial scale. In other words, we have been told the profile of the earth with its mountains and oceans was already established when man first appeared.

"In the last few decades, however, numerous facts have emerged from mountains and valleys that tell a different story. In Kashmir, Helmut de Terra discovered sedimentary deposits of an ancient sea bottom that was elevated at places to an altitude of 5,000 feet or more and tilted at an angle of as much as 40 degrees; the basin was dragged up by the rise of the mountain. But this was entirely unexpected: [According to Arnold Heim and August Gausser, *The Throne of the Gods, an Account of the First Swiss Expedition to the Himalayas,* (1939), p. 218], 'These deposits contain Paleolithic fossils' and this according to...Heim, Swiss geologists, would make it plausible that the mountain passes in the Himalayas may have risen, in the age of man, three thousand feet or more, 'however, fantastic changes so extensive may seem to a modern geologist.'

"*Studies on the Ice Age in India and Associated Human Cultures,* published in 1939 by de Terra, working for the Carnegie Institution, with the assistance of Professor T.T. Paterson of Harvard University, is one long argument and demonstration that the Himalayas were rising during the Glacial Age and reached their present heights only after the end of the Glacial Age, and actually in historical times. From other mountain ridges came similar reports.

"De Terra divided the Ice Age of the Kashmir slopes of the Himalayas into Lower Pleistocene (embracing the first glacial and interglacial stages), Middle Pleistocene...and Upper Pleistocene (comprising the last to glaciations and an interglacial stage).

"[According to de Terra and Paterson, p. 223], 'The scenery which this region presented at the beginning of the Pleistocene must have differed greatly from that of our time... The Kashmir valley was less elevated, and its southern rampart, the Pri Panjal, lacked that Alpine grandeur that enchants the traveler today...' Then various formation groups [they add] moved, 'both horizontally and vertically, resulting in a southward displacement of older rocks upon foreland sediments accompanied by uplift of the mobile belt.'

"[They continue on page 225 that], 'The main Himalayas suffered sharp uplift in consequence of which the Kashmir lake beds were compressed and dragged upward on the slope of the most mobile range... Uplift was accompanied by a southward shifting of the Pri Panjal block toward the foreland of northwestern India.' The Pri Panjal massif that was pushed toward India is at present 15,000 feet high.

"In the beginning of this period the fauna was greatly impoverished, but thereafter, judging from remains, large cats, elephants, true horses, pigs and hippopotami occupied the area.

"'In the Middle Pleistocene, or Ice Age, there was a 'continued uplift.' 'The archaeological record prove that early paleolithic man inhabited the adjourning plains.' De Terra refers to 'abundance of paleolithic sites.' Man used stone implements of 'flake' form, like those found in the Corner forest-bed in England.

"Then once more, the Himalayas were pushed upward. [de Terra and Paterson state on page 222], 'Tilting of terraces and lacustrine [lake] beds' indicates a 'continued uplift of the entire Himalayan tract' during the last phases of the Ice Age.

"In the last stages of the Ice Age, when man worked stone in the mountains, he might have been living in the Bronze stage down in the valleys. It has been repeatedly admitted by various authorities—quoted subsequently in this book [*Earth in Upheaval*]—that the end of the glacial epoch may have been contemporaneous with the time of the rise of the great cultures of antiquity, of Egypt and Sumeria and it, follows, also of India and China. The Stone Age in some regions could have been contemporaneous with the Bronze Age in others. Even now there are numerous tribes in Africa, Australia, and Tierra del Fuego, the southern tip of the American, still living in the Stone Age, and many other regions of the modern world would have remained in the Stone Age had it not been for the importation of iron from more advanced regions. The aborigines of Tasmania never got so far as to produce a polished— neolithic—stone implement, and in fact, barely entered the crudest stone age. This large island south of Australia was discovered in 1642 by Abel Tasman; the last Tasmanian died in exile in 1876, and the race became extinct.

"The more recent uplifts in the Himalayas took place also in the age of modern man. [de Terra and Paterson state on page 223], 'The postglacial terrace record suggests that there was at least one prominent postglacial advance [of ice],' and this, in the eyes of de Terra and Paterson, is indicative of a diastrophic movement of the mountains. [They add], 'We must be emphatic on one particular feature—namely the dependence of Pleistocene glaciation on the diastrophic [upward motion] character of a mobile mountain belt. This relationship, we feel, has not been sufficiently recognized in other glaciated regions, such as Central Asia and the Alps, where similar if not identical, conditions are found.'

"It has been generally assumed that loess—thin windblown dust that is built into clays—is a product of a glacial age. However, in the Himalayas, de Terra reported finding neolithic, or polished stone, implements in loess and commented: 'Of importance for us is the fact that loess forma- tion was not restricted to the glacial age, but that it continued…into postglacial times.' In China and in Europe, too, the presence of polished stone artifacts in loess prompted a similar revision. The neolithic stage that began, according to the accepted scheme, at the end of the Ice Age, still persisted in Europe and in many other places at the time when, in the centers of civilization, the Bronze Age was already flourishing.

"R. Finsterwalder, exploring the Nanga Parbat massif in the western Himalayas (26,600 feet high), dated the Himalayan glaciation as post- glacial; in other words, the expansion of the glaciers in the Himalayas took place much closer to our time then had previously assumed. Great uplifts of the Himalayas took place in part after the time designated as the Ice Age, or only a few thousand years ago. [See R. Finsterwalder, "*Zeitschrift Der Gesellschaft Fur Erdkunde Zu Berlin*" (1936) p. 32ff.]

"Heim, investigating the mountain ranges of western China, adjacent to Tibet and east to the Himalayas, came to the conclusion (1930) that they had been elevated *since* the Glacial Age. [See J.S. Lee, *The Geology of China*, p. 207]

"The great massif of the Himalayas rose to its present height in the age of modern, actually historical man [Heim and Gausser, *Throne of the Gods*, p. 220 state], 'The highest mountains in the world are also the youngest.' With their topmost peaks the mountains have shattered the entire scheme of the geology of the 'long, long ago.'

"The Siwalik Hills are in the foothills of the Himalayas north of Delhi; they extend for several hundred miles and are 2,000 to 3,000 feet high. In the nineteenth century their unusually rich fossil beds drew the attention of scientists. Animal bones of species and genre, living and extinct, were found there in most amazing profusion. Some of the animals looked as though nature had conducted an abortive experiment with them and had discarded the species as not fit for life. The carapace of a tortoise twenty feet long was found there, how could such an animal have moved on hilly terrain? [See D. Wadia, *Geology of India* (2nd ed.; 1939) p. 268]. The *Elephas ganesa,* an elephant species found in the Siwalik Hills, had tusks about fourteen feet long and over three feet in circumference. One author [J.T. Wheeler, *The Zonal Belt Hypothesis,* (1908) p. 68] says of them: 'It is a mystery how these animals ever carried them, owing to their enormous size and leverage.'

"The Siwalik fossil beds are stocked with animals of so many and such varied species that the animal world of today seems impoverished by comparison. It looks as though all these animals invaded the world at one time: 'This sudden bursting on the stage of such a varied population of herbivores, carnivores, rodents and primates, the highest order of the mammals, must be regarded as a most remarkable instance of rapid evolution of species,' writes D. Wadia in his *Geology of India* [p. 268]. The hippopotamus, which 'generally is a climatically specialized type' (de Terra), pigs, rhinoceroses, apes, oxen, filled the interior of the hills almost to bursting. A.R. Wallace, who shares with Darwin the honor of being the originator of the theory of natural selection, was among the first to draw attention, in terms of astonishment, to the Siwalik extinction.

"Many of the genre that comprised a wealth of species were extinguished to the last one; some are still represented, but by only a few species. Of nearly thirty species of elephants found in the Siwalik bed, only one species has survived in India. [According to Wadia, p. 279], 'The sudden and widespread reduction by extinction of the Siwalik mammals is a most startling event for the geologist as well as the biologist. The great carnivores, the varied races of elephants belonging to no less than 25 to 30 species...the numerous tribes of large and highly specialized ungulates [hoofed animals] which found such suitable habitats in the Siwalik jungles of the Pleistocene epoch, are to be seen no more in an immediately succeeding age.' It used to be assumed that the advance of the Ice Age killed them, but subsequently it has been recognized that great destructions took place in the age of man, much closer to our day.

"The older geologists thought that Siwalik deposits were alluvial in their nature, that they were debris carried down by the torrential Himalayan streams. But it was realized that this explanation 'does not appear to be tenable on the ground of the remarkable homogeneity that the deposits possess' and a 'uniformity of lithologic composition' in a multitude of isolated basins, at considerable distance from one another.

[Wadia, p. 270] There must have been some agent that carried these animals and deposited them at the feet of the Himalayas, and after the passage of a geologic age...[are] signs of more than one destruction. There was also a movement of the ground: [Wadia states, p. 264], 'The disrupted part of the fold has slipped bodily over for long distances, thus thrusting the older pre-Siwalik rock of the inner ranges of the mountains over the younger rocks of the outer ranges.'

"If the cause of these paroxysms and destruction was not local, it must have produced similar effects at the other end of the Himalayas and beyond that range. Thirteen hundred miles from the Siwalik Hills, in central Burma, the deposits cut by the Irrawaddy River [according to Wadia, pages 274-275] 'may reach 10,000 feet.' 'Two fossiliferous horizons occur in this series separated by about 4,000 feet of sands.' The upper horizon (bed), characterized by mastodon, hippopotamus, and ox, is similar to one of the beds in Siwalik. [Wadia adds], 'The sediments are remarkable for the large quantities of fossil-wood associated with them... Hundreds and thousands of entire trunks of silicified trees and huge logs lying in the sandstones' suggest the denudation of 'thickly forested' areas. Animals met death and extinction by the elementary forces of nature which also uprooted forests and from Kashmir to Indo-China threw sand over species and genre in mountains thousands of feet high.

"In the Andes, at 16° 22′ south latitude, a megalithic city was found at an elevation of 12,500 feet, in a region where corn will not ripen. The term 'megalithic' fits the dead city only in regard to the great size of the stones in its walls, some of which are flattened and joined with precision. It is situated on the Altiplano, the elevated plain between the Western and Eastern Cordilleras, not far from Lake Titicaca, the largest lake in South America and the highest navigable lake in the world, on the border of Bolivia and Peru.

"[Clemens Markham, in *The Incas of Peru*, (1910), p. 21 states], 'There is a mystery still unsolved on the plateau of Lake Titicaca, which if stones could speak, would reveal a story of deepest interest. Much of the difficulty in the solution of this mystery is caused by the nature of the region, in the present day, where the enigma still defies explanation...' [Markham also states, p. 23], 'Such a region is only capable of sustaining a scanty population of hardy mountaineers and laborers. The mystery consists in the existence of ruins of a great city at the southern side of the lake, the builders being entirely unknown. The city covered a large area, built by highly skilled masons, and with the use of enormous stones.'

"When the author of the quoted passages posed his question to the scholarly world, Leonard Darwin, the president of the Royal Geological Society, offered the surmise that the mountain had risen considerable after the city had been built.

"'Is such an idea beyond the bond of possibility?' asked Sir Clemens. [p. 23]; Under the assumption that the Andes were once some two to three thousand feet lower than they are now, 'maize would then ripen in the basin of Lake Titicaca, and the site of the ruins of Tiahuanacu could support the necessary population. If the megalithic builders were living under these conditions, the problem is solved. If this is geologically impossible, the mystery remains unexplained.' [Markham p. 23]

"Several years ago another authority, A. Posnansky, [in *Tiahuanacu, The Cradle of American Man,* (1945), p. 15], wrote in a similar vein: 'At the present time, the plateau of the Andes is inhospitable and almost sterile. With the present climate, it would not have been suitable in any period as the asylum for great human masses' of the 'most important prehistoric center of the world.' [On pages 1 and 39 he writes], 'Endless agriculture terraces' of the people who lived in this region in pre-Inca days can still be recognized. 'Today, this region is at a very great height above sea level. In remote periods, it was lower.'

"The terraces rise to a height of 15,000 feet, twenty-five hundred feet above Tiahuanacu, and still higher, up to 18,400 feet above sea level, or to the present line of eternal snow on Illimani.

"The conservative view among evolutionists and geologists is that mountain making is a slow process...there never could have been spontaneous uplifting on a large scale. In the case of Tiahuanacu, however, the change in altitude apparently occurred after the city was built, and this could not have been the result of a slow process that required hundreds of thousands of years to produce visible alteration.

"Once Tiahuanacu was at the [lake] water's edge; then Lake Titicaca was ninety feet higher, as its old strand line discloses. But this strand line is tilted and in other places it is more than 360 feet above the present level of the lake. There are numerous raised beaches; and stress was put on 'the freshness of many of the strandlines and the modern character of such fossils as occur.' [See H.P. Moon, "The Geology and Physiography of the Altiplano of Peru and Bolivia," *The Transacts of the Linnean Society of London,* 3rd Series, Vol. I, Pt. 1 (1939), p. 32]

"Further investigation into the topography of the Andes and the fauna of Lake Titicaca, together with a chemical analysis of this lake and others on the same plateau, established that the plateau was at one time at sea level, or 12,500 feet lower than it is today. [According to Posnansky, *Tiahuanacu,* p. 23], 'Titicaca and Poopo, lake and salt bed of Coipagna, salt beds of Uyuni—several of these lakes and salt beds have chemical compositions similar to those of the ocean.' As long ago as 1875, Alexander Agassiz demonstrated [in *Proceedings of the American Academy of Arts and Sciences,* 1876] the existence of a marine crustaceous fauna in Lake Titicaca. At a higher elevation, the sediment of an enormous dried-up lake, whose waters were almost potable [according to Posnansky, *Tiahuanacu,* p. 23], 'is full of characteristic [sea] mollusks, such as Paludestrina and Ancylus, which shows that it is, geologically speaking, of relatively modern origin.'

"Sometimes in the remote past the entire Altiplano with its lake, rose from the bottom of the ocean. At some other time point a city was built there and terraces were laid out on the elevation around it; then in another disturbance, the mountains were thrust up and the area became uninhabitable.

"The barrier of the Cordilleras that separates the Altiplano from the valley to the east was torn apart and gigantic blocks were thrown down into the chasm. Lyell, combating the idea of a universal flood, offered the theory that the bursting of the Sierra barrier opened the way for a large lake on the Altiplano, which cascaded down into the valley and caused the aborigines to create the myth of a universal flood. [See Lyell, *Principles of Geology*, I, p. 89, p. 270]

"Not so long ago an explanation of the mystery of Lake Titicaca and of the fortress Tiahuanacu on its shores was put forward in the light of Hörbiger's theory: A moon circled very close to the earth, pulling the waters of the oceans toward the equator; by its gravitational pull, the moon held day and night, the water of the ocean at the altitude of Tiahuanacu. [H.S. Bellamy, *Built Before the Flood: The Problem of Tiahuanacu Ruins,* (1947), p. 14, states], 'The level of the ocean must have been at least 13,000 feet higher.' Then the moon crashed into the earth, and the oceans receded to the poles, leaving the island with its megalithic city as a mountain above the sea bottom, now the continent of the tropical and subtropical Americas. All this happened millions of years before our moon was caught by the earth, and thus the ruins of the megalithic city Tiahuanacu are millions of years old, that is, the city must have been built long 'before the Flood.'

"This theory is bizarre. The geological record indicates a late elevation of the Andes, and the time of its origin is brought ever closer to our time. Archaeological and radiocarbon analyses indicate that the age of the Andean culture and of the city is not much older than four thousand years. [According to F.C. Hibben, *Treasure in the Dust,* (1951), p. 56] Not only the 'built before the Flood' theory collapses; so does the belief that the last elevation of the Andes was in the Tertiary, or more than a million years ago.

"Sometime in the remote past the Altiplano was at, or below sea level, so that originally its lakes were part of a sea gulf. The last upheaval, however, took place in an early historical period after the city of Tiahuanacu had been built; the lakes were dragged up, and the Altiplano and the entire chain of the Andes rose to their present height.

"The ancient stronghold of Ollantaytambo in Peru is built on top of an elevation; it is constructed of blocks of stone twelve to eighteen feet high. [According to Don Ternel in *Travel,* April 1945], 'These Cyclopean stones were hewn from a quarry seven miles away… How the stones were carried down to the river in the valley, shipped on rafts and carried up to the site of the fortress remains a mystery archaeologists cannot solve.'

"Another fortress or monastery, Ollantayparubo, in the Urubamba Valley in Peru, northwest of Lake Titicaca, [according to Bellany's *Built Before the Flood,* p. 63], 'perches upon a tiny plateau some 13,000 feet above sea-level, in an uninhabitable region of precipices, chasms and gorges.' It is built on red porphyry blocks. The blocks must have been brought [Bellamy continues] 'from a considerable distance...down steep slopes, across swift and turbulent rivers, and up precipitous rocks—faces which hardly allow a foot-hold.' It has been suggested that the transportation of the building blocks was feasible only if the topography of these localities was different at the time of the construction. However, definite proof in this connection is lacking, and changes in topography must be deduced from abandoned terraces, from [sea] mollusks of the dried up lakes, from tilted shorelines, and from other similar indications.

"Charles Darwin, on his travels in South America in 1834-35 was impressed by the raised beaches at Valparaiso, Chile at the foot of the Andes. He found that the former surf line was at an altitude of 1,300 feet. He was impressed even more by the fact that sea shells found at this altitude were still undecayed, to him a clear indication that the land had risen 1,300 feet from the Pacific Ocean in a very recent period. [In his *Geological Observations on the Volcanic Islands and Parts of South America*. Pt. II, Chap. 15, Darwin states], 'within the period during which upraised shells remained undecayed on the surface.' And since only a few intermediary surf lines can be detected, the elevation could not have proceeded little by little.

"Darwin also observed [on the same page] that 'the excessively disturbed condition of the strata in the Cordillera, so far from indicating single periods of extreme violence, presents insuperable difficulties, except on the admission that the masses of once liquefied rocks of the axes were repeatedly injected with intervals sufficiently long for their successive cooling and consolidation.'

"At present, it is the common view that the Andes were created, not so much by compression of the strata, as by magma, or molten rock invading the strata and lifting them. The Andes also abound in volcanoes, some exceedingly high and enormously large.

"The foothills of the Andes hide numerous deserted town and abandoned terraces, monuments...that go up the slope of the Andes and reach the eternal snow line and continue under the snow to some unidentified altitude prove that it was not a conqueror nor a plague that put the seal of death on gardens and towns. In Peru [E. Huntington in *Hyluings-Skrift*, (1935), p. 578 states], 'aerial surveys in the dry belt west of the Andes have shown an unexpected number of old ruins and an almost incredible number of terraces of cultivation.'

"When Darwin mounted the Uspallata Range, 7,000 feet high in the Andes, and looked down on the plain of Argentina from a little forest of petrified trees broken off a few feet above the ground, he wrote in his *Journal [Of Researches...During the Voyage of H.M.S. Beagle* entry of March 30, 1835], 'It required little geological practice to interpret the marvelous story which this scene at once unfolded; though I confess I was at first so much astonished that I could scarcely believe the plainest evidence. I saw the spot where a cluster of fine trees once waved their branches on the shores of the Atlantic, when the ocean—now driven back 700 miles—

came to the foot of the Andes. I saw that they had sprung from volcanic soil which had been raised above the level of the sea, and that subsequently this dry land, with its upright trees, had been let down into the depths of the ocean. In these depths, the formerly dry land was covered by sedimentary beds, and these again by enormous streams of submarine lava—one such mass attaining the thickness of a thousand feet; and these deluges of molten stone and aqueous deposits five times alternately had been spread out. The ocean which received such thick masses must have been profoundly deep; but again the subterranean forces exerted themselves and now I beheld the bed of that ocean, forming a chain of mountains more than seven thousand feet in height... Vast and scarcely comprehensible as such changes must ever appear, yet they have all occurred within a period, recent when compared with the history of the Cordillera; and the Cordillera itself is absolutely modern as compared with many of the fossiliferous strata of Europe and America.' But how extremely young the Cordillera of the Andes is, only the research of recent years has brought out." [18]

Henry H. Horworth in *Nature,* Vol. 5, (March 28, 1872), pp. 420-422 cites Darwin respecting the height of the raised elevation when Darwin was discussing Hacienda of Quintero, in central Chile. Darwin claims, "The proofs of the elevation of this whole line of coast is unequivocal..." and speaking of northern Chile states, "I have convincing proofs that this part of the continent of South America has been elevated near the coast at least 400 to 500 feet, and in some parts 1,000 to 1,300 feet, since the epoch of existing shells, and further inland the rise may have been greater." [From Darwin's *Naturalist's Voyage*]

When did the Andes mountains rise? According to the *New York Times,* (Oct. 3, 1989), pp. C1 and C14, "Archaeologists working in Peru have unearthed stunning evidence that monumental architecture, complex societies and planned developments first appeared and flowered in the New World *between 5,000 and 3,500 years ago.*" [emphasis added] The author of the article, William K. Stevens goes on to say, "Around 4,000 to 3,700 years ago, activity abruptly shifted and irrigated agriculture replaced fishing as the main economic resource." Why would a civilized people leave a thriving, hospitable environment to go live inland in an inhospitable region? Stevens states,

"It is something of a mystery to archaeologists why any major civilization would develop in the Andean valleys and on the Peruvian coast. The region's altitude and aridity make it 'grossly hostile' said Dr. [Michael] Moseley [an archaeologist at the University of Florida who has long worked in the region] who added: 'That anyone ever lived there is a bit of a surprise.'"

People do not move from more comfortable regions to ones that are 'grossly hostile.' They move instead to regions that are distinctly more conducive to life. In all texts that deal with Egyptian, Hindu, Chinese, and Mesopotamian civilizations, the authorities claim that these civilizations moved into hospitable river valleys from arid regions, not the other way around. Stevens adds, "The emerging picture of this earliest American civilization is that of a people tied initially to the sea, but then moving abruptly—no one knows why—into the Andes highlands to build a flourishing economy based on irrigated agriculture that prospered in spite of the harsh, cold, arid climate at altitudes around 10,000 feet."

But it is much more reasonable and probable that these cities were built at lower elevations and were uplifted with the Andes about 3,500 years ago when civilization there declined.

The evidence presented by Velikovsky is highly conclusive. Why then do Sagan and his colleagues hide from it? Macbeth informs us that the Alps, Himalayas and Andes

> "Apparently all...have risen extensively since men moved in, and much of the upthrusting has occurred in the short period since the retreat of the glaciers. It is impossible to express this precisely in years, but the span of time is almost infinitesimal when compared to the figures commonly used by geologists. Needless to say, the upthrusting was not a quite everyday event. Checking a couple of college textbooks used by my children, I found that practically nothing was said about mountain building and that the subject seems to baffle the scholars."[19]

What is most baffling is Sagan acting as if this information is non-existent. His explanation of this evidence of the very recent rise of the mountain ranges to support *his* assertion would be most welcome. One can only suspect that he does not deal with it because it probably cannot be explained away to fit his uniformitarian time table.

MAMMOTH BONES C14

Sagan states,

> "The idea that mammoths were deep frozen by a rapid movement of the Earth's geographical pole a few thousand years ago can be tested—for example, by carbon-14 or amino acid racemization dating. I should be very surprised if a very recent age results from such tests."[20]

The use of amino acid racemization which Sagan suggests to carry out these tests has been found to be of little worth with porous bones. According to Gregory Heisler, "The Dating Game," in *Discover*, (September, 1992), p. 82,

> "In the 1970's a flush of excitement over a technique called amino acid racemization led workers to believe that another continent—North

America—had been occupied by humans fully 70,000 years ago. Further testing at the same American sites proved that the magical new method failed by one complete goose egg. The real age of the sites was closer to 7,000 years."

On page 83 Heisler adds that, "amino acid racemization scorned for the last fifteen years, thanks to the discovery that the technique, unreliable when applied to porous bone, is quite accurate when used on hard ostrich egg shells." To date mammoth bones requires dating of its porous materials, and therefore, the mechanism for dating of mammoths suggested by Sagan is of no value. With respect to Carbon-14 that is a different story.

What can one say then except that on page 275 of Charles Hapgood's *The Path of the Poles,* we learn that "Radiocarbon" *American Journal of Science,* No. II, p. 43, had dated mammoth's bone, M774, found in Santa Isabel Iztapan, State of Mexico, that was found associated with stone implements. Its radiocarbon date is 2,640 years old, plus or minus 200 years. This dates it quite close to Velikovsky's last catastrophe of 2,700 years ago. In *Radiocarbon,* (Vol. 15, No. 1), a mammoth tusk from Bavaria, Germany, on the basis of averaging three separate dates is dated around 1900 B.C. Jim Hester's "Late Pleistocene Extinction and Radio-carbon Dating," in *American Antiquity,* (Vol. XXVI, 1960), pp. 57-77 has presented data that mammoth bones found in Florida mixed with other extinct animals and human artifacts was found to be 2000 years old based on radio carbon dating. These dates are simply dismissed because of a 10,000 year credo.

L. Krishtalka in *Nature,* Vol. 317 (1984) pp. 225-226 makes the explicit point that dating the extinction of the mammoths is used by theorists to maintain this 10,000 year credo. He states, "their selective acceptance of only 'good' dates—those that fit the model...for example, dates for human beings in North America no older than 12,000 yr B.C., and those for mammoths younger than 10,000 B.P.— may play fast and loose with evidence that doesn't fit." these assumed dates. What Krishtalka is claiming is that theorists will not accept when found, nor publish when found, dates of mammoth bones that are younger than 10,000 years. Only a few of these dates have somehow escaped detection and were printed. In our discussion of Carbon-14 dating, we will demonstrate that this selective acceptance of only dates that fit the accepted theory is common.

However, B. Bower reports that "Dwarf mammoths outlived last Ice Age" in *Science News,* (March 27, 1993), p. 197, "Woolly mammoths, those icons of the ice age that most paleontologists assumed died out around 9,500 years ago survived in miniature form...until about 4,000 years ago on a Arctic Ocean island according to new findings. Mammoth teeth found in 1991 on Wrangel Island located 120 miles off the coast of northeast Siberia range from approximately 7,000 to 4,000 years old, report Andrei V. Sher and Vadim E. Garutt, paleontologists at the Russian Academy of Sciences in Moscow." According to *Discover* (January 1994), p. 54, "The tusk and bone fragments [of the Wrangel pigmy elephants] were younger still, just 3,730 years old.

Even Carl Sagan has come to the conclusion that certain Ice Age members of the elephant family survived well into historical times. In his book *Comet,* p. 273 Sagan suggests, "The last mastodon died perhaps as recently as 2000 B.C." But if the mastodon survived into historical times is it really improbable that its cousin, the mammoth did not? Surely, if one form of elephant survived longer than assumed is it reasonable and plausible that the other forms did not, because dates for mammoth survival do exist in spite of the suppression of these findings.

WHEN WAS THE MOON LAST MOLTEN?
Sagan then writes,

> "Velikovsky believes that the Moon, not immune to the catastrophes which befell the Earth, had similar tectonic events occur on its surface a few thousand years ago, and that many of its craters were formed then... There are some problems with this idea as well: samples returned from the Moon in the *Apollo* mission show no rocks melted more recently than a few hundred million years ago." [21]

It seems apparent that Sagan is quite sure about his evidence even though he states his position without citing where the evidence is found. The Moon formed with the Earth about 4.5 billion years ago and was heated then by its radioactive elements. Thus, the Moon was molten then and atomic dating processes will prove this. Let us see how old the Moon is by these dating processes to determine when it supposedly was last molten. Ian T. Taylor informs us that,

> "The oldest rocks on earth have a reported age of 3.8 billions years. However, it was realized that the moon would have crusted over at about the same time as the earth; since there is no wind or water to cause erosion, it was believed moon rocks would provide a direct radiometric age for the earth. Sure enough, after retrieval of the moon rock samples in the *Apollo* program, Holme's estimation was claimed to be exactly confirmed, and the age of the earth confidently stated in the popular press and textbooks to be 4.5 billion years. However, the official reports and scientific journals, in which actual results of the radiometric determinations were given, showed that the ages of the moon rock samples varied between 2 and 28 billion years [according to Whitcome

and DeYoung who cite *earth and Planetary Science Letters* for 1972-77; *Science,* 1970, Vol. 167, pages 462-555]. Quite evidently, the data for public consumption had been selected to confirm the theory."[22]

Based on modern theories, the universe was born about 20 billion years ago. If the Moon was molten when it was born, as Sagan seems to imply, it was molten 28 billion years ago or before the universe was born. Apparently Sagan does not wish to discuss this information or other relevant data about the Moon that contradicts his assertion about the age of the Moon's last melting.

DUST

Sagan, in his book *Comet,* states,

> "In its annual voyage around the Sun, the Earth runs into particles of the zodiacal [dust] cloud, mainly in the dawn hemisphere; more rarely, faster moving debris catches up with the Earth in its twilight hemisphere. The total being accumulated all over the planet is about a thousand tons of dust a day. The number of fine particles collected by stratospheric aircraft —U-2's and others—is about what we would expect for particles in the zodiacal cloud captured by the moving Earth."[23]

Ian T. Taylor informs us that,

> "Petterson (1960) of the Swedish Oceanographic Institute, working on high mountain tops filtered measured quantities of air and analyzed the particles he found. Since the meteorites that have survived contain an average of 2.5 percent nickel, then the nickel content of the dust extracted, represented that which came from meteors rather than from terrestrial sources. From a knowledge of the total volume of the earth's atmosphere, Petterson reckoned that 14 million tons of meteoric dust settled on the Earth's surface each year; however, because of some variability in results, he concluded with a more conservative figure of five million tons... Isaac Asimov, the popular science writer, took the more liberal figure and concluded that at that rate, the dust piles up to about ten-millionths of an inch per year. This is certainly not much to get excited about. However, he then pointed out that over nearly five billion years, this would add up, if undisturbed to a layer of fifty-four feet deep over the entire surface of the earth. [See Isaac Asimov in *Science Digest,* January 1959.] Recalling that this dust is mostly iron and nickel oxides, it will be evident that no such layer or any trace of it is to be found; then, of course, it is argued that wind and water carried it all away and it is now in the ocean sediments.
>
> "Asimov, writing at about the time the *Apollo* moon landing was being planned, was reflecting a concern among scientists that in the absence of wind and rain, a similar depth of dust would have accumulated on the moon's surface... There was before them the prospect that the Apollo lunar module would land only to disappear by slowly sinking into the moon dust! To avoid this very possibility, the lunar module was

equipped with large pad feet. On 21 July 1969, more than 600 million people watched as television transmitted mankind's first footstep onto the Moon's surface. Neil Armstrong's reply to CBS interviewer Walter Cronkite is worth quoting since the opening dialogue, reported by Wilford of *The New York Times*, (21 July 1969, page 1), concerned the depth of the dust: 'The surface is fine and powdery. I can pick it up loosely with my toe. It does adhere in fine layers like powdered charcoal to the sole and sides of my boots. I only go in a small fraction of an inch, maybe an eighth of an inch.' As if to confirm this, astronauts Armstrong and Aldrin had great difficulty planting the American flag into the rocky and virtually dust-free ground, yet not one comment was made on the significance of the absence of the great depth of dust."[24]

Furthermore, meteorites and comets also strike the lunar surface. These bodies are not slowed by an atmosphere, like that which covers the Earth. Upon impact, these bodies are traveling at such great velocity that they vaporize much of the impact area and throw dust in all directions that would fall back on the lunar surface. Donald Goldsmith in his book *Nemesis the Death Star and Other Theories of Mass Extinction,* (NY 1985), p. 25, informs us that a meteorite upon impact produces a crater, "five to ten times the object's diameter, and would eject a vast quantity of matter upward and horizontally...most of the material ejected by the impact would be dust and grit." In addition, passing comets would occasionally approach the Earth–Moon system and the dust from their tails sweeping over the Moon would cover the lunar surface with more dust. Kenneth Hsu in his book *The Great Dying,* (NY 1986), pp. 189-190, states that, "A large comet need not even hit the earth to produce dust; a near miss would leave enough debris in the earth's atmosphere to produce a complete blackout..." of the Sun over the entire globe. The same condition would most certainly also occur to the Moon even though the Moon lacks an atmosphere. Over the history of the solar system, a great many such close encounters must have occurred and left cometary dust on the lunar surface. In combination, the influx of interplanetary dust with the dust created by cometary and meteorite impacts and the close passages of comets that give off great amounts of dust, would produce a considerable depth of dust on the Moon. It is inconceivable that the dust layer on the Moon over a 4.6 billion period would only be one - eighth of an inch deep.

Harold Slusher, in *Age of the Cosmos,* (San Diego 1980) has a chapter titled "Cosmic Dust Influx". On page 41 he states,

"The Moon moves through the same region of space that the Earth does, and consequently, should have about the same influx of cosmic dust on its surface as on the Earth. Astronomers had been concerned that a lunar spaceship upon landing would sink into the supposed huge amount of dust that should have accumulated on the surface of the Moon in about 4.5 billion years of assumed time. The rocket would be stuck in layers of 'mud' and not be able to leave the Moon. Also, in the 'sea' [maria] areas, where lunar ships landed, there should have accumulated more dust than elsewhere on the Moon. Yet, the amount of dust is amazingly small. What could have happened to all the dust, assuming a 4.5 billion year old Moon?"

According to Bevan M. French in *The Moon Book,* (NY 1977), p. 144, there is only "1 to 2 percent of meteoric material [that] has been mixed with the ground-up lunar rock." This presents a great problem. Based on the age of the Moon, its soil or regolith, as some astronomers refer to this debris on the lunar surface, should have a much greater percent of meteoric material. After all, the Moon has supposedly been bombarded by large and small meteors for 4.5 billion years. There is no process by which meteoric material that was pulverized by impact with the Moon could be removed. The present influx of meteors from meteor showers has supposedly been occurring periodically over these same billions of years. There is no plate tectonic motion on the Moon to subduct this material deep below the surface. The lunar soil is several meters or yards thick. Again, based on any reasonable analysis, the lunar soil should have a great deal more meteoric material than 1 or 2 percent.

Patrick Moore in *Star and Sky,* Vol. 1, p. 10 for April 1979 remarks that on the Moon, "meteoric material does seem to be in strangely short supply, and one of the world's leading authorities on meteorites, G.J.H. McCall has even asked plaintively, 'Where have all the mete-orites gone?' if the Moon is 4.5 billion years old. The materials that should have collected over the long history of the Moon, namely dust and meteorites, are in extremely short supply." Sagan does not discuss this; but perhaps he should explain this discrepancy. Now, should the layer of dust be far less deep by employing a very conservative estimate of it, nevertheless, it is not there in anything like the depth required to fit the notion that the Moon has been acquiring dust for 4.5 billion years. However, if much of the Moon with its surface dust melted because it had a catastrophic encounter with a huge comet-like body recently, then the enigma of its thin dust layer and the small meteoric content of its soil is in accordance with what is known.

In fact, Sagan in his book *Comet,* tells us that,

"If this cometary dust had fallen on the Earth over its entire history at the same rate that it does today, and if nothing destroyed it after landfall, there would be a dark, powdery layer about a meter thick everywhere on Earth. [And then Sagan adds]... If a single large comet were pulverized, and all its debris were spread smoothly over the Earth, a layer about a centimeter thick would result."[25]

A centimeter is somewhat less than half an inch; therefore, if only a part of the dust of the passing comet fell on the Moon, as is clearly seen, the dust layer appears to be explained. Bevan M. French in *The Moon Book,* (NY 1978), pp. 203-204, states,

"The exposed surface of a lunar rock is a natural particle counter. The tiny, fast moving particles of cosmic dust that strike the rock produce tiny, glass-lined microcraters. If the exposure age of the rock is known, we need only count the craters on the exposed surface to calculate the rate at which the rock has been bombarded...

"So far as we know, the impact rate of small dust particles onto the moon has been constant throughout the past few million years. However, some recent studies have raised questions about this assumption. Scientists who make a complete study of the microcraters on a large *Apollo 16* sample (60015)...were able to date the formation ages of single microcraters by studying the solar flare particle tracks preserved in the glass linings of each crater. The rock had been exposed on the lunar surface for about 80,000 years, but the data indicates that more microcraters had formed during the last 10,000 years than in the earlier stages of the rock's exposure.

"Such a recent increase in microcrater formation would imply a sudden influx of dust size particles into the earth-moon region about 10,000 years ago. [Around the time of Venus' birth from Jupiter] One possible explanation is the arrival of a new comet which began to shed dust as it passed close to the sun."

THERMOLUMINESCENCE TESTS

This then brings us back to the question of when the Moon was last molten. NASA was indeed quite interested in this question as was Velikovsky who sent a letter to the *New York Times,* on August 5, 1971 in which he requested that core samples of lunar soil be extracted from a depth of three feet and be subjected to *thermoluminescence testing.* Velikovsky contended this would show that the Moon was molten in historical times. NASA did have such tests performed of which Sagan should be aware. The tests were carried out by R. Walker of Washington University, St. Louis, as reported in *Kronos,* (Winter-1977), pp. 37-38. However, the cores were from only six inch soil depths and not three feet as Velikovsky suggested. Thus, the material would be affected by the long 14 day lunar day period. Since the Moon lacks an atmosphere like the Earth's, it is subject to greater solar radiation. In spite of all this, the shallow lunar core sample tested by Walker showed that the Moon had been molten less than 10,000 years ago—not hundreds of millions of years ago as Sagan had represented.

How reliable is a thermoluminescence test? From *New Scientist*, we learn about a priest, Father Eugene Stockton, who discovered tools that apparently were made by Paleolithic man of Australia. He brought his find to the University of Woolongong where it was subjected to tests.

> "Initially, the investigators at the university dated the tools as being 27,000 years old. But when Gerald Nanson and Bob Young reanalyzed the tools last year (1986) *with a more reliable technique called thermoluminescence, they established their true age.*"[26] [emphasis added]

The science writers of *New Scientist* say that "thermoluminescence tests" are "reliable". The space scientists of NASA are certainly aware of the reliability of this dating method. That is clearly why they undertook testing of lunar soil by Walker. Thus, once again, when Sagan claims that the Moon was last molten millions of years ago, he was ignoring plain evidence. Furthermore, the fit of the small amount of dust on the Moon taken together with the date indicated by the thermoluminescence test indicates the Moon was molten only a few thousand years ago.

If indeed the Moon was heated by tidal effects 3,500 to 2,700 years ago, it should still retain some of this residual heat. Velikovsky claimed in the *New York Times,* for July 21, 1969, a few feet under the lunar surface a steep thermal gradient would be discovered. The space scientists held that the Moon was a cold body and that there would be hardly any heat in the gradient. *Time* magazine, p. 67, reported that the lunar heat gradient was "surprisingly high."[27] Bevan M. French in *The Moon Book,* (NY 1978), p. 103, states,

> "The *Apollo 15* astronauts drilled two holes 1 to 2 meters deep and about 2 centimeters in diameter into the lunar soil. Into each hole they lowered a long probe with sensitive electrical thermometers placed along its entire length... Data from the thermometers revealed that the temperature was 1 degree Celsius higher at the bottom of the hole than at the top."

Some scientists were so upset by these high readings that they "suggested...instrument malfunctions."[28] This corroborates the thermoluminescence tests and the dust layer evidence as well.

Further, Velikovsky writing in *Pensée,* Vol. 2, No. 2 (1972), p. 19, "When Was the Lunar Surface Last Molten?" cites Thomas Gold's "Apollo II Observations of a Remarkable Glazing Phenomenon on the Lunar Surface", *Science,* Vol. 165 (1969).

> "Gold looking for a cause of the glazing assumed 'a giant solar outburst in geologically recent times'... How recent? 'The glazing occurred less

than 30,000 years ago: otherwise, the glaze would have been eroded and dusted over by slow bombardment of the moon by cosmic dust...the event must have taken place some thousands of years ago...to allow enough time for the metal-plating process to coat the glass."

Thus, the glass patches lying on the lunar soil give evidence of very recent large thermal changes on the Moon.

DUST AGAIN

There are additional points regarding the age of the dust found in interplanetary space. First, J.P. Bradley, et. al., in a discussion of dust in interplanetary space published in *Science,* Vol. 226 (1984), p. 1432, states,

"Nuclear tracks have been identified in interplanetary dust particles (IDP's) collected from the stratosphere. The presence of tracks unambiguously confirms the extraterrestrial nature of IDP's and the high track densities (10^{10} to 10^{11} per square centimeter) suggest an exposure age of approximately 10^4 years [10,000 years] within the inner solar system."

Thus, even the age of interplanetary dust can be no older than 10,000 years. Although some scientists believe the dust may be replenished by short-period comets, some others such as Bradley feel comets represent an inadequate source. Nevertheless, the large quantities of youthful dust strongly imply very recent catastrophism in the solar system. The dust should show a variety of ages going back at least 30,000 years!

Bradley's view is now accepted as correct. *Sky and Telescope,* March 1989, p. 243, informs us that, "For some time, astronomers have known that comets do not [at present] produce enough dust to maintain the zodiacal dust clouds." Research indicates forty percent of this dust could come from asteroid collisions.

Second, the question is, where is this dust found distributed in the solar system? According to Velikovsky, the catastrophic events he described all occurred inside and between the orbits of Jupiter and Mercury. Therefore, interplanetary dust should be located inside this space. In giving birth to Venus, an enormous amount of dust would have been produced inside the orbit of Jupiter, and that is what is found. Mitchell M. Waldrop and Richard Kerr in an article titled "IRAS Science Briefing" in *Science,* Vol. 222, (1983), p. 916, claim that the Infrared Astronomy Satellite found a vast ring of dust between the orbits of Mars and Jupiter. Velikovsky also pointed out in *Earth in Upheaval,* p. 288, back in 1955 that,

"The zodiacal light, or the glow seen in the evening sky after sunset, stretching in the path of the sun and other planets (ecliptic), the mysterious origin of which has for a long time occupied the minds of astronomers, has been explained in recent years as the reflection of the solar light from two rings of dust particles one following the orbit of Venus."

Hence, the rings of dust are located where the cataclysms described by Velikovsky also occurred. If, as Velikovsky claimed, proto-planet Venus was a comet whose orbit decayed and changed from an elliptical one which brought it close to Earth to the one it currently follows only a few thousand years ago, then it should have also left zodiacal dust as a ring between the Earth and itself. Over a few thousand years this dust ring should have migrated so that its materials are distributed inward, that is, closer to the Sun. It should be remembered that the dust tails of comets point away from the Sun.

C. Leinert, et. al., have discovered this torus of zodiacal dust and in their article, "The Zodiacal Light From 1.0 to 0.3 AU as Observed by the Helios Space Probes", *Astron. Astrophys.,* Vol. 103, (1981), pp. 177-188. Their paper discusses several years of analyzing zodiacal light by the Helios Space Probes. They report that both probes show a torus of zodiacal dust between the orbit of the Earth and 0.3 AU. Furthermore, the dust thins out over this distance. Interestingly, the authors admit that they can find no explanation for this unusual discovery.

This dust evidence again supports Velikovsky's thesis respecting the changing orbit of Venus.

Third, according to Waldrop and Kerr, the dust is swept into the Sun by the Poynting-Robertson effect in at most 20 to 30 thousand years. However, the measured age of the planetary dust particles, based directly on counting nuclear particle tracks, is only about 10,000 years. Both results imply that these dust clouds in orbit about the Sun are extremely young.

The evidence of the interplanetary dust is surprisingly in full accord with Velikovsky's hypothesis. It is young; it is found at the places where celestial catastrophes are described to have occurred, and it has not been removed by the solar wind and/or drawn into the Sun.

On the surface of the Moon, on the ice cap in Greenland and in interplanetary space, the evidence from celestial dust is in full accord with Velikovsky's thesis.

CRATERS ON EARTH

Sagan then states,

"Furthermore, if lunar craters were to have formed abundantly 2,700 or 3,500 years ago, there must have been a similar production at the same time of terrestrial craters larger than a kilometer across. Erosion on the Earth's surface is inadequate to remove any crater of this size in 2,700 years. There are not large numbers of terrestrial craters of this size and age, indeed, there is not a single one."[29]

In *Earth in Upheaval*, Velikovsky devotes a chapter to this topic titled "The Carolina Bays" in which we find,

"Peculiar elliptical depressions or 'oval craters' locally called 'bays', are thickly scattered over the Carolina coast of the United States and more sparsely over the entire Atlantic coastal plain from southern New Jersey to northeastern Florida. The marshy depressions are numbered in the tens of thousands and according to the latest estimate, their number may reach half a million [See Douglas Johnson, The Origin of the Carolina Bays, (1942) W.F. Prouty "Carolina Bays and Their Origin", *Bulletin of the Geological Society of America,* LXIII, (1952) pages 167-224].

"Measurements made on more prominent ones seaward from Darlington, show that the larger bays average 2,200 feet [2/3 of a kilometer] in length and in single cases exceed 8,000 feet. [Over 2 1/2 kilometers or more than a mile and a half.] A remarkable feature of these depressions is their parallelism: the long axis of each of them extends from northwest to southeast, and the precision of the parallelism is 'striking.' Around the bays are rims of earth, invariably elevated at the southeastern end. These oval depressions may be seen especially well in aerial photographs. Any theory of their origin must explain their form, the ellipticity of which increases with the size of the bays; their parallel alignment; and the elevated rims at their southeastern ends.

"In 1933 a theory was presented by Melton and Schriever of the University of Oklahoma, [See F.A. Melton and A.W. Schriever, 'The Carolina Bays—Are They Meteorite Scars?', *Journal of Geology* XLI, 1933], according to which the bays are scars left by a 'meteoric shower or colliding comet.' Since then the majority of authors who have dealt with the problem have accepted this view, and it has found its way into textbooks as the usual interpretation. [See D. Johnson above, *The Origin of the Carolina Bays,* p. 4.] The authors of the theory stress the fact that 'Since the origin of the bays apparently cannot be explained by the well-known types of geological activity, an extraordinary process must be found. Such a process is suggested by the elliptical shape, the parallel alignment, and the systematic arrangement of elevated rims.'

[According to Melton and Schriever above, *Journal of Geology,* XLI p. 56], The comet must have struck the northwest, 'If the cosmic masses approached this region from the northwest, the major axes would have the desired alignment.' The time when the catastrophe took place was estimated at sometime during the Ice Age. ...The deposition of sand and silt, a process which doubtlessly occurred while the region was covered by the sea during the terrace-forming marine invasion of the Pleistocene [glacial] period. But the possibility was also envisaged that 'the collision took place' through 'the shallow ocean water during the marine invasion' [after the end of the Ice Age]. The swarm of meteorites must have been large enough to hit an area from Florida to New Jersey.

"Some critics disagree with the idea that the bays originated in the Ice Age or 'are relatively ancient,' and place their origin in more recent time. [See D. Johnson, *Origin of the Carolina Bays,* p. 5] The craters were produced by meteoric impact, either by direct hits or by explosion in the air close to the ground, thus causing the formation of vast numbers of depressions. Some of the bays, it is assumed, are on the bottom of the ocean. It was stressed [by C.P. Olivier, *Meteors,* (1925), p. 240] that 'a very large number of meteorites have been discovered in the southern Appalachian region, in Virginia, North and South Carolina, Georgia, Alabama, Kentucky, and Tennessee.'"[30]

It should also be noted that this coastal region is subjected repeatedly by seasonal hurricanes, to weathering and erosion. But these storms of wind and rain have not erased these craters; hence their fresh circular rims indicate that they are only a few thousand years old.

A very short lived geological phenomenon, for example, is a small lake. Over a few thousand years, detritus flowing into lakes fills them with sediment and they disappear. This is made explicit by geologist Ronald B. Parker of the University of California at Berkeley in *The Tenth Muse,* (NY 1986), p. 76 wherein he states,

"As any geologist will tell you, lakes are temporary features in the long term...be assured, that in the time frame of geologists, a life span of a few hundred or a thousand years is temporary. The fact is that lakes do become filled in with sediment and accumulated organic matter over periods of many years to form rich agricultural land in temperate climates, dry playas in desert regions... [etc.]"

In this regard, an article in the *Journal of Geology,* Vol. 57 (1949), titled "Oriented Lakes of Northern Alaska" described lakes quite similar in structure to the Carolina Bays. The distribution of the Alaskan lakes is spread out over an area of 25,000 square miles of the Arctic coastal plane which is equivalent in the area to the Carolina Bays. There are, according to the authors, tens of thousands of lakes or lake basins. The sizes of the lakes are from a few tens of feet; the largest lakes are 9 miles long and 3 miles wide, many with averages between 1 and 3 miles long and 1/2 mile wide. Their shapes are generally elliptical, cigar shaped, rectangular, ovoid or egg shaped, while others are deformed by ice action. The lakes are parallel to each other, intersect or overlap one another just as the Carolina Bays. Often the lakes are arranged in rows also just as the Carolina Bays.[31]

In *New Scientist,* evidence of cratering on the ocean bottom is reported. We are informed that,

"The topography of the sea floor around Britain, like that of its land area, has formed over many thousands of years and results from many well understood processes. So it is surprising that recent studies have

discovered a wide expanse of a sea bed in the middle of the North Sea—
15,000 and 20,000 square kilometers in area -which appears on the sonar
pictures to have a topography much like a miniature lunar
landscape..."[32]

The craters are between 33 and 330 feet in diameter. The authors of
the article go on to say that the craters are generally elliptical and tend
to have a common direction of elongation. This is like the Carolina
Bays and Alaskan lakes and, like those craters, the North Sea craters
tend to form lines in the same direction. Small meteorites falling into
the sea would lose much of their force and thus, produce shallower
and smaller craters in general. This is what is observed in the North
Sea.

However, a swarm of larger meteors falling at a higher angle to the
surface of the Earth would create more highly circular craters. In this
respect, it is interesting that Commander Jacques-Yves Cousteau, the
famous French sea explorer, discovered such circular craters in the
Caribbean Sea. His discovery is reported by Janet Gregory in *FSR*.
She relates that Cousteau, after returning home to his base in Monaco
in the south of France, told of strings of circular blue holes that he and
scientists aboard the *Calypso* observed on the floor of the Caribbean.
These circular blue holes, when they were explored, were discovered
to be highly circular depressions, each about 300 yards (900 feet) in
diameter, but only a few feet deep. Apparently, some formed lines
about 25 miles long. This would extrapolate to between 300 to 400
shallow craters for each linear arrangement. Thus, there would be
thousands or tens of thousands and possibly a few hundred thousand of
these craters in the sea bed of the Caribbean. What Cousteau claimed
was that these craters were scooped out of rocks in the sea bed. This
implies explosive force.[33]

I.N. Lancaster, in the *Geographical Journal*, Vol. 144, (1978), p. 81,
states of oriented lakes in the Kalahari Desert that,

"The pans or small dry or ephemeral lakes, of the southern Kalahari in
Botswana, are contained in shallow, sub-circular to sub-elliptical,
enclosed depressions in the surface sandwich mantle of the region."

No one knows the precise cause of these oriented lakes, but S.J. Shand
in the *Scientific Monthly*, Vol. 62, (1946), p. 95 describes one of these
lakes which is 1,500 yard long and 1,000 yards wide and is surrounded
by a sandy rampart 30 to 50 feet high. He claims that there are
thousands of similar such lakes on the coastal plain of South Africa,
with many more in the dry interior. The closeness in similarity to the
Carolina Bays is found to be quite strong, implying a similar origin.
No such lakes are being formed anywhere on the Earth at present.

L.P. Killigrew and R.J. Gilkes in *Nature,* Vol. 247, (1974), p. 454, describes playa lakes of southwestern Australia. These lakes number in the thousands and range in size from 4 thousandths to over 100 square kilometers. Large numbers are elliptical and exhibit a slightly west to north orientation. From aerial photographs in the article, these lakes also show many features similar to the Carolina Bays. In "Oriented Lakes and Lineaments of Northeastern Bolivia", George Plafker writes in the *Geological Society of American Bulletin,* Vol. 75, (1964), p. 503, stating that these,

> "Oriented lakes occur over an area of roughly 45,000 square miles in the Beni basin where the water table is at or near the surface. They extend from the margin of the Brazilian shield outcrop, westward to within 20 miles of the sub-Andean zone. In the area of detailed study there were 104 oriented and dry lakes more than 1 km in length, and almost an equal number of lakes less than 1 km in maximum dimension. Lakes constitute 3 percent of the total area with roughly one lake more than 1 km long per 40 square miles... Most of the lakes in the Beni basin have axes or long straight segments of shoreline that tend within 10 degrees of N. 45 degrees E. or 45 degrees W. Lake shorelines in the mapped area either oriented in one or both of these two directions or are completely unoriented. The lakes range from about 1,000 feet square to 12.4 miles long by 5.4 miles wide."

Plafker also tells us about lakes he observed on the Old Crow Plain of Canada in the Yukon Territory which are also oriented and appear to be similar to those he described in Bolivia. It is quite probable that all these oriented lakes were formed around the same time by meteoric explosions and then affected by the wind, water and ice if they are located in areas with cold climates. But it should be remembered that lakes have lifetimes of only a few thousand years and thus, these lakes are quite young. Sagan claimed that there was not a single recent crater over a kilometer in diameter to be found on the surface of the Earth. In a sense he is correct because there is not one, but several thousand such recent craters. In this sense he is also quite wrong. Thus, when discussing the question of recent craters, he writes, "On these questions [of terrestrial cratering] Velikovsky seems to have ignored critical evidence."[34] When the evidence is examined, it strongly contraindicates Sagan's assertion.

As to more accurately dated craters of recent origin, Lewis M. Greenberg states,

> "As it happens, the Barringer crater in Arizona (1.2 km in diameter) whose presently accepted age is 50,000 years, once had it age fixed at 2,000 to 3,000 years by Barringer and from 5,000 to 10,000 years by Tilghman. (See E.L. Krinov, *Giant Meteorites,* Oxford, (1966), p. 104)

The Indians who settled in this district are well acquainted with the crater's existence. 'They have a legend that at one time one of their gods descended from Heaven in blazing magnificence to find rest beneath the ground.' (Krinov pp. 82–83) Since archaeologists estimate that man did not appear in this region until 20–25,000 years ago, the Indian legend has proven to be somewhat disconcerting. It has even caused Nininger to note 'the possibility of a discrepancy in the estimates of the crater's age.' The New Quebec Crater (3.2 km in diameter) has been estimated by Meen to be between 3,000 and 15,000 years [old] (Krinov, p. 52). I would also refer Sagan to the work of F. Dachille ("Interactions of the Earth with Very Large Meteorites, *Bulletin of South Carolina Academy of Sciences,* Vol. 24, 1962), who mentions the discovery of a possible crater basin, 240 km in diameter and 5–10,000 years [old] in date, under the Antarctic ice near Long. 14 degrees E. Lat. 70 degrees S.

"Of more than passing interest is a 'proven' meteorite impact crater at Kaalijarvi, Estonia SSR (150 m) which has been dated to ca. 710–580 B.C. (See *Radiocarbon,* Vol. 8, (1966), p. 436). Independently, Krinov (op. cit., p. 40) considers that the age of this crater is approximately 4,000 –5,000 years."[35]

GLOBAL FLOODING

Sagan states,

"Velikovsky believes that the close passage of Venus or Mars to the Earth would have produced tides miles high [that is]…hundreds of miles high. This is easily calculated… To the best of my knowledge, there is no geological evidence for a global inundation of all parts of the world at any time between the sixth and fifteenth centuries B.C. If such floods had occurred, even if they were brief, they should have left some clear trace in the geological record."[36]

As was stated earlier, Velikovsky's theory is that the Ice Age ended by a catastrophe which among other events caused oceanic flooding of much of the Earth's land masses. Thus, let us examine the evidence. Stephen J. Gould informs us that,

"In 1923, J. Harlan Bretz proposed a striking and unorthodox explanation for the channeled scablands of Eastern Washington. This peculiar topography is developed within a series of elongated basins called coulees… They are traceable up gradient to the southern extent of the last glaciation and down gradient to the Snake or Columbia rivers…the peculiar features [of the area suggest]…that the channels were once filled with water to a great height; the deep gouging of basalt within the channels…[and other evidence] does not look like the ordinary work of rivers…

"Bretz concluded, therefore, that the coulees had been carved by a single gigantic flood that had filled them to a depth of more than 1,000 feet, had cut through hundreds of feet of basalt in places and had ended in a matter of days. He envisioned the scope of the event as follows:

> 'Fully 3,000 square miles of the Columbia plateau were swept by the glacial flood, and the loess and silt cover removed. More than 2,000 square miles of this area were left as bare, eroded, rock cut channel floors, now the scablands, and nearly 1,000 square miles carry gravel deposits derived from the eroded basalt. It *was* a debacle [catastrophe] which swept the Columbia Plateau. (Bretz 1923, p. 649).'

"Bretz's hypothesis evoked from the geological establishment a flood of commentary, nearly all of it negative. The common theme running through all of this criticism was the rejection of his ideas in favor of gradualistic explanations, often on *a priori* grounds."[37]

Today the truth of a gigantic flood is accepted by the scientific community, but only on the *a priori* ground that the flood was caused when an ice dam broke and a huge glacial lake emptied. However, a tidal wave sweeping the glacier would produce the same effect. In fact, W.C. Hunt rejects the suggestion that an ice dam could hold water estimated to be 2,100 feet deep. He writes in *Bulletin of Canadian Petroleum Geology*, (1977), Vol. 25, p. 468, that the depth of water in Lake Missoula held by an ice dam "is an impossibility" and goes on to state, "When one considers that modern engineering employs bedrock grouting for securing the footings of 500 foot (150 m) dams, it must strike any reader as virtually frivolous to suggest that chance emplacement of glacial ice might have dammed Clark Fork across a 7 mile (11 km) span lacking in intermediate abutments, and then retained water at four times the pressure of modern engineered concrete dams!" Hunt proved that the dam was actually emplaced elsewhere and "would have had an unsupported length of approximately 50 mi (80 km)." Such a length of dam made of chance-placed ice is incredible. Thus, it is highly probable that the flood was not caused by a dammed lake. To flood this area, the tidal wave must have come either from the Pacific Ocean or the Gulf of Mexico which were not covered by the ice sheet during the last Ice Age.

In this respect we return to the thesis of Warren C. Hunt published in the *Bulletin of Canadian Petroleum Geology*. Hunt presents just this hypothesis, that a tidal wave swept over North America from the ocean at the end of the Ice Age. He claims that the erratic stones of Alberta, Canada lie on materials that are radiocarbon dated to be quite young. Therefore, the erratic boulders, some of them huge, which lie upon the young detritus could not have been pushed by the ice into this area after the detritus was laid down because the ice had melted.

Hunt maintains that the only way to produce this phenomenon is to have an oceanic tidal wave do the job. He writes,

> "If such a tide had come in...and inundated the Athabaska Valley, up to 5,000 feet (1823 meters) or so above sea level, it would have floated the glacier in the 50 or so miles (80 km) comprising the probable rock source. That such a phenomenon could have happened and, in fact, probably did happen, is the thesis of this paper.
>
> "Violent earthquaking would have been the first event, perhaps due to proximal passage of a cosmic body. Seismicity is conceived to have caused collapse of mountains and ridges with consequent avalanching out upon the surface of the Athabaska Valley glacier. That event was followed by a great tide from the Gulf of Mexico."[38]

Hunt then gives evidence of relic beaches and evaporate deposits of Lake Bonneville to reinforce his thesis. In summation Hunt states that, "The concept of a celestial body interfering with Earth gravity to cause great tides is not alien to the theory of uniformitarianism."[39]

Velikovsky states in the chapter titled "Sea and Land Change Places" in *Earth in Upheaval*,

> "The most renowned naturalist to come from the generation of the French Revolution and the Napoleonic Wars was Georges Cuvier. He was the founder of vertebrate paleontology, or the science of fossil bones, and the science of extinct animals. Studying the finds made in the gypsum formation of Montmarte in Paris and those elsewhere in France and the European continent in general, he came to the conclusion that in the midst of even the oldest strata of marine formations there are other strata replete with animals or plant remains of terrestrial or fresh-water form; and that among the most recent strata, or those that are nearest the surface, there are land animals buried under heaps of marine sediments." [Cuvier wrote] 'It has frequently happened that lands which had been laid dry, have been again covered by the waters, in consequence either of their being engulfed in the abyss, or of the sea having merely risen over them... These repeated irruptions and retreats of the sea have neither all been slow nor gradual; on the contrary, most of the catastrophes which had occasioned them have been sudden; and this is especially easy to be proven, with regard to the last of these catastrophes, that which, by a twofold motion, had inundated, and afterwards laid dry, our present continents, or at least a part of the land which forms them at the present day.'"[40] [See Georges Cuvier, *Essay on the Theory of the Earth,* (5 ed.; 1827), pp. 13-14 English Translation]

Cuvier contends that he is not speaking of Europe only, but that the flood inundated "our present continents." Velikovsky presents a great deal of evidence to support this flood episode. Now, if indeed the continents were flooded, then the level of the oceans would have been lowered. Is there evidence that there was a drop in the ocean level 3,500 years ago? In fact, Velikovsky presents a chapter, "Dropped Ocean Level" in *Earth in Upheaval,* in which we find:

"In many places of the world the seacoast shows either submerged or raised beaches. The previous surf line is seen on the rock of raised beaches; where the coast became submerged, the earlier water life is found chiseled by the surf in the rock below the present level of the sea. Some beaches were raised to a height of many hundred feet, as in the case of the Pacific coast of Chile, where Charles Darwin observed that the beach must have risen 1,300 feet only recently—'within the period during which upraised shells have remained undecayed on the surface.' He thought also that the 'most probable' explanation would be that the coast level, with 'whole and perfectly preserved shells,' was 'at one blow uplifted above the future reach of the sea,' following an earthquake. In the Hawaiian Islands there is a 1,200 foot raised beach. On Espiritu Santo Island in the New Hebrides in the southern Pacific, corals are found 1,200 feet above sea level...

"In numerous instances evidences of submergence and emergence are seen on the same rock. One such case we have...the Rock of Gibraltar. To a lesser degree the phenomenon is repeated in Bermuda...

"These changes date from different ages, but common to all of them is the absence of intermediate surf lines; if the emergence or submergence had been gradual, intermediate surf lines would be seen in the rock.

"R.A. Daly observed that in a great many places all around the world there is a uniform emergence of the shore line of eighteen to twenty feet. [That is, the sea level dropped 18 to 20 feet] In the southwest Pacific, on the islands of Tutuila, Tau and Ofu and on Rose atoll, all belonging to the Samoan group but spread over two hundred miles, the same emergence is evident. In Daly's opinion, this uniformity indicates that the rise was due to 'something else than crustal warping.' A force pushing from inside would not be 'so uniform throughout a stretch 200 miles long.' [See R.A. Daly, *Our Mobile Earth*, p. 177] Nearly halfway around the world, at St. Helena in the South Atlantic, the lava is punctuated by dry sea caves, the floors of which are covered with water-worn pebbles, 'now dusty because untouched by the surf.' The emergence here is also twenty feet. At the Cape of Good Hope caves and beaches 'also prove recent and sensibly uniform emergence to the extent of about 20 feet.'

"Daly proceeds: 'Marine terraces, indicating similar emergence, are found along the Atlantic coast from New York to the Gulf of Mexico; for at least 1,000 miles along the coast of eastern Australia; along the coasts of Brazil, southwest Africa, and many islands of the Pacific, Atlantic and Indian Oceans; and in all these and other published cases, the emergence is recent as well as the order of magnitude. Judging from the condition of the beaches, terraces and caves, the emergence seems to have been simultaneous on every shore.' [See *Our Mobile Earth*, p. 178]

"Of course Daly also found many places where the change in the position of the shoreline was of a different magnitude, but 'these local exceptions prove the rule.' In his opinion, the cause of the world-wide emergence of the shore lies in the sinking of the level of all seas of the globe, 'a recent world-wide sinking of ocean level' which could have been caused by water being drawn from the oceans to build the ice caps of Antarctica and Greenland. Alternately, Daly thinks it could also have resulted from deepening of the oceans or from an increase of their area.

"P.H. Kuenen of Leyden University, in his *Marine Geology,* finds Daly's claims confirmed: [in *Marine Geology,* (1950), p. 538, Kuenen writes], 'In thirty-odd years following Daly's first paper many further instances have been recorded by a number of investigators the world over, so that this recent shift is now well established.'

"Whatever was the cause of the phenomenon observed, it was not the result of a slow change; in such a case we would have intermediate shore lines between the present surf line and the twenty-foot line on the same beaches, but there were none.

"Of special interest is the time of change. According to Daly [in *Our Mobile Earth,* p. 179], 'This increase of the ice cap or caps has been tentatively referred to late Neolithic time, about 3,500 years ago. At that approximate date, there was some chilling of the northern hemisphere at least, following a prolonged period when the world climate was distinctly warmer than now. Late-Neolithic man lived in Europe 3,500 years back.'

"As to the date of the sudden drop of oceanic level, Kuenen writes [in *Marine Geology,* p. 538], 'The time of the movement was estimated by Daly to be probably some 3,000 to 4,000 years ago. Detailed field work in the Netherlands and in eastern England has shown a recent eustatic depression of the same order of magnitude as deduced by Daly. Here the time can be fixed as roughly 3,000 to 3,500 years ago.' Thus, the work in the Netherlands and in England confirmed not only Daly's finding, but also his dating. The ocean level dropped, of course, all over the world. It was not a slow subsidence of the bottom, or a slow spread of the ocean over land, or a slow evaporation of oceanic water: whatever it was, it was sudden and therefore, catastrophic."[41]

Some scientists have challenged Daly's thesis. R.T. Walcott, in *Quaternary Research,* Vol. 2 for 1972, claims that the beaches should be understood as the result of the rebound of the Earth's crust because of postglacial rebound. This would only make sense if all the continents were covered by the ice cap or that there were many intermediate beaches. There is no such evidence for this conclusion. In addition, the Samoan Islands and Marquesas in the Pacific Ocean and other tropical islands of the Atlantic and Indian Oceans are not part of the continents, but show the same oceanic change. To maintain that both glaciated and non-glaciated shore lines emerged because of postglacial rebound on all continents and that this same phenomenon produced the same evidence on islands at great distances from the continents is without a mechanism and thus, lacks foundation.

Let us examine this from another view point. If there were enormous tidal waves on the oceans, would they not annihilate human inhabitants that probably lived on the islands of the oceans, and therefore, human habitation of these islands would date only from after 3,500 years B.P.? In this respect, Brian M. Fagan in *The Great Journey,* (London 1987), p. 75, tells us that,

> "The tropical routes [from Asia to America] are thousands of miles longer [than the Aleutian route] and involve unpredictable wind patterns, prolonged calms, and weeks-on-end out of sight of land. Whatever Pacific route might have been followed, prehistoric migrants would have to possess a technology capable of building strong boats to use on the open sea. And here crucially we can consider the evidence of the first colonization of Polynesia, the vast Pacific region that includes New Zealand, Hawaii and Easter Island. Countless excavations have shown that not until 3,500 years ago, with the advent of the outrigger canoe, did this whole area begin to be colonized. Before then, there were simply no human inhabitants at all."

Again, the date 3,500 years ago corresponds to the evidence that Velikovsky developed.

This concept that ancient man could not travel the seas prior to 3500 B.P. is now known to be in error. It is contradicted by what we know of the Indian Archaic of North America who lived 8,000 to about 4,000 years ago, and then suddenly disappeared. They are known to have been able to travel on the oceans in large ocean going canoes. Thus, the appearance of humans in the Pacific indicates that the Pacific was swept by tidal waves around 3,500 years ago and within a century ancient man began to recolonize the islands. The oceans all dropped about 20 feet 3,500 years ago and this is observed by mass extinctions found on the continents. Thus, Sagan's statement that "there is no geological evidence for a global inundation of all parts of the world either in the eighth or in the fifteenth centuries B.C."[42] shows once again, his great reluctance to approach the evidence produced by fellow scientists that supports Velikovsky.

FAUNAL EXTINCTIONS

Sagan wishes to know, "Where are the extensive faunal extinctions..." if such a catastrophe as Velikovsky describes in *Worlds in Collision* had taken place.[43] Critics of Velikovsky's theory would argue that the fossil remains of animals that became extinct when the Ice Age ended are at least 10,000 years old and not 3,500 years old.

To discuss this argument scientifically, evidence of the recentness or lateness of the mass extinction that accompanied the end of the Ice Age is necessary. In this respect, an article from *Science News Letter,* sheds some light.

> "Radical changes in our ideas of the course of events in recent geological time—say the last half million years or so—may be brought about by the discovery in Utah of the unfossilized skull of an extinct camel, with a bit of dried flesh still clinging to the bone. The relatively fresh condition of the specimen argues that its one-time possessor died only a few centuries or millennia ago; present ideas hold that this particular sort of camel did become extinct a half-million years ago. If this camel really died so long ago, the bone should have been largely or wholly replaced by stone, and there should have been no flesh on it at all.
>
> "The find was reported by Prof. Alfred S. Romer of the University of Chicago. The skull was sent to him by Prof. A.L. Mathews of the University of Utah for examination.
>
> "Prof. Romer's first guess was that it might be a relic of a herd of dromedaries imported into the Southwest during the 1870's as an experiment which terminated unsuccessfully. But a critical examination of its anatomical details showed many points of close resemblance to the skulls of very ancient extinct American camels, and marked differences from those of existing Asiatic and African forms. In his opinion, the animal belonging to the genus *Camelops* which is supposed to have been extinct for at least half a million years.
>
> "Prof. Romer's tentative answer to the riddle is not that the skull has remained unfossilized, yet undestroyed, for half a million years, but that the species did not become extinct then, surviving instead until comparatively recent times.
>
> "Such an answer, he points out, would also help to settle the conflict over the antiquity of man in America. Many scientists refuse to accepts as authentic the occasional finds made on the continent of stone or bone implements associated with the remains of animals supposed to have been extinct for hundreds of thousands of years. Prof. Romer states that other recently discovered remains of camels, lions and other animals in the West also hint at a longer survival of these extinct beasts than has hitherto been supposed."[44]

C14 DATING THE EXTINCTIONS

What is, furthermore, pertinent to the evidence is radiocarbon dating. W. Dort published results in the *Antarctic Journal of the United States* that freshly slaughtered seals when subjected to radiocarbon analysis are dated to be 1,300 years old.[45] B. Hubner published results in *The Physiology of Forest Trees* that wood from a growing tree when radiocarbon dated was found to be 10,000 years old.[46] And M. Keith and G. Anderson published results in *Science* that the shells of living mollusks when radiocarbon dated was found to be 2,300 years old.[47] Thus, the dating by radiocarbon, though much respected by the scientific community, is not without serious problems. It is believed that such a method is perhaps accurate back no more than 50,000 years. However, dinosaur bones, coal and oil that are supposed to be millions of years old have yielded radiocarbon dates.[48] This is supposed to be impossible.

In fact, J. Ogden of a Carbon-14 testing laboratory wrote an article titled, "The Use and Abuse of Radiocarbon" in the *Annals of the New York Academy of Science,* where we learn that an investigator that brings a specimen for testing is actually to supply the date that he will accept for the age of the material. Then if a date is found that is close to the figure that has been requested, it is published with plus and minus tolerance levels that appear to make the test appear honest. Ogden states, "It may come as a shock to some, but fewer than 50 percent of the radiocarbon dates from geological and archaeological samples in northeastern North American have been adopted as 'acceptable' by investigators."[49] R.E. Lee in an article titled "Radiocarbon: Ages in Error", in *Anthropological Journal of Canada,* (1981) 19 (3):9 p. 25, states,

"The necessity for calibration over the last 7,000 years is well recognized and attended to, while the probable error in older dates receives no practical consideration at all. At a range of 20,000 to 30,000 years, it is true, one can only guess at the full extent of the problem. But one can be reasonably sure about its trend: *too young.*" [Lee's emphasis]

He goes on to say on page 27, "*Radiocarbon method is still not capable of yielding accurate and reliable results. There are gross discrepancies,* the chronology is uneven and relative, and the accepted dates are actually selected dates." Lee then cites R. Stuchenrath in the *Annals of the New York Academy of Sciences,* (1977), Vol. 288, p. 188, who informs us that "This whole blessed thing is nothing but 13th century alchemy and it all depends upon which funny paper you read."

These statements from workers in the field of radiocarbon dating show that such evidence is not to be relied on as many textbooks have claimed.

Frank C. Hibben in *The Lost Americans,* revised and updated (NY 1968), in his chapter "Radioactive Times" discussed the process of Carbon-14 dating. After outlining several of the problems associated with using this method, he states on pp. 139-140, "Even with these drawbacks and pitfalls...archaeologists and laboratory technicians began to hammer out the exact history of the earliest Americans. *The dates badly out of line were disregarded.*" [emphasis added] What we are told is Carbon-14 dates show extinct animals and trees may well have lived into historical times; but because these dates do not fit the preconceived theory that the Ice Age ended far more recently than the Carbon-14 evidence shows, these dates are simply disregarded.

How common is the activity of disregarding Carbon-14 dates that contradict the chronological expectations? Apparently it is quite common. Ron Willis in *Info Journal*, Vol. 3, (1973), pp. 1–7, also deals with this dating process. In discussing this he states, "There are anomalous dates in the series which do not fit. This is common in the C-14 process. *Like any good archaeologist, I will ignore the dates that do not fit.*" [emphasis added] Thus, the archaeologists corroborate the statements cited above that the relics brought for dating are not accepted if they do not agree with the preconceived assumptions regarding their age by the archaeologists.

Hibben, cited above, page 140 then informs us that, "The early workers were amazingly accurate in their guesses as to the antiquity of the earliest American hunters." This accuracy was only obtained by throwing away all the dates that were "badly out of line." These statements clearly indicate that "all good archaeologists" organize the evidence so the *a priori* expectations are validated. Therefore, whenever Carbon-14 dates show that the Ice Age ended far more recently than theory allows, it is bad contaminated evidence and is disregarded. On the other hand, when this same method gives dates that fit the expected chronology, it is hailed as "amazingly accurate," because this is good uncontaminated evidence. The major criteria for the distinction between the good, accurate, acceptable evidence and the bad, inaccurate, unacceptable evidence, we are told by the archaeologists, and paleontologists is whether or not it fits the expectations of the chronology.

This approach to evidence is no different than the methods employed by the astronomers when they find ancient documents that describe the lunar month as being of a different length than expected. In both cases, when evidence contradicts the expectations, the inconvenient data is ignored. Finagling with scientific evidence is called "culling" in the parlance of the scientists. As can be clearly seen, the uniformitarian researchers have no qualms about culling their data because their scientific philosophy cannot be in error; only the evidence may be. Thus, the textbooks that support the present chronology as accurate, do so with "culled" data. Carbon-14 dating is not so much 13th century alchemy as it is 20th century scientific fraud.

In fact, it was reported in *New Scientist* for September 30, 1989, p. 26 in an article titled "Unexpected errors affect dating techniques," by Andy Coghlan that, "Britain's Science and Engineering Research

Council...commissioned a trial that compared the accuracy with which 38 laboratories around the world dated artifacts of known age [by Carbon-14 dating]. Of the 38, only seven produced results that the organizers of the trial considered to be satisfactory." The fact of the matter was that not one testing laboratory got the exact date and 31 of the 38 world's leading Carbon-14 testing laboratories were off by hundreds to thousands of years. This clearly indicates that the dates offered for past events based on Carbon-14 dating is a myth. On the basis of this evidence we find that if 7 out of 38 laboratories produced acceptable results than only 18.5 percent were acceptable while 81.5 percent were erroneous. Would anyone be willing to accept a dating methodology that was shown in error four times greater than the acceptable results. Such a finding would be the death knell to any scientific technique.

Therefore, a more acceptable method of dating would be the examination of human fossils and human artifacts that are known to be very recent. However, if there was a great flood that produced the extinctions, the fossils should show evidence of this immense catastrophic flooding. This then brings us to Velikovsky's evidence.

To produce all of it in Velikovsky's words would require almost one-third of his book. It seems best that only the most germane material be cited.

ART AND DRAWINGS OF EXTINCT ANIMALS

If many types of fauna actually lived into historical times and were destroyed by a catastrophe, then ancient man of Mesopotamia, Egypt and even of Mexican civilization should have left a record of their existence. Geologists of Sagan's persuasion maintain these organisms became extinct at least 10,000 years or more ago. But what do the archaeologists say? Velikovsky states in *Earth in Upheaval*, p. 187 that,

"K.S. Sandford, writing of the conflict of views between geologists and archaeologists in England says [in *Nature* (Dec. 2, 1933)], 'The difference of opinion in some instances is so complete that one or the other must assuredly be wrong.' Those who measure the time in terms of cultural or physical anthropology and archaeology stand in very definite opposition to all estimates [of the time of extinction] based on a geological or on a paleontological time scale.

"As an additional argument, the archaeologist points to pictures of extinct animals in Babylonian and Egyptian bas-reliefs..."

On page 87 of *Earth in Upheaval*, Velikovsky cites L. Frobenius and Douglas C. Fox's, *Prehistoric Rock Pictures in Europe and Africa*, (Museum of Modern Art, 1937), p. 38 respecting,

"Orientalists of the last century...[that] decided that [rock] drawings in the Sahara were the work of Phoenicians. It was likewise observed that on the drawings discovered by Barth, the cattle wore discs between their horns just as in Egyptian drawings. Also, the Egyptian god Set was found pictured on the rocks. And there were rock paintings of war chariots drawn by horses, 'in an area where these animals could not survive two days without extraordinary precautions.' [P. LeCler, *Sahara*, (1954), p. 46]

"The extinct animals in the drawings suggest that these pictures were made sometime during the Ice Age; but the Egyptian motifs made in the very same drawings suggest that they were made in historical times."

Thus, there are *bas reliefs* in Egypt and Babylonia of extinct animals and rock drawings made by Saharans who knew the Egyptian god Set and the war chariot which had to be of an historical period also drawing pictures of extinct animals.

The animals depicted on the walls of Babylonian and Egyptian buildings and Saharan rock drawings had, according to the geologists, been extinct for at least 8,000 years at the time of those early civilizations. The bones of extinct animals are rarely found complete or properly assembled in any location. Yet somehow the stone sculptors of these early times and rock painters were able to accurately represent their forms. Similar evidence of elephants is found in the Americas, that is, depictions of extinct elephants. These forms had been extinct at least 8,000 years as well. Yet, they are represented by ancient Mexican artisans. How can such a state of affairs exist? If Sagan is correct that these faunal forms died out over 10,000 years ago, then one must assume that in Babylonia, Egypt and Mexico universities existed where artisans went to study paleontology—the science of reconstruction of extinct organisms from only a few bones. Or perhaps the Babylonians, Egyptians and Mexicans traveled to the ancient caves of Europe to sketch the drawings of stone age man that were already 8,000 years older at that period. What is apparent is that these animals lived into historical times and early civilized man had observed them and sculptured and drew them accurately.

UNIFORMITARIAN CAUSES

One of the hypotheses for the extinction is that early man hunted these animals to extinction. But if man did so in America, where the human population was relatively small and man did so in Europe and Asia where the human population was greater, why did the elephant survive in Africa and India? Were the Africans and Indians inferior hunters or did they have inferior appetites compared to their American, European and North Asian cousins? If early man hunted the horse in America to extinction, which can run swiftly for great distances, why didn't he do so with the musk-ox which is slower and when frightened forms a standing circle to stand off attackers? Standing still in a circle, the musk-ox was a perfect target for ancient man. Furthermore, several species of birds also became extinct. It is difficult to understand how these avians were hunted to extinction. In *Science News,* Vol. 132, (1987), p. 285, Paul S. Martin of the University of Arizona-Tucson states that, "...overkill is unlikely to have killed 10 classes of North American birds at the end of the last Ice Age." Martin "contends that nearly all of the bird extinctions occurred among scavengers that feed on carnivore leftovers" But, since not all the classes were scavengers, what could have killed the birds? Not climate changes, since the climate became warmer. Martin goes on, "Radiocarbon chronologies are bad in North America, worse in Europe." He states that, "There are some species extinctions that will probably never be reliably dated."

The geologists' other extinction hypotheses tells us that the climate changed. But the extinctions occurred not only in the high latitudes toward the poles, but also in the tropical latitudes around the equator.

The concept that animals were hunted to extinction can be shown to be incorrect by the example of the horse. In *Earth in Upheaval,* Velikovsky writes on page 208,

> "Fossil bones of horses indicate that this was a very common animal in the New World in the Ice Age. But when the soldiers of Cortes, arriving at the shores of America, rode their horses which they had brought from the Old World, the natives thought that gods had come to their country. They had never seen a horse.
>
> "Of the horses the Spaniards brought to America some went astray, became wild, and filled the prairies, traveling in herds; the land and its vegetation and its climate proved to be exceedingly well suited for the propagation of this animal."

If the horse had been hunted to extinction in earlier times, why didn't the American Indians hunt it to extinction once these herds of horses spread over the continent? If they could have hunted these animals to extinction in an earlier period, they most certainly could have hunted them to extinction soon after these animals were reintroduced into the region. Thus, the hunting hypothesis for extinction appears to be meaningless. Furthermore, the very ability of the horse to thrive in North America makes the climate hypothesis for extinction also meaningless. If the climate change was responsible for the extinction of the horse, why was the horse, when reintroduced into the same climate that supposedly destroyed it, able to thrive and multiply so fruitfully? Thus, there is no explanation of the mass extinction that ended the Ice Age. However, if a global catastrophe occurred and the bones of animals and trees along with human artifacts and bones are found together in great masses all around the globe, we can expect that huge tidal waves swept over the lands.

ALASKA AND CANADA

Thus, Velikovsky describes muck deposits filled with millions upon millions of broken bones of extinct mammoth, mastodon, super bison and horse found throughout the lower reaches of the Yukon and according to F. Rainey in *American Antiquity*, 1940, Volume 5,

> "may be considered to extend in greater or lesser thickness over all unglaciated areas of the northern peninsula" [of Alaska and into northern Canada. The] "millions upon millions of animals torn limb from limb [are]...mingled with uprooted trees."[50]

Throughout these masses of shattered bones and trees is volcanic ash. And the depth of these masses of bones, trees and ash is great, "as much as 140 feet." F.C. Hibben of the University of New Mexico ["Evidence of Early Man in Alaska", *American Antiquity*, VIII (1943), p. 256] states that,

> "Although the formation of deposits of muck is not clear, there is ample evidence that at least portions of this material were deposited under catastrophic conditions. Mammal remains are for the most part, dismembered and disarticulated, even though some fragments yet retain, in their frozen state, portions of ligaments, skin, hair and flesh. Twisted and torn trees are piled in splintered masses... At least four considerable layers of volcanic ash may be traced in these deposits, although they are extremely warped and distorted."
>
> Velikovsky then informs us that, "In various levels of the muck, stone artifacts were found, 'frozen *in situ*' at great depths and in apparent association, with the Ice Age fauna, which implies that 'men were contemporary with extinct animals in Alaska.' [See Rainey, *American Antiquity*, V, p. 307] Worked flints, characteristically shaped, called Yuma points, were repeatedly found in the Alaskan muck, one hundred and more feet below the surface. One such spear point [according to Hibben, *American Antiquity*, VIII, p. 257] was found there between a lion's jaw and a mammoth's tusk. Similar weapons were used only a few generations ago by the Indians of the Athapascan tribe, who camped in the upper Tanana Valley. [See Rainey, *American Antiquity*, VI, p. 301] [And Hibben in *American Antiquity*, VIII, p. 256 writes] 'It has also been suggested that even modern Eskimo points are remarkably Yuma like,' all of which indicates that the multitudes of torn animals and splintered forests date from a time not many thousands of years ago."[51]

William N. Irving and C.R. Harington in *Science*, report having found the jaw bone of a child perhaps eleven or twelve years old in the graveyard of the permafrost of the Yukon.[52]

SIBERIA

In Siberia on the other side of the Arctic Ocean are found mass graveyards of mammoth bones by the millions upon millions as well as on the New Siberian Islands.

> "In the stomachs and between the teeth of the mammoths were found plants and grasses that do not grow now in northern Siberia. 'The contents of the stomachs have been carefully examined,' [according to Whitney in the *Journal of the Philosophical Society of Great Britain*, XII (1910), p. 56] they showed the undigested food, leaves of trees now found in Southern Siberia, but a long way from the existing deposits of ivory. Microscopic examination of the skin showed red blood corpuscles, which was a proof not only of a sudden death, but that death was due to suffocation either by gases or water, evidently the latter in this case. But the puzzle remained to account for the sudden freezing up of this large mass of flesh so as to preserve it for future ages.
>
> "What could have caused a sudden change in the temperature of the region? Today the country does not provide food for large quadrupeds, the soil is barren and produces only moss and fungi a few months in the year; at that time the animals fed on plants. And not only mammoths pastured in northern Siberia and on the islands of the Arctic Ocean. On Kotelnoi [according to Whitney above, p. 50] 'neither trees, nor shrubs, nor bushes exist...and yet the bones of elephants, rhinoceroses, buffaloes, and horses are found in the icy wilderness in numbers which defy all calculation.'"[53]

Interestingly, at the Bird's Eye Division of General Foods Corporation, New York, experiments were carried out with carnation and gladioli flowers that were placed in a solution of stomach acid,

> "to find out the minimum rate of temperature decrease required in order to preserve parts of the flowers such that they would still be identifiable. Then, from the dimensions of the mammoth and the known rates of heat transfer (heat loss) through fur, skin, fat, flesh, etc., an outside temperature was computed that would reduce the stomach temperature at the previously determined rate in order to preserve the buttercup flowers. The staggering conclusion was that the mammoth, and presumably all the tens of thousands of other frozen animals in the north was overcome in mid-summer by a cold blast with temperatures lower than minus 150 degrees Fahrenheit... The lowest recorded temperatures on earth have never reached this extreme, while the temperatures in these polar regions today have never since recovered to the point where buttercups will grow again."[54]

But Velikovsky informs us that,

> "On Maloi, one of the group of Liakhov Islands [in the Arctic Ocean above the Arctic Circle] Toll found bones of mammoths and other animals together with the trunks of fossil trees, with leaves and cones.

[Whitney, above, p. 50 states] 'This striking discovery proves that in the days when the mammoth and rhinoceroses lived in northern Siberia, these desolate islands were covered with great forests, and bore a luxuriant vegetation.'"[55]

The point is that large trees should not be able to grow on these Arctic islands nor on the tundra of Alaska and Siberia. According to Ivan T. Sanderson, *The Dynasty of ABU,* (NY 1962), p. 80, "pieces of large treetrunks of types that do not—and can not—live at those latitudes today for purely biological reasons [were found]. The same goes for huge areas in Siberia."

How is it possible to freeze mammoths in mid-summer with a drop in temperature to -150 degrees Fahrenheit? Siberia is known to have a short, but very hot summer. In this respect, Sir Fred Hoyle, the noted British astronomer, tells us,

> "Some years ago, I had reason to work out mathematically which place on Earth receives the most solar radiation at mid-summer. I found to my astonishment it was the pole. It is obviously not the pole that receives the most radiation at midday, but because the Sun never sets at mid-summer for latitudes inside the Arctic Circle, sunlight is received there for the full twenty-four hours."[56]

Velikovsky contends that the mechanism responsible for what we find in Alaska and Siberia was a shift of the Earth's pole by the close passage of a stupendous comet.

IPIUTAK

Before leaving the Arctic region one may ask, if the climatic conditions were so much warmer in the Arctic area, would not human settlements be found there? Velikovsky informs us that,

> "It is assumed here that in historical times, neither northeastern Siberia nor western Alaska were in the polar regions, but that as a result of the catastrophes of the eighth and seventh centuries [B.C.] this area moved into that region. This assumption implies that these lands, to the extent that they were not covered by the sea, were most probably places of human habitation...
>
> "In 1939 and 1940, 'one of the most startling and important finds of the century' (E. Stefansson) was made at Point Hope in Alaska, on the shores of Bering Strait: an ancient city of about eight hundred houses, whose population had been larger than that of the modern city of Fairbanks [in 1950] was discovered there, north of 68 degrees, about 130 miles within the Arctic Circle.
>
> "Ipiutak, as the location of this ancient city is called by the present Eskimos, must have been built before the Christian era; two thousand years is thought a conservative estimate of its age. The excavations have yielded beautiful ivory carvings unlike any known Eskimo or other

American Indian culture of the northern regions. Fashioned of logs, the strange tombs gave up skeletons which stared up at the excavators with artificial eyeballs carved of ivory and inlaid with jet... Numerous delicately made and engraved implements also found in the graves, resembled some of those produced in North China two or three thousand years ago; others resemble carvings of the Ainu people in northern Japan and the natives of the Amur River in Siberia. The material culture of these people was not a simple one, of the kind usually found in the Arctic, but elaborate and that of a sophisticated people, in this sense more advanced than any known Eskimos and clearly derived from eastern Asia."[57]

[The above citation is from Evelyn Stefansson, *Here in Alaska*, (1943) pp. 138ff] Altogether there were some 800 buildings uncovered at Ipiutak. Thus, its population was 4,000 inhabitants or more. How did they survive in such numbers during the six months of winter darkness? Unless the arrangement of the polar axis were different so that the climate was milder, it appears impossible that these early city dwellers could survive the long, dark, cold winter. The people who excelled the Eskimo "did not have seal oil lamps, sleds or slate tools" we are told by *National Geographic*.[58] Without seal oil lamps, how did they supply light for themselves for six months of night? Without sleds, how did they transport logs for the graves? How did they travel? Clearly what is found strongly points to a different climate and position of the polar axis.

THE BRONZE AGE IN SIBERIA

What about on the other side of the Arctic Ocean? Velikovsky informs us that,

"In recent years, Russian archaeologists have discovered abundant remains of human culture in northeastern Siberia in the frozen taiga where frozen bodies of mammoths are found and where nobody suspected human abodes in ages past. There was human population in northeastern Siberia in paleolithic time, in neolithic time and in the bronze time too.

"Paleolithic artifacts were found in Yakutia; rock drawings very similar to the Paleolithic drawings on the rocks and in the caverns of France and Spain, were found in the valley of the Lena, near the village Shishkino. [A.P. Oklidnikov, "Excavation in the North", published in the Soviet Union, English title *Vestiges of Ancient Culture*, (1951) states] 'In the neolithic age, *about two to three millennia before our era*, neolithic races, descendants of earlier inhabitants of Yakutia...spread to the very coast of the Arctic Ocean in the north and the Kolyma in the east...'

"On the lower Lena, [river] north of the confluence with Viliy, inside the polar circle, monuments are found of a characteristic culture; outstanding finds were made near the lake Yolba, not far from Jigansk.

"As soon as the archaeologists started a methodic investigation of the area, in Yakutsk itself, was found a workshop of an ancient metallurgist in which, at the end of the second millennium before the present era, he made bronze axes similar to the axes manufactured about that time in the Near East and in Europe.

[Oklidnikov continues] "In the Yakutsk taiga two and a half [or three] thousand years ago, there already lived artisans in metals who were able to extract copper from ore, to melt it and pour it into forms, and to make axes, beautiful bronze tips for the spears, knives and even swords."

"These relics of a civilization in the taiga of northeastern Siberia imply that the climate changed there in the age of advanced man. Before the ice froze the region, voracious members of the elephant family roamed there in large herds."[59]

Furthermore, Hapgood writes,

"The Soviet publication *Sputnik*, in its issue of November, 1968, reported the discovery of evidence of human occupation of the New Siberian Islands, as well as of Spitzbergen, during the ice age. Both archipelagoes are virtually uninhabitable now, especially the New Siberian Islands, which lie only 10 degrees, or about 600 miles from the [north] pole. *Sputnik* gives the source of information as the newspaper *Kommunist Tajikistana,* and says, 'Archaeologists have discovered traces of a Stone Age settlement on the Novosibirsk Islands (New Siberian Islands)... They have found bone implements and arrowheads as well as needles and axes skillfully fashioned from mammoth tusks.'

"Spitzbergen [a group of islands about 800 miles from the North Pole] was once inhabited too. Proof of this can be seen in the fragments of prehistoric cliff drawings found near the present day settlement of Ny Alesund. On the rock face are well-preserved incised outlines of whales and deer."[60]

LAKES OF THE GREAT BASIN

Velikovsky discusses the lakes of the Great Basin in the western part of the United States. These lakes do not have outlets and thus, the salt and mineral content of their waters can be measured against the feeding sources to determine their age. Velikovsky writes,

"Abert and Summer Lakes in southern Oregon have no outlets. They are regarded as remnants of a once large glacial lake, Chewaucan. W. van Winkle of the United States Geological Survey investigated the saline content of these two lakes and wrote [in U.S. Geological Survey, Water Supply Paper 363. (Washington 1914) that], 'A conservative estimate of the age of Summer and Abert Lakes, based on the concentration and area, the composition of the influent waters, and the rate of evaporation, is 4,000 years.'"[61]

Owen Lake, east of Mount Whitney in California also has no outlet. Velikovsky reports that,

"H.S. Gale analyzed the water of the lake and of the river for chlorine and sodium and came to the conclusion that the [Owens] river required 4,200 years to supply the chlorine present in the lake and 3,500 years to supply its sodium. Ellsworth Huntington of Yale found these figures too high, because no allowance was made for greater rainfall and 'freshening of the lake' in the past and consequently he reduced the age of the lake to 2,500 years [in *Quaternary Climates,* monographs by J. Claude Jones, Ernst Antevs, and Ellsworth Huntington (Carnegie Institution of Washington, 1925) p. 200] which would place its origin not far from the middle of the first millennium before the present era."[62]

Lake Lahontan in the Great Basin in Nevada also lacks an outlet. As it dried up, it split into several smaller lakes. Velikovsky states,

"More recently, Lahontan and its residual lakes were explored anew by J. Claude Jones, and the results of his work were published as 'Geological History of Lake Lahontan', by [I. Russell of] the Carnegie Institution of Washington. [Monograph 11 (1886)] He investigated the saline content of [residual lakes], Pyramid and Winnemucca...and of the Truckee River that feeds them. He found that the river could have supplied the entire content of chlorine of these two lakes in 3,881 years. 'A similar calculation, using sodium instead of chlorine gave 2,447 years necessary.'"[63]

However, I. Russell "found bones of Ice Age animals in the deposits of the ancient lakes. Bones of horses, elephants and camels, animals that became extinct in the Americas, were found in the Lahontan sediments, as well as spear points of human manufacture."[64]

LA BREA TAR PIT

In the La Brea tar pit on the western outskirts of Los Angeles are found in the asphalt, clay and sand according to J.C. Merriam in *Memoirs of California,* Vol. I, No. 2 (1911), "a most remarkable mass of skeletal material." Velikovsky informs us that,

"these fossils were regarded as representing the fauna of the late Tertiary (Pliocene) or early Pleistocene (Ice Age)... The animal remains are crowded together in an unbelievable agglomeration... Among...animals unearthed in this pit were bison, horses, camels, sloths, mammoths, mastodons and birds including peacocks [and saber-toothed tigers]... Separate bones of a human skeleton were also discovered in the asphalt of La Brea... However, it does not show any deviation from the normal skulls of Indians."[65]

J.C. Merriam, cited above, informs us that the skeletal remains of *Felix atrox,* a species of lion found in La Brea asphalt as well as horse and a camel are so alike to those found in the sediments of Lake Lahontan that they must have been contemporaneous. Antevs and Jones, also cited above, "On the basis of his analysis...came to the conclusion that the extinct animals lived in North America into historical times."[66]

FLORIDA

"On the Atlantic coast of Florida, at Vero in the Indian River region, in 1915 and 1916, human remains were found in association with the bones of Ice Age (Pleistocene) animals, many of which either became extinct, like the saber-toothed tiger, or have disappeared from the Americas, like the camel... W.H. Holmes, head curator of the Department of Anthropology of the United States National Museum, who investigated the pottery...from Vero [said that the bowls found] 'such as were in common use among Indian tribes of Florida.' When compared with vessels from Florida earth mounds, 'no significant distinction can be made; in material, thickness of walls, finish of rim, surface finish, color, state of preservation, and size and shape,' the vessels 'are identical'... But the bones of man and his artifacts (pottery) were found among extinct animals... E.H. Sellards, state geologist of Florida, and a very capable paleontologist, wrote in the debate that ensued: 'That the human bones are fossils normal to this stratum and contemporaneous with the associated vertebrates is determined by their place in the formation, their manner of occurrence, their intimate relation to the bones of other [extinct] animals, and the degree of mineralization of the bones.' This 'degree of mineralization of the human bones is identical with that of the associated bones of the other [extinct] animals.'... In 1923-1929, thirty-three miles north of Vero, in Melbourne, Florida, another such association of human remains and extinct animals was found, 'a remarkably rich assemblage of animal bones, many of which represent species which became extinct at or after the close of the Pleistocene [Ice Age] epoch.' The discoverer, J.W. Gidley, of the United States National Museum [in the *Bulletin of the Geological Society of America*, Vol. XL, pp. 491-502], established unequivocally that in Melbourne—as in Vero—the human bones were of the same stratum and in the same state of fossilization as the bones of the extinct animals. And again human artifacts were found with the bones... I. Rouse [in *Transactions of the New York Academy of Sciences*, Ser. II, Vol. 12 (1950), pp. 224ff] a recent analyst of the much debated fossils of Florida [states] that, 'the Vero and Melbourne man should have been in existence between 2000 B.C. and the year zero A.D.'"[67]

William R. Corliss in *Science Frontiers*, (Glen Arm, MD 1994), p. 217 describes another bone bed in Florida,

"A new bone bed has been discovered south of Tampa. Paleontologists say it is one of the richest fossil deposits ever found in the United States. It has yielded the bones of more than 70 species of animals, birds, and aquatic creatures. About 80% of the bones belong to plains animals, such as camels, horses, mammoths, etc. Bears, wolves, large cats, and a bird with an estimated 30-foot wing span are also represented. Mixed in with all the land animals are sharks' teeth, turtle shells, and the bones of fresh and salt water fish. The bones are all smashed and jumbled together, as if by some catastrophe. The big question is how bones from such different ecological niches—plains, forests, ocean—came together in the same place?"

AGATE SPRING QUARRY, NEBRASKA

"In Sioux County, Nebraska, on the south side of the Niobrara River, in Agate Spring Quarry, is a fossil-bearing deposit up to twenty inches thick. The state of the bones indicate a long and violent transportation before they reached their final resting place... The animals found there were mammals...small, twin horned rhinoceros...(Moropus) [a horse with legs and claws] like those of carnivorous animals; and bones of a giant swine that stood six feet high... A few miles to the east, in another quarry were found skeletons of an animal which, because of its similarity to two extinct species, is called a gazelle camel."[68]

"A.L. Kroeber [in *The Maya and Their Neighbors,* (1940), p. 476] sees no easy way to avoid the conclusion that 'some of the associations of human artifacts with extinct animals may be no more than three thousand years old.'"[69]

ELEPHANTS

Not only that, but the American Indians also made representations of elephants. Ignatius Donnelly in his book *Atlantis,* (NY 1985), pp. 169–170 states,

"There are in Wisconsin a number of mounds of earth representing different animals—men, birds and quadrupeds. Among the latter is a mound representing an elephant, 'so perfect in its proportions, and complete in its representation of an elephant, that its builders must have been well acquainted with all the physical characteristics of the animal which they delineated.'"

"On a farm in Louisa County, Iowa, a [smoking] pipe was ploughed up which also represented an elephant. We are indebted to the valuable work of John T. Short (*The North Americans of Antiquity,* p. 530) for a picture of this singular object. It was found in a section where the ancient mounds were very abundant and rich in relics. The pipe is of sandstone, of the ordinary mound builder's type and had every appearance of age and usage. There can be no doubt of its genuineness. The finder had no conception of its archaeological value.

"In the ruined city of Palenque we find, in one of the palaces, a stucco bass-relief of a priest. His elaborate head-dress or helmet represents very faithfully the head of an elephant."

Thus, like the *bas-reliefs* of Egypt and Babylonia, the elephant which supposedly had been extinct at least 10,000 years, is faithfully depicted.

Again we are asked to believe that for many thousands of years American Indians kept the image of the elephant in their memory. But the bones of elephants are rarely found complete or properly assembled in any location. The implication is clear. Their forms were *not* extinct in historical times and the early American Indians had observed them and copied them accurately. And there is other evidence as well.

Ivan T. Sanderson in *The Dynasty of ABU,* (NY 1962), pp. 125-26, discusses stellas discovered in Copan in Central America.

> "The two top and dominant figures on each edge of one stele are most perfectly and naturalistically represented heads of elephants—not loxodonts, mammoths, or mastodons but obviously elephants... [These were presented and discussed by G. Elliott Smith, "Pre-Columbian Representations of Elephants in America," *Nature,* Vol. 15, (Nov. 1915), pp. 340-341, and *Nature,* Vol. 16, (Dec. 1915), p. 425]
>
> "To summarize the arguments, we may say that because elephants in ancient Central America did not then, and still do not fit into the prescribed scheme of history, but because these carving were undeniably authentic, everything possible and impossible was immediately put forward to 'explain them away.' They were not elephants at all, said some of the learned, but the enlarged heads of the giant parrot like bird of that country known as the macaw; or representations of turtles, or of a 'bat-god' wearing a symbolic headdress, or such forth. All these things are well known from Mayan carvings, but each is invariably quite distinct, for the Mayas were very accurate in their animal representations. Anything seemed acceptable as long as it was not an elephant.
>
> "The Copan carvings are of elephants and they are perfectly executed with their trunks slightly curled to one side and backward, but nobody was prepared to admit the fact. What clinched the matter, however, was a careful search for and reappraisal of the extant original Mayan codices—hieroglyphic texts on scrolls, done in many colors, a few of which survived the wholesale burning by the Church in the early days. Brought to light were several dozen quite obvious elephants, elephant symbols and figures of men wearing elephant headdresses..."

Other evidence of elephants living into recent times is reported by Neill J. Harris, "The Riddle of America's Elephant Slabs'" *Science Digest,* Vol. 69, (March 1971), pp. 74-77.

> "Commonly known Elephant Slabs, the carved rocks were found in 800 year-old Indian ruins located on the south side of the Animas River just opposite Flora Vista. According to a famous archeologist, the late Earl Halstead Morris, the small boy sold the carvings to archeologist Charles Avery Amsden, who reportedly took Morris to where the boy had found the slabs. Morris, an expert on primitive Southwestern artifacts dated these Indian ruins at about A.D. 1200, based on potsherds found lying on the surface of the Flora Vista site.

"'I can see no reason to doubt the authenticity of these specimens,' wrote Morris concerning the Elephant Slabs 'but how to explain them I would not say. In all my experience I have seen nothing similar'...

"The principle slab measures six inches wide, six inches long; a deep groove on the left-hand side shows where it was probably broken off from a larger stone. The unknown petroglypher meticulously chiseled 55 signs and pictures into exceeding hard stone and left no obvious traces of tool slippage or overcrossing lines.

"The second slab measures six inches wide, 14 inches long, and bears only ten faintly incised signs including outlines of an elephant—the surprising fact that gave the slabs their names—a bird, and what looks like a mountain lion. [The first slab contains glyphs of two birds and two elephants]. Both slabs, particularly the larger one, closely resemble the stone hoes or knives of the Southwest..."

George F. Carter's "A Note On The Elephant In America," *Epigraphic Society, Occasional Publications,* (Vol. 18), 1989 p. 90 informs us that,

"A very late kill of a mastodon has been recorded for Equador. The creature was mired in a gully, and attacked by man and killed... The key element in this case, the mastodon, was cooked in place by heaping earth over it, and then building huge fires to cook the creature in an earth oven effect. The earth [that] heaped the carcass contained potsherds... This places the kill within the pottery period of that area and a very early date would be 3000 B.C. Since some of the pottery was said to be decorated, a later date is quite possible. The sherds were destroyed in a fire in the museum where they were housed and so the exact age cannot be determined. The excavation was by a well known archeologist and equally well known paleontologist. There is no reason to question any of the facts."

Why are the facts of recently dated finds of elephants dismissed? Carter in "The Mammoth In American Epigraphy", *Epigraphic Society, Occasional Publications,* (Vol. 18), 1989, p. 213 tells us,

"We have a case of basic assumptions ruling the conclusion. It is assumed as established fact that the mammoth died out 10,000 years ago... If the mammoth became extinct 10,000 years ago, how does one explain the Eskimo, the Algonquin and other eastern United States Indians' vivid memory of the mammoth? The Indian descriptions are very graphic. They surely were describing elephants. (mammoths, mastodons and elephants are lumped together) One legend says that the agricultural Indians complained bitterly to the Great Spirit against the mammoth damaging their corn fields. The Great Spirit then killed them off by hurling thunderbolts at them. This was told to Thomas Jefferson in Washington and Cotton Mather in New England, and that is just what is portrayed [on relics found] in the Bucks County and Holly Oaks specimens. Now why would the Indians invent a tale like that? And how could they describe the elephant as to size, trunk, tusks and all if they had never seen one?"

SOUTH AMERICA

If the extinctions are so prevalent in North America should not evidence of like nature also exist in South America? Charles Hapgood in his chapter titled, "The Pleistocene Graveyards of South America" writes,

"The discoveries of vast quantities of animal remains in almost every part of South America have invariably been made in recent [within the last 10,000 years] formations. As long ago as 1887 Sir Henry H. Horworth in his monumental work, *The Mammoth and the Flood*...summarized our knowledge of these beds as follows, 'In South America, the Pleistocene beds are developed on a very large scale. They cover plains of the Argentine Republic, in the form of modified lehm or loess, to which the name Pampas mud was given by Darwin...'

"According to Burmeister, they are richest in organic remains in the province of Buenos Aires, becoming less rich as we travel westward and northward. Rich deposits of this [Pleistocene] age have also been found in the Banda Oriental, at various points on the river Parana and at Berrero in Patagonia.

"Burmeister says, 'the diluvial deposit containing bones of animals of this [Pleistocene] age extends over the whole Brazilian plains from the flanks of the Cordilleras to the borders of the Atlantic.' They have also been found abundantly in Bolivia on the great plateau; and also west of the mountains in Peru and Chile.

"From Caracas in the north, to the Sierra of Tandel in Patagonia in the south, they have, in fact, occurred in more or less abundance over the whole continent...

"That the surface beds of the Pampas and the deposits in the caves were synchronous, is admitted by all explorers...

"Nor is there any doubt that both sets of beds date from the same horizon as the mammoth beds of other countries.

"The fauna of the Pleistocene beds of the southern states of North America, is, in fact, largely identical with that from the beds we are now discussing."[70]

Were human remains found associated in the same manner as those found in North America? Hapgood states,

"In a limestone cavern on the borders of the Lagoa do Sumidouro, some three leagues from Santa Lucia, Dr. P.W. Lund excavated the bones of more than thirty individuals (human) of both sexes and various ages. The skeletons lay buried in hard clay overlying the original red soil forming the floor of the cave and were found mixed together in such great confusion—not only with one another, but with the remains of *Megatherium* and other Pleistocene mammals—as to preclude the idea that they had been entombed by the hand of man. All the bones, whether human or animal, showed evidence of having been contemporary with one another.

"In other caves investigated by Lund, bones of ancient man were found alongside those of the formidable *Smilodon*, a giant feline which became extinct during the last Pleistocene times. Referring to the evidence from these and other Brazilian fossiliferous caves, the Marquis de Nadaillac wrote:

'...Doubtless these men and animals lived together and perished together, common victims of catastrophes.'[71]

Hapgood gives several such examples and adds,

"the groups or caches of animal fossils unearthed at widely separated South American localities in which incongruous animal types (carnivores and herbivores) are mixed promiscuously with human bones [are widespread]. These are found not only in the Pampas formation, but also in Brazilian caves and in volcanic ash at Punin and elsewhere."[72]

A. d'Orbigny maintains that,

"It would seem that one cause destroyed the terrestrial animals of South America, and that this cause is to be found in great dislocations of the ground caused by the upheaval of the Cordilleras.

"If not, it is difficult to conceive on the one hand the sudden and fortuitous destruction of the great animals which inhabited the American continents, and on the other the vast deposit of Pampan mud.

"I argue that this destruction was caused by an invasion of the continent by water, a view which is completely *en rapport* with the facts presented by the great Pampan deposit, which was clearly laid down by water.

"How, otherwise, can we account for this complete destruction and the homogeneity of the Pampas deposits containing bones? I find an evident proof of this in the immense number of bones and of entire animals whose numbers are greatest at the outlets of the valleys, as Mr. Darwin shows.

"He found the greatest number of the remains at Bahia Blanca, at Bajada, also on the coast, and on effluents of the Rio Negro, also at the outlet of the valley [exactly as the deposits of Alaska and Siberia]. This proves that the animals were floated, and hence, were chiefly carried to the coast.

"The hypothesis necessitates that the Pampas mud was deposited suddenly as the result of violent floods of water, which carried off the soil and other superfluous debris, and mingled them together. This homogeneity of the soil in all parts of the Pampas, even in places 200 leagues apart is very remarkable.

"These are not different strata differently coloured, but a homogeneous mass, which is more or less porous, and shows no signs of distinct stratification. The deposit is also of one uniform colour, as if it had been mixed in one muddy flood slightly tinted by oxide of iron.

"The bones again are only found isolated in the lower strata, while entire animals occur on the circumference or the upper part of the basin.

"Thus, they are very rare at Buenos Aires, while they abound in the Banda Oriental and in the White Bay. Mr. Darwin says they are heaped up in the latter place, which again supports the contention.

"Another argument may be drawn from the fact that the Pampas mud is identical in colour and appearance with the earth in which the fossil remains occur in the caverns and fissures of Minaes Geraes in Brazil, and the fragments brought by M. Claussen are completely like the others in colour and texture.

"My final conclusion from the geological facts I have observed in America is, that there was a perfect coincidence between the upheaval of the Cordilleras, the destruction of the great race of animals and the great deposit of Pampas mud."[73]

How fresh are these South American fossils? Charles Darwin in *Voyage of the Beagle,* (1876 edition), p. 181, states, "...it is difficult to believe that they have lain buried for ages underground. The bones contain so much animal matter, that when heated in the flame of a spirit lamp, it not only exhales a very strong animal odor, but likewise burns with a slight flame." Very ancient fossil bones do not burn with a flame.

Charles H. Hapgood's *Earth's Shifting Crust,* (NY 1958), p. 225 cites A. Geike who discussed Charles Darwin's analysis of extremely young ocean beaches found high in the Andes Mountains stating,

"On the west coast of South America, lines of raised terraces containing recent [sea] shells have been traced by Darwin as proof of a great upheaval of the part of the globe in modern geological times. The terraces are not quite horizontal, but rise to the south. On the frontier of Bolivia, they occur from 60 to 80 feet above the existing sea level, but nearer the higher mass of the Chilean Andes they are found at one thousand, and near Valparaiso at 1,300 feet. That some of these ancient sea margins belong to the human period was shown by Mr. Darwin's discovery of shells with bones of birds, ears of maize, plaited reeds and cotton thread..."

If the Andes Mountains took many millions of years to rise to their present height, as Sagan assures us, why haven't these ancient beaches with all their detritus been eroded away? The only way such sea shore terraces could exist and not have been eaten away by erosional processes is that they were only uplifted recently. And, if the beaches were raised gradually, as Sagan suggests, we should find numerous intermediate beaches between the highest levels and the sea. Where are these numerous beaches? One cannot suggest that they were nearly all eroded away, but that the highest beach miraculously avoided being weathered out of existence since it is the oldest beach. Again this points to the sudden recent uplifting of the Andes Mountains.

Another form of evidence comes in that of plant fossils. Edward W. Berry's "The Age of the Bolivian Andes", in the *National Academy of Science Proceedings,* Vol. 3, for 1917, provided evidence that the Andes were lifted to their present height in Pleistocene times. He reported the finding of fossil plants at the site of Corocoro, which is some 13,000 feet above sea level and also at Potosi, which is at 14,000 feet. Significantly among the flora were found tropical fern trees similar to those of the Amazon basin. Berry (p. 283) states, "The sea deposited a part of these strata [high in the Bolivian Andes] in late Tertiary or Pleistocene time, and since that time there have been differential vertical movements amounting to a minimum of 13,500 feet." Berry's conclusion is that, "There is then, definite evidence that parts of the high plateau and the eastern Cordillera stood at sea level in the late Tertiary."

ARCTIC MUCK

How similar is the mud of the Pampas to the muck of the Arctic coast? The implication of this question is that both were laid down by a catastrophe of enormous flooding over wide expanses of the land; but the Pampas material is not frozen, while the Arctic muck is. The catastrophe for each appears to be an enormous flood. In *Pursuit,* for October 1969, appeared this article titled, "Much About Muck."

"In a fine report on a highly informative talk given by a Mr. E.M. Benson, Vice President of the North American Producing Division of the Atlantic Richfield Oil Company, to the Long Beach Petroleum Club of California on the new oil field in northern Alaska, there appeared a rather noteworthy quote. This read: 'Drilling down through the 1,000-foot thick frozen earth can produce some surprises. One of our wells brought up an 18 inch long chunk of tree trunk almost 1,000 feet below the surface. It wasn't petrified—just frozen,' the oil company executive said. The reason this statement is noteworthy is not because the reporter seems to have been impressed, but rather that a man of Mr. Benson's experience—and he started as a worker in the fields—should use the word 'surprise' in this case.

"We are going to hear a lot about this frozen earth or 'muck' from now on because of this vast oil strike on the Arctic shores of the Alaskan peninsula. It is indeed full of surprises; but a tree trunk in it, and even at a depth of a thousand feet, is not at all surprising. What surprise there was on this occasion was probably due to the fact that it came to light in an area devoid of trees today and hundreds of miles from any forest growth. The nature of muck is not generally understood, and the theories on its origin are even less widely known...

"A world map of the distribution of this frozen soil and muck reveals several very interesting things, the most outstanding aspect being that it lies low, level plains or tablelands. Unless it was caused by some cosmic forces that we have not yet detected, it would appear to be a subaerial [flood] deposit derived from massive erosion of higher grounds and with steeper slopes [as the mud of the Pampas was carried from the detritus of the higher ground of the Cordilleras]. However, its depth in some places, and over enormous areas, has always caused even the most open-minded geologists to boggle. The Russians, who own the major land areas covered by this substance have conducted prolonged studies on it for half a century and have in some places drilled down to over 4,000 feet, but still without reaching solid rock. The conundrum is, of course, how do you get that thickness of what is manifestly surface-derived material if it is the result of mere run-off? To this there would appear to be but one answer.

"First, the lands now blanketed with this material must at one time have been much higher above sealevel, so that stuff could be deposited upon them, rather than running on beyond and out into the sea. Alternatively, the sealevel would have been much below than that of today; but in this case we are asked to suppose that universal sealevel was not too long ago, geologically speaking, more than 4,000 feet lower. If neither of these situations pertained when the first, and lowest layers of this muck were laid down, just what were the conditions, since no such strata could be laid down even under shallow, tranquil coastal seas? To suggest that the uplands from which this stuff came were once much higher and had a steeper run-off is begging the question, and doesn't help at all. Yet, there is the bloody muck lying all over the lot and to enormous depths. It has to be accounted for...

"[The tree trunk 1,000 feet down] comes as no surprise at all to geologists who have specialized in the surface constitution of the Arctic regions. A mere section of tree trunk is a mild relief compared to some of the things that the muck has yielded. In the New Siberian Island, for instance whole trees have turned up; the trees of the family that includes the plums; and with their leaves and fruits. No such hardwood trees grow today anywhere within two thousand miles of those islands. Therefore the climate must have been very much different when they got buried; and, please note, they could not have been buried in frozen muck which is [as hard as] hard rock, nor could they have retained their foliage if they were washed far north by currents from warmer climates. They must have grown thereabouts, and the climate must have been not only warm enough, but have had a long enough growing period of summer sunlight for them to have leafed and fruited.

"Ergo, either what is now the Arctic was at the time as warm as Oregon, or the land that now lies therein was at that time elsewhere. Geophysicists don't go for an overall warming of this planet to allow such growth at 72 degrees north; otherwise everything in the tropics would have boiled! Thus, we are left with the notion that either the whole earth's crust [or the pole] has shifted, or bits of it have drifted about. But then comes another problem—the Time Factor.

"Along with the plum trees, and other non-Arctic vegetation there are found associated animal remains of many kinds. One of these is the famous mammoth. Now everybody has somehow got the totally erroneous idea that these great hairy beasts are found in ice. Not one has ever been found in ice: they are all in this frozen earth of muck. Then, just because of their layer of fat and their covering of long hair, everybody likewise thinks that they were Arctic types. A moments consideration will disclose just how ridiculous an idea this is. A large elephantine needs some half a ton minimum of fresh green food a day to maintain itself, and there were apparently (at least according to the number of their bones and bodies that have [thus far] been found in the muck) hundreds of thousands of them up till *only a few thousand years ago*. For a minimum of eight months out of the year there is nothing for such large animals to eat north of the tree line in the Arctic, though some Barren Ground Caribou and a few Muskox get along by scratching through the shallow snow to get at tundra moss and lichens. Therefore, these elephantines must have migrated far south for the winter or the climate must have been much milder than it is today, or the land they lived in was elsewhere.

"But not even this pinpoints the reason for the muck or explains just how all the junk that is found in it, even down to thousands of feet, got there."[74] [emphasis added]

Thus, the muck of the Arctic could well be the result of a shift of the polar axis which gave rise to an enormous flood that did in the Arctic region what d'Orbigny claims occurred in South America to form the mud of the Pampas.

What then of the middle latitudes? Are there deposits of material found in these regions that are thick and extensive and uniform in nature just like the Pampas mud and the Arctic muck, and are also filled with the bones of Ice Age mammals? Such material has been known to exist for a long time; it is called *loess*. The great problem, like the Pampas mud and the Arctic muck is, how to account for these enormous loess deposits. Two major explanations for the loess have been presented by the geologists. One is that it was deposited by wind action. But if this were indeed the case, the grains of loess blown from one region to another, over the surface of the Earth, would cause the grains to collide and rub against each other and the surface rock. The many abrasions would have produced grains of matter that were highly rounded in shape. Loess, by contrast, is everywhere found to be highly angular in shape and this categorically denies its eolian depositional origin.

On the other hand, it is argued by the geologist that the loess could not have been deposited by water because under such conditions it would show a layering or stratified structure which it distinctly lacks. However, this analysis is based strictly on the uniformitarian supposition that the loess was deposited gradually over many thousands of years. But if the loess was deposited as a huge flood of slurry mud, its nature would be in full accord with the phenomena attributed to it.

The loess is uniform in color, texture and composition like the Pampas mud and Arctic muck.

In places it is 1,500 feet or more in thickness and covers extensive areas.

It has not changed the terrain upon which it lies, but simply fills in all valleys between hills.

It contains very large quantities of the bones of mammals among which is the mammoth and other Ice Age fauna and also man and his artifacts.

It contains innumerable tubes and capillaries running down through the loess, through which the water in it drained.

The major objection of the geologists to the concept that the loess was deposited by water is that this hypothesis can only be congruent with the known facts if the loess was deposited by a catastrophic flood. Nevertheless, its similarity to the Pampas mud and the Arctic muck strongly suggest its origin was also of a similar and catastrophic nature.

However, Sagan need only read Velikovsky's book to gather the evidence for how great an extinction occurred at the close of the Pleistocene. Velikovsky cites L.C. Eiseley regarding "species... believed to have been destroyed to the last specimen" in the closing Ice Age.

> "Animals strong and vigorous, suddenly die out without leaving a survivor. The end came, not in the course of the struggle for existence— with the survival of the fittest. Fit and unfit, and mostly fit, old and young, with sharp teeth, with strong muscles, with fleet legs, with plenty of food around, all perished. These facts as I have already quoted [according to Eiseley] drive 'the biologist to despair as he surveys the extinction of so many species and genres in the closing Pleistocene [Ice Age].'"[75] [See L.C. Eiseley, *American Anthropologist*, Vol. XLVIII, (1946), p. 54.]

Frank C. Hibben in his book *The Lost Americans*, gives a description of the extinction. Hibben writes:

> "The Pleistocene period ended in death. This is no ordinary extinction of a vague geological period which fizzled to an uncertain end. This death was catastrophic and all inclusive... The large animals that have

given their name to the period became extinct. Their death marked the end of an era.

"But how did they die? What caused the extinction of forty million animals? This mystery forms one of the oldest detective stories in the world. A good detective story involves humans and death. The conditions are met at the end of the Pleistocene. In this particular case, the death was of such colossal proportions as to be too staggering to contemplate...

"The 'corpus delecti' of the deceased in this mystery may be found almost everywhere...the animals of the period wandered into every corner of the New World not actually covered by the ice sheets. Their bones lie bleaching on the sands of Florida and in the gravels of New Jersey. They weather out of the dry terraces of Texas and protrude from the sticky ooze of the tar pits of Wilshire Boulevard in Los Angeles. Thousands of these remains have been encountered in Mexico and even in South America. The bodies lie as articulated skeletons revealed by dust storms, or as isolated bone and fragments in ditches or canals. The bodies of the victims are everywhere in evidence.

"It might at first appear that these great animals died a natural death; that is, the remains that we find in the Pleistocene strata over the continent represent the normal death that ends the ordinary life cycle. However, where we can study these animals in some detail such as in the great bone pits of Nebraska, we find literally thousands of these remains together. The young die with the old, foal with dam and calf with cow. Whole herds of animals were apparently killed together, overcome by some common power.

"We have already seen that the muck pits of Alaska are filled with the evidence of universal death. Mingled in these frozen masses are the remains of many thousands of animals killed in their prime... The evidences of violence, they are as obvious as in the horror camps of Germany. Such piles of bodies of animals or men simply do not occur by an ordinary natural means."[76]

WHALES

If indeed the sea in a gigantic tidal wave covered vast areas of the Earth, should there not be evidence of oceanic life that would be tossed on top of the land and that died there? Velikovsky tells us,

"In bogs covering glacial deposits in Michigan, skeletons of two whales were discovered. Whales are marine animals. How did they come to Michigan in the post glacial epoch? Whales do not travel by land. Glaciers do not carry whales, and the ice sheet could not have brought them to the middle of a continent. Besides, the whale bones are found in post glacial deposits. Was there a sea in Michigan *after* the glacial epoch, only a few thousand years ago?

"In order to account for whales in Michigan, it was conjectured that in the post glacial epoch, the Great Lakes were part of an arm of the sea. At present the surface of Lake Michigan is 582 feet above sea level.

"Bones of whale have been found 440 feet above sea level, north of Lake Ontario; a skeleton of another whale was discovered in Vermont, more than 500 feet above sea level, [See Dana, *Manual of Geology*, p. 983] and still another in the Montreal-Quebec area, about 600 feet above sea level. [See C. Dunbar, *Historical Geology*, p. 453]

"Although the Humphrey whale and beluga occasionally enter the mouth of the St. Lawrence, they do not climb hills. To account for the presence of whales in the hills of Vermont and Montreal, at elevations of 500 to 600 feet, requires the lowering of the land to [below] that extent. Another solution would be for an ocean tide, carrying the whales, to have trespassed upon the land. In either case herculean force would have been required to push mountains below sea level or to cause the sea to irrupt, but the latter explanation is clearly catastrophic. Therefore the accepted theory is that the land in the region of Montreal and Vermont was depressed more than 600 feet by the weight of ice and kept in this position for a while after the ice melted. [But then salt water should have entered the lakes with the whales and the Great Lakes should be saline and not fresh water lakes.]

"But along the coast of Nova Scotia and New England, stumps of trees stand in water, telling of once forested country that became submerged. And opposite the mouths of the St. Lawrence and the Hudson rivers are deep canyons stretching for hundreds of miles into the ocean. These indicate that the land became sea, being depressed in post-glacial times. Then did both processes go on simultaneously in neighboring areas here [in Michigan and Vermont and Montreal] up, there [in Nova Scotia, the St. Lawrence and the Hudson river areas] down?

"A species of Tertiary whale, *Zeuglodon,* left its bones in great numbers in Alabama and other Gulf states. The bones of these creatures covered the fields in such abundance and were [according to George McCready Price, *Common-Sense Geology*, (1946), pp. 204-5] 'so much of a nuisance on the top of the ground that farmers piled them up to make fences.' There was no ice cover in the Gulf states; then what had caused the submergence and emergence of the land there?"[77]

It was also reported in *The New York Times* for March 12, 1987, p. A22, that scientists had found whale bones in the high Andes Mountains, "Scientists have found fossil whales and other marine animals in mountain sediments in the Andes, indicating that the South American mountain chain rose very rapidly from the sea." Since the bones of whales were found at an altitude of 5,000 feet, it is assumed that the whales died, then settled to the sea floor 15 million years ago and were covered by marine sediments. Then they were gradually raised to their present height. However, this analysis is denied by the evidence. A great many whales and other sea animals had to all have died in the very same area of the ocean at the same time; most would have become bloated with gas and floated to the surface. To settle on the same seabed location they would have to have drifted over days or weeks together and not be eaten. But according to the article,

> "Nearly all of the fossils were embedded in surface rock and easy to pick up... Best of all, despite weathering, many of the smallest fossils were remarkably intact and will be easy to study... Assemblages comparable to these are virtually unknown in the Andes *since geological upthrusting generally destroys fossil beds.*" [emphasis added]

If the uniformitarian analysis is employed, the gradual upthrusting of the Andes from the bottom of the sea would obliterate nearly all the fossils and would not have left such a large assemblage of bones intact. However, a huge tidal wave would throw the whales on the land and buried them in a light covering of sediment. The evidence shows that these whales were found at the surface with many other sea animals. It was also reported that in the ground were found strata that contained land animals such as camel and rhinoceros.

In *Discover*, (May 1991), pp. 45-48, James Trefil discusses large numbers of whales found in a valley one hundred miles southwest of Cairo, Egypt in the Sahara desert, in which 243 fossil whale skeletons has thus far been discovered. The fossils are *Zeuglodon,* like those discovered all over the southern parts of the United States. The whale bones are scattered among sand dunes and the whales are just falling out of the rocks or are buried in the sand. When the wind exposes parts of the whale bones, the paleontologists dig them out as rapidly as possible because the windborne sand erodes away the bones that are exposed. Thus, it is probable many other whales have been exposed and destroyed by erosion over time. As is accepted, the Sahara was covered by sand only after the Ice Age ended. Therefore, the whales had to somehow have gotten into the sand after the Ice Age ended and the Sahara became a desert.

Allan O. Kelly and Frank Dachille in *Target Earth,* (Carlsbad, CA 1953), p. 168 describe,

> "deluge lakes...found in other parts of the world. In Tunisia and Algeria there are a number of salt lakes, some of them below sea level, which show elevated shore lines of Pleistocene age. The University of California's Expedition of 1947-48 found the bones of whales along with those of present day mammals in these old lake beds."

In *Nature,* Vol. 38 for 1888, p. 134 is a discussion on whale bones 330 feet above sea level in Sweden which states,

"At a recent meeting of the Scientific Society of Upsala, Dr. C. Aurivillius read a paper on the skeleton of the so-called Swedenborg whale (Eubalana svedenborgii, Lillj), discovered last November in the province of Halland, in a layer of Marl 50 feet above the sea. Remains of this species of whale have only been found once before, viz. early last century, when some parts of one were discovered in the province of Western Gothland, 330 feet above the sea, and *70 miles inland*... The skeleton has been presented to the Upsala Museum." [emphasis added]

In *Nature*, Vol. 183, 1959, p. 272 D.L. Dineley and P.A. Garrett write,

"The preservation of Pleistocene or recent land mammals in the Siberian permafrost has long been known, but the literature does not appear to include mention of marine mammals preserved in ice. Particular interest, therefore, is attached to the discovery in 1958 of part of a whale carcass entombed in the ice cored moraine of Sveabreen, Ekmanfjord, in Vestspitsbergen..."

"This question of how and when the animal became entombed is a difficult one. One would expect a dead whale to float and hence, decompose during the summer months, even if it died in the winter... As long as it remained in the permafrost, the animal would not decompose. It is surprising that the body remained intact during the movement. (Of the ice cored moraine) The estimate of the event is thought to be 2,500 years ago, but radiocarbon dating may narrow the time when the whale died."

Stephen Jay Gould in *Hen's Teeth and Horses Toes,* (NY 1983), p. 323, states, "Falling sea level has accompanied nearly every mass extinction that the Earth has suffered; this correlation is about the only aspect of mass extinction that evokes general agreement among geologists."

Thus, the evidence speaks clearly of an enormous extinction a few thousand years ago notwithstanding Sagan's statement that he cannot find this evidence. There is a great deal more information in *Earth in Upheaval,* that deals with evidence from the other continents, the oceans and recent geological publications. The small amount of information presented above will give some idea of what is available in the literature in greater depth and detail. The evidence from the other continents, islands and oceans shows that there is indeed plenty of scientific data to support the thesis that a catastrophe of world-wide dimensions shook the globe only a few thousand years ago.

ARCHAEOLOGY

Sagan asks, "And what of the archaeological...evidence?"[78] The reader will recall that earlier, Velikovsky claimed that there were four world ages or four world-wide catastrophes of varying intensity. In *Worlds in Collision,* he cites Hesiod in his chapter titled "The World

Ages". "Hesiod, one of the earliest Greek authors, wrote about four ages." In Tibet, Velikovsky writes,

"Analogous traditions of four expired ages persist on the shores of the Bengal Sea and in the highlands of Tibet." In the Americas, Velikovsky cites Brasseur's *Histoire des Nations Civilisées de Mexique* (1857-1859), Vol. I p. 53, which states, "The ancients [of Mexico] knew that before the present sky and earth were formed, man was already created and life had manifested itself four times."[79]

Therefore, if the world experienced four catastrophes, there should be evidence of at least these four events and even of earlier ones as well.

Claude F.A. Schaeffer, the greatest archaeologist of the twentieth century, carried out an archaeological survey across a vast area of the Middle East. Velikovsky, in *Earth in Upheaval,* in a chapter titled "The Ruins of the East" summarizes Professor Schaeffer's work.

"In the ruins of excavated sites throughout all lands of the ancient East, signs are seen of great destruction that only nature could have inflicted Claude Schaeffer, in his great recent work, [Stratigraphie comparée et chronology de l'Asie occidentale, Oxford Univ. Press, (1948)] discerned six separate upheavals. All of these catastrophes of earthquake and fire were of such encompassing extent that Asia Minor, Mesopotamia, the Caucasus, the Iranian plateau, Syria, Palestine, Cyprus and Egypt were simultaneously overwhelmed. And some of these catastrophes were in addition, of such violence, that they closed great ages in the history of ancient civilizations." [emphasis added].

Velikovsky continues,

"The enumerated countries were the subject of Schaeffer's detailed inquiry; and recognizing the magnitude of the catastrophes that have no parallels in modern annals or in the concepts of seismology, he became convinced that these countries, the ancient sites of which he studied, represented only a fraction of the area that was gripped by the shocks.

"The most ancient catastrophe of which Schaeffer discerned vestiges took place between 2,400 and 2,300 before the present era. It spread ruin from Troy to the valley of the Nile. In it, the Old Bronze Age found its end. Laid waste were cities of Anatolia...Tarsus, Aligar and those of Syria, like Ugarit, Byblos, Chagar, Bazar, Tell Brak, Tepe Gawra, and of Palestine, like Beth Shan and Ai; and of Persia, and of the Caucasus. Destroyed were the civilizations of Mesopotamia and Cyprus, and the Old Kingdom in Egypt came to an end, a great and splendid age. In all cities, walls were thrown from their foundations, and the population markedly decreased. [Schaeffer states], 'It was an all encompassing catastrophe. Ethnic migrations were, no doubt, the consequence of the manifestation of nature. The initial and real causes must be looked for in some cataclysm over which man had no control.' It was sudden and simultaneous in all places investigated. [The above and the following quotations are from Schaeffer, *Stratigraphie Comparee,* pp. 534-567]

"In a few centuries, migrating and multiplying themselves, the descendants of the survivors of the ruined world built new civilizations: the Middle Bronze Age. In Egypt it was the time of the Middle Kingdom, a short, but glorious resurrection of Egyptian civilization and might. Literature reached its perfection, political might its apogee. Then came shock that in a single day made of the empire a ruin, of its art debris, of its population corpses. Again it was the entire ancient Near East, to its uttermost frontiers; that fell prostrate; nature which knows no boundaries, threw all countries into a tremor and covered the land with ashes.

"This brilliant period of the Middle Kingdom in Egypt, during which flourished the art of the Middle Kingdom in Egypt and the exquisite art and industry of the Middle Minoan Age [on Crete], and in the course of which great centers of trade like Ugarit in Syria enjoyed remarkable prosperity, was suddenly terminated..."

"The great activity of international trade which, during the Middle Bronze Age, had been characteristic of the eastern Mediterranean and most of the lands of the Fertile Crescent, suddenly stopped in all this vast area... In all the sites in Western Asia examined up to now a hiatus or a period of extreme poverty broke the stratigraphic and chronological sequence of the strata... In most countries the population suffered great reduction in numbers; in others settled living was replaced by a nomadic existence." [Thus writes Schaeffer.]

Velikovsky continues,

"In Asia Minor the end of the Middle Bronze came suddenly, and a rupture between that age and the Late Bronze is evident [according to Schaeffer] in 'all sites that were stratigraphically examined.' Troy, Boghazkoi, Tarsus, Alisaar, present the same picture of life vanishing with the end of the Middle Bronze.

"In Tarsus, between the strata of the 'brilliantly developed civilization' of the Middle Bronze and those of the Late Bronze, a layer of earth five feet thick was found without a sign of habitation—a 'hiatus.' [This is similar to the mud of the Pampas or Muck of the Arctic, but on a much smaller scale.] At Alaca Huyuk the transition from Middle Bronze to Late Bronze was marked by upheaval and destruction and the same may be said of every excavated site in Asia Minor.

"On the Syrian coast and in the interior [Schaeffer writes], 'we find a stratigraphic and chronological rupture between the strata of the Middle Bronze and Late Bronze at Qalaat-er-Rouss, Tell Simiriyan, Byblos, and in the necropoles [graveyards] of Kafer-Djarra, Oraye, Majdalouna.' All the necropoles examined in the upper valley of the Orontes creased to be used and habitation of the great site of Hama was interrupted at the moment the Middle Kingdom in Egypt went down. Also, in Ras Shamra, there is a marked gap between the horizons of the Middle and Late Bronze.

"In Palestine, at Beth Mirsim, there was an interruption in the habitation of the site after the fall of the Middle Kingdom in Egypt. In Beth-Shan, between the layers of the Middle Bronze and Late Bronze, the excavators came upon an accumulation of debris a meter thick. [Schaeffer states], 'It indicates that the transition from the Middle Bronze to the Late Bronze was accompanied by an upheaval that broke the chronological and stratigraphical sequence of the site.' A similar situation was found at Tell el Hesy by Bliss. Earth tremors played havoc also with Jericho, Megiddo, Beth-Shemesh, Lachich, Ascalon, Tell Taanak. The excavator of Jericho found that the city had been repeatedly destroyed. The great wall surrounding it fell in an earthquake shortly after the end of the Middle Kingdom. [See J. Garstang and G.B.E. Garstang, *The Story of Jericho,* 1940]

"Concussions devastated the entire land of the Double Stream. The Russian-Persian borderland also shows that there was no continuity between the Middle Bronze and Late Bronze. In the Caucasus, not an archaeological vestige was found of the centuries between these two ages.

"A sea tide broke onto the land on the coast of Ras Shamra bringing further destruction in its wake.

"It appears also that the end of the Middle Kingdom was marked by volcanic eruptions and lava flows. On the Sinai Peninsula, at an early and undisclosed date, a flow of basaltic lava from the fissured group—the Sinai massif is not a volcano—burned down forests, leaving a desert behind. [This according to Flinders Petrie, *Ancient Egypt,* 1915] In Palestine, lava erupted, filling the Jezreel Valley. Early in this century a Phoenician vase was found embedded in lava. Geologists have asserted that volcanic activity in Palestine ceased in prehistoric times. [H. Gressman, in *Palastinas Erdgeruch In der Israelitischen Religion,* (1909), pp. 74-75 states], 'The assertion of the geologists thus becomes very questionable'... The vase found in lava proves volcanic activity there [according to Gressman, p. 75] 'in historical times.' The verdict of the archaeologists is that [according to A. Lods, *Israel,* (1932), p. 31, That] the vase 'dates from the fifteenth century before the present era.'"[80]

Regarding archaeological evidence from the Americas, we have already discussed Tiahuanacu in the Andes of South America and Ipiutak in Alaska. But what of the civilization of Mexico?

Byron Cummings' article "Ruins of Cuicuilco May Revolutionize Our History of Ancient America", appeared in *National Geographic Magazine,* Vol. 44, for 1923, pp. 202-220. In it Cummings describes a pyramid and possibly a city lying in the central Mexican valley on the great plateau near Mexico City. It is considered one of the oldest structures in the Americas and some archaeologists suggest that it may even pre-date the pyramids of Egypt.

All around the pyramid and the surrounding area is, what may be called Pedregal lava. Cummings' team dug down through this lava which is 25 feet thick and then uncovered several layers of archaeological occupation that were interspersed with thick layers of volcanic ash. That is, each level of culture was destroyed and buried beneath volcanic material upon which the next developed.

Below the Pedregal lava was an archaeological stratum that is currently classified as Early Classical, which roughly spans the period between 200 to 400 A.D. Beneath this stratum lay a thick layer of ash below which was found the remnants of an earlier culture which was quite different than that above it. It did possess some characteristics of the latter society; but the fact that there were far greater differences than the few superficial similarities, shows that the earlier culture was terminated by violent destruction and buried under volcanic ash which when removed, showed archaeological artifacts of a very primitive society.

The Carbon-14 dating of the materials carried out many years later on biological artifacts also showed sharp breaks between the major cultural layers. But what is most significant is that the hiatuses are dated closely to the two catastrophes described in *Worlds in Collision.* Sometime after around 1870 to 1770 B.C., there is a complete break in the stratum and another between 610 to 710 B.C. These dates come relatively close to 1500 and 689 B.C. dates as suggested by Velikovsky. This indicates that in the America, great catastrophes closed these periods of human development in the same manner as Schaeffer described in the Near East.

Furthermore, Cummings also found fine water deposited silt between the facing stones of the pyramid. He estimated that the pyramid was immersed in as much as six feet of water in the ancient past, giving evidence of an inundation in the heart of the Mexican plateau.

The dating of these layers is further corroborated by K.V. Flannery, et. al. in *Science,* Vol. 58 for 1967, pp. 445ff in an article which deals with Mesoamerican Zatopec culture. Flannery gives the starting dates of the major periods in Mesoamerica as 8000, 1500 and 600 B.C. C.J. Ransom who first reported Flannery's work in *The Age of Velikovsky,* p. 181, also points out that,

"The timing may not be exact, but largely because of carbon dating it is now apparent that the general periods suggested by Velikovsky are reasonable, and the legends generated during these periods correlate to those of other countries. Frank Waters is well-known for his writing about the history and myths of Native Americans and Mesoamericans. In 1975 he published *Mexico Mystique,* which is divided into two parts and describes the history and then the myths of the Aztecs, Olmecs, Mayas, Toltecs, and other groups in Mexico. In his analysis of the mythologies of these cultures, Waters reviewed the work of Velikovsky which is relevant to these areas. He concluded that although the timing of some events still creates some problems, '...Velikovsky's theory runs parallel to Mesoamerican myth in general outline.'"

Thus, from Alaska in the north to the Andes in South America, and the Valley of Mexico, is archaeological evidence that apparently supports Velikovsky.

THE COLUMBIA PLATEAU

Norman Macbeth in *Darwin Retried*, (Ipswich MA, 1971), pp. 111-112 describes Velikovsky's evidence of the Columbia Plateau thus,

"Something like 200,000 square miles in Idaho and eastern Washington and Oregon are covered with lava, which in many places is 5,000 feet or more in depth. All the volcanoes in the world, working at their present paltry scale and tempo through any period of time, could never produce such quantities of lava; hence this is a direct challenge to the uniformitarian theory. To make matters worse, much of the lava seems to be fresh and a figurine of baked clay was found at a depth of 320 feet. If men were present and making figurines before the eruptions ceased, the eruptions must have been very recent.

"I found this report fully confirmed by Ruth Moore, [in *The Earth We Live On*), (Knopf 1956), Chapter 17] an able popularizer in the earth sciences. Miss Moore actually gives far more astonishing details on the lava than Velikovsky does, but she never mentions the figurine or the freshness. She is an ardent admirer of Lyell and maintains a conviction in the uniformitarian theory that allows her to say [on p. 345]... 'nothing has ever indicated that Lyell was wrong about the general uniformity of the earth's behavior.'"

By ignoring evidence like this, Sagan can claim that there is no evidence.

Thus, we are brought full circle. The geologists maintain the disruption that ended the Ice Age was slow and happened 10,000 years ago. But in Egypt and Babylonia, ancient men saw and depicted supposedly long extinct fauna on the walls of their buildings. The geologists say volcanic activity did not occur in Palestine in historical times; but pottery 3,500 years old—or of historical time is found embedded in lava there. And there is other archaeological evidence some of which, like Ipiutak in Alaska or depictions of extinct elephants by American Indians or Tiahuanacu in the Andes or Cuicuilco in Mexico or a bronze workshop in the muck of the Arctic tundra of Siberia, or stone-age relics of ancient man in Spitzbergen above the Arctic Circle, or caves with relics of stone-age man high in the Alps mountains many miles inside a glacial ice field that we have discussed. But there is more in Velikovsky's book *Earth in Upheaval.*

TIDAL DESTRUCTION OF THE MOON

Finally Sagan asks, "Where is the evidence of extensive melting in these centuries near where the tidal distortion is greatest?"[81] In *Worlds in Collision* Velikovsky writes,

> "The great seas of dried lava and the great craters on the dead planet [the Moon] devoid of air and water bespeak the dreadful devastations, even death itself, that interplanetary contacts can leave in their wake. The great formations of craters, mountain rifts, and plains of lava on the moon were formed not only in the upheavals described in this book, but also in those which took place in earlier times. The moon Moonis a great unmarked cemetery flying around our Earth, a reminder of what can happen to a planet."[82]

If the Moon suffered great catastrophes 3,500 and 2,700 years ago, and was shaken by large meteorite impacts and powerful perturbations, it would still be vibrating from these events. This, indeed, is now known to be the case. Sagan, in his book *Cosmos,* pp. 85-86 discusses this lunar wobbling or libration stating,

> "When an object impacts the Moon at high speeds, it sets the Moon slightly wobbling. Eventually the vibrations die down, but not in...[a] short period... Such a quivering can be studied by laser techniques... Such measurements performed over a period of years reveal the Moon to be librating or quivering with a period (about three years) and amplitude (about three meters)..."

Sagan assumes the wobbling was caused by another later event than that suggested by Velikovsky. Nevertheless, Velikovsky's catastrophes would certainly create just such a libration of the Moon.

Now, if the Moon suffered during these catastrophes, wouldn't the Earth also have undergone violent perturbations; and shouldn't it also be librating or wobbling as the Moon does? This also is indeed the case. The discoverer in the last century of this strange wobble, Seth Carlo Chandler, has had this phenomenon named after him—it is called the Chandler Wobble. The Earth's poles wobble back and forth in a period of about 14 months over a distance of a few meters.

Velikovsky, in *Earth in Upheaval,* p. 113 stated that, "Simon Newcomb, foremost American mathematical astronomer in his paper, "On the Periodic Variation of Latitude" [*Astronomical Journal,* XI (1891); Cf. idem, in *Monthly Notices of the Royal Astronomical Society,* LII (1892), No. 35] wrote,

> "'Chandler's remarkable discovery that the apparent variation in terrestrial latitudes may be accounted for by supposing a revolution of the

axis of rotation of the earth around that figure...is in such discord with the received theory of the earth's rotation, that, at first, I was disposed to doubt its possibility.' However, on reconsideration he found a theoretical justification: 'Theory then shows that the axis of rotation will revolve around that of a figure, in a period of 306 days and in a direction from west to east.'"

"G.V. Schiaparelli, the Italian astronomer, in his research, *De la rotation de la terre sous l'influence des actions geologicques* (1889), pointed out that in the case of displacement the pole of inertia (or of figure) and the new pole of rotation would describe circles around each other, and the earth would be in a state of strain. 'The earth is at present in this [wobbling] condition and as a result the pole of rotation described a small circle in 304 days, known as the Eulerian circle.' This phenomenon of wobbling points to a displacement of the terrestrial poles sometime in the past."

If the Chandler Wobble produces a strain on the Earth's rotation, it should be damping [getting smaller and smaller] because over time, strains on the Earth will disappear unless some force can account to keep the Earth wobbling. But, as a matter of fact, measurements over time show that the Chandler Wobble is damping. The *Encyclopedia Britannica, Macropedia,* Vol. 17, (London 1986), p. 555 informs us that, "Chandler's Wobble, an oscillation of the Earth's axis with a period of about 14 months, appears to be damping with an amplitude reduction factor of $1/a$ about 15 years. There are problems in assessing the damping accuracy and determining this nutation [irregular motion] accuracy." Since the very nature of the Earth situated as it is in space, tends to damp this polar motion, the Chandler Wobble cannot be generated by the present internal nor external conditions of the Earth. John Wahr, in "The Earth's Inconstant Motion", *Sky and Telescope* for June 1986, p. 549 admits, "The Chandler Wobble is a good example of our great (though diminishing) ignorance. In truth, we don't know much more about its origin than we did in Chandler's time a century ago." However, the Chandler Wobble, like the wobble of the Moon, can both be explained by a catastrophe to both in the recent past and Velikovsky's theory does explain the Moon and the Earth's wobbling motions.

Velikovsky claims the great maria of the Moon are the result of immense gravitational stresses that heated the surface and actually melted it. Therefore, if this is the case, one side of the Moon would show this phenomenon much more than the other. This we showed earlier to be the case. Thus, what other scientific evidence supports this view?

If the Moon was tidally melted by a large passing body, then the maria should show this by a pattern that runs in a great circle arc across the Moon's surface. R.J. Malcuit et. al.; in an article titled "The

Great Circle Pattern of Large Circular Maria" in *The Moon*, (1975) Vol. 12, p. 55, which states,

"The circular maria—Orientale, Imbrium, Serenitiatis, Crisium, Smythii, and Tsiolkovsky—lie nearly on a lunar great circle. This pattern can be considered the result of a very close, non-capture encounter between the Moon and the Earth early in solar system history."

However, earlier we showed that dust on the Moon indicates its craters are extremely young; therefore, the arcurate pattern of large maria must be of recent tidal melting. What is also significant is the interpretation that these large lunar basins were produced by tidal melting and not by impact.

If a large body crashed into the Moon when it first formed, it would find the Moon melted from radioactive elements and thus, it too would melt. If, on the other hand, it crashed into the Moon after the Moon had cooled and solidified it, would on impact, have been so hot that it would have vaporized. However, if a planetary sized body approached the Moon, it would have distorted the shape of the Moon on one side, pulled the dense material of the lunar core toward the surface and where this hot material of the core came to the surface together with tidal forces of the passing body would have created huge seas of melted material. Thus, the side of the Moon with these maria which faces the Earth should show all these phenomena and more.

If the catastrophes described by Velikovsky occurred, the last event should cause the Moon to reverberate, that is, to experience earthquakes. There are two types of moonquakes, one type that recurs every 28 days or about once every lunar month. However, the location of this type of quake is not random; according to Bevan M. French in *The Moon Book*, (Westford, MA 1978), p. 228,

"In addition to being deep, (about 600 to 800 kilometers below the surface) most moonquakes are also localized; that is, they occur again and again at specific places on the moon. At present, 40 such 'centers' of moonquake activity have been identified, all but one on the near side of the moon. The centers are not distributed at random. With only a few exceptions, they lie along two great belts that are 100 to 300 kilometers wide and run for about 2,000 kilometers across the moon."

These belts of weakness are precisely what one would expect if the Moon had recently undergone enormous catastrophic stresses. What makes these deep seated moonquakes so difficult for the geophysicists to explain is that they should not exist. Moonquakes located 600 to 800 kilometers below the lunar surface cannot occur at that depth because of the high pressure and high lunar temperature gradient. In *Scientific American*, Vol. 260, for Jan. 1989, Cliff Frohlich's article "Deep Earthquakes", p. 48, discusses the problem related to deep-focus earthquakes. The reason on the Earth that such deep-focus earthquakes cannot occur is that at a depth of 60 kilometers, rocks should be so hot that they become ductile. Under stress, the rocks do not break violently but flow to deform. A similar condition pertains to the Moon. At some point below the surface, perhaps 200 km, its rocks will be so hot that they will deform and not break catastrophically. Thus, it is probable that the Moon recently experienced an enormous catastrophe that created unevenly distributed pressure centers that are correcting themselves after the event. G. Latham et. al., in *Science,* Vol. 174, (1971), pp. 687-692 admit, "The moonquakes appear to be releasing internal strain of unknown origin."

The same should also hold true for the Earth. Frohlich shows that since 1964, over 60,000 earthquakes have originated well below the Earth's surface in the same deep regions in which such events are considered impossible. This also implies that both the Earth and the Moon experienced a large catastrophic recent event to leave these pressure signatures of the event that are relaxing.

French goes on to state, pp. 229-230 that,

> "not all moonquakes are regular or related to tides. The lunar seismome-
> ters have also detected another kind of moonquake that occurs in groups
> at apparently random times, apparently unrelated to the moon's motions
> around the earth. These moonquakes are similar to the swarms of small,
> shallow earthquakes that often accompany the eruption of volcanic lavas
> on the earth."

But, magma in the upper regions of the Moon would have cooled billions of years ago based on present theory. French states, p. 237 that there are numerous observations reported by astronomers for hundreds of years, "such as glows, hazes, brief color changes and temporary obscurations of lunar surface features. Many of the observations, made in recent times by careful observers, are convincing, although no adequate explanation has yet been proposed..." to explain these volcanic gas emissions. This being the case, there must still be volcanic activity occurring near the lunar surface. However, the Moon cannot be ancient and still possess magma at or near the surface. French states, p. 237, "Present-day volcanic eruptions are hard to reconcile with the ancient ages of all the lunar lavas." However, volcanic activity near the Moon's surface is exactly what is expected based on Velikovsky's theory. If the Moon recently experienced a cosmic catastrophe, it would have created pockets of magma below the lunar surface which it seems to possess. The shallow moonquakes reflect volcanism.

Related to this is the finding by the astronomers of several types of volcanic features primarily on the near side of the Moon where volcanic activity has been observed as glows, hazes, etc. Nicholas M.

Short, in *Planetary Geology*, (Englewood Cliffs, NJ 1975), p. 109, writes,

> "Most of the more than four hundred lunar domes, the dark-halo craters, rings, wrinkled ridges, straight and sinuous rilles, and other endogenic features of volcanic origin are associated with mare lavas, volcanic fillings in large craters or, in some instances, the smooth plains... The majority of these features are concentrated on the near side of the Moon, as would be expected from the prevalence of the maria on [that] face."

It has been known for a considerable time that the side of the Moon that contains the major maria has a slight bulge. It seems the lunar core has been pulled by some force toward the side facing the Earth. This cannot be explained only by the fact that this side of the Moon faces the Earth; and that the pull of the Earth's gravity created this bulge.

The basic problem with the Moon's bulge is explained by the fact that the Moon possesses a high thermal gradient; that is, it becomes hotter with depth. Based on rheology—the science of the deformation of solids caused by gravity—bulges on large bodies will sink. S.K. Runcorn in *Scientific American*, Vol. 257 in an article titled "The Moon's Ancient Magnetism", for Dec. 1987, p. 63 explains,

> "What caused the [lunar] bulge? Laplace supposed that strains might have developed when the Moon initially cooled from a molten state, and the shape resulting from these strains had been retained after solidification. Later, Sir Harold Jeffreys suggested that the strain could have been caused by the tidal action of the Earth. By this explanation, the bulge would have been frozen in when the Moon was about 40 percent of its current distance from the Earth.
>
> "Both Laplace and Jeffreys assumed the rigidity of the Moon would have guaranteed that an early distortion (such as a bulge) would remain to the present day. Yet, if solid state creep occurred in the Moon at one-trillionth the rate at which we now know it occurs in laboratory materials at modest temperatures, such a primeval bulge would have disappeared long ago."

Dr. Runcorn has his own catastrophic theory for the Moon's bulge. Velikovsky's hypothesis that the Moon suffered a tidal interaction with proto-planet Venus some 3,500 years ago would also produce a lunar bulge. It is clear that the Moon's bulge was caused by a relatively recent and, quite probably, a very recent event.

However, the side of the Moon with the major maria also possesses "Mascons" or large concentrations of mass below the surface. M. Zeilik tells us, "The fact that almost all maria have associated mascons implies that some common process produced large amounts of dense material under the maria."[83]

Now, Carl Sagan believes the Moon formed billions of years ago. But if the Moon was then internally heated by radioactive elements, the large bodies that would impact to form the mascons would have melted. We know from previous discussion of the Moon that it still possesses a high thermal gradient. Therefore, a large body that entered the Moon's interior must have melted billions of years ago. According to Allen L. Hammond in an article in *Science,* for 1972,

> "[the] heat flux from the moon of 3.3 microwatts per square centimeter (about half of the heat flux from the Earth), a value that is considerably higher than was expected. The result is also higher than that predicted by thermal calculations based on geochemical models of the moon's interior, and, if sustained by further measurements...it might lead to a revision in estimates of the amount of material (whose decay is thought to be the heat source) present within the moon."[84]

Thus, the mascons, even with a lower thermal gradient, are not explained by the accepted evolutionary model. And Peter Cadogan in *The Moon—Our Sister Planet,* (Cambridge 1981), p. 271, states, "Perhaps the most remarkable aspect of mascons is their survival."

However, it is a well-known geological fact that the youngest rocks on Earth are also the most radioactive. Therefore, the maria that were formed most recently should contain the highest concentration of radioactive materials. Hammond goes on to say, about these nearside areas,

> "The region around the Imbrium basin, however, appears to have some unusual features. Observations of the gamma rays given off by radioactive materials showed much higher concentrations of uranium, thorium and potassium in Mare Imbrium and in the neighboring lava flows of Oceanus Procellarum than elsewhere on the moon. [These are the largest maria on the moon and would, therefore, contain the deepest radioactive materials in their surface lavas.] The observations were made from lunar orbit, and the...*Apollo 15* spacecraft allowed about 15 percent of the moon's surface to be mapped. The results of the experiment, conducted by a team headed by James Arnold of the University of California at San Diego, indicate concentrations of about 10 parts per million of thorium in the Mare Imbrium soil, compared with 1...ppm in the eastern part of the moon...why the Imbrium and Procellarum regions should be the overwhelming source of the radioactive elements on the moon's surface, as they appear to be, has been difficult for geochemists to explain."[85]

It is only difficult to explain if one believes the Moon's surface has not recently undergone tidal distortion which brought radioactive material from inside the Moon to its surface. Thus, the near side of the Moon exhibits arcs of maria, mascons, moonquake zone belts, domes, dark halo craters, rings, wrinkled ridges, straight and sinuous rills, high radio activity all associated with recent volcanic activity which still lingers as clouds and hazes.

Further, if the Earth, during the catastrophe, was not subjected to a strong magnetic field, should not the Moon also show a strong remanent magnetism in its rocks? Hammond, cited above, tells us,

> "Even more disturbing to many geochemists is evidence that the moon had a magnetic field, and hence, presumably a molten iron core throughout most of its first 1.5 billion years. Residual magnetism has been found in many of the lunar rock samples. In an experiment conducted by Paul Coleman and his colleagues at the University of California in Los Angeles, a magnetometer in a small satellite launched into lunar orbit by the *Apollo 15* crew has now recorded measurable amounts of residual magnetism over much of the moon, an indication that the magnetized samples are not isolated phenomena. Coleman believes that the residual field is due to a magnetized crust, which would imply that the moon was immersed in a strong magnetic field at the time the crust was formed.
>
> "Neither the magnetic field associated with the solar nebula nor the dipole magnetic field of the Earth, according to S. Runcorn of the University of Newcastle-upon-Tyne in England, could have magnetized the moon's crust because of the orientation of the field in the solar wind and the length of period during which the moon would have had to be very close to the earth...
>
> "The magnetic evidence, therefore leads to a picture of the Moon's evolution that conflicts with [uniformitarian] models based on geochemical considerations."[86]

Thus, neither the Earth's nor the Sun's magnetic fields are responsible for the Moon's magnetic field. According to Peter Cadogan in *The Moon—Our Sister Planet*, (Cambridge, England 1981), p. 312,

> "there is simply not enough metallic iron inside the Moon to carry the required magnetism. Even with 4.8% [iron] a magnetic field as high as 75 oersted would be necessary, and it is difficult to see how this could have been generated in the early solar system. For any reasonable solar field, the amounts of metallic iron necessary would be incompatible with the Moon's measured density, moment of inertia and magnetic permeability."

This simply means the Moon could never have generated its magnetic field in its early history.

If neither the Earth nor the Sun nor the Moon could generate the magnetic field where did the field come from? According to Cadogan, p. 314, "the rocks of the lunar crust have been magnetized externally, rather than internally." For this to occur, a body had to come close enough to the Moon to melt its rocks and at the same time immerse the Moon's surface in a strong magnetic field. Venus is described as expelling planetary thunderbolts. Hence, it was at that time, a highly electric and magnetic body which could melt and magnetize the Moon's rocks.

All the evidence discovered on the Moon supports Velikovsky's hypothesis and contradicts the views that Sagan supports.

Lastly, the Moon, as discussed, still seems to have a good deal of radioactive elements. This evidence produces a problem for the Moon's craters being ancient. The theory of Earth's and other terrestrial bodies formation currently accepted by most astronomers is the gradual condensation and accretion process. According to Raymond Siever in *The Solar System,* (San Francisco 1975) a *Scientific American* book, p. 59,

> "As the planet grew it began to heat up as a result of the combined effect of the gravitational infall of its mass, the impact of meteorites and the heat of radioactive decay of uranium, thorium and potassium... Eventually the interior became molten. The consequence of the melting was what has been called the iron catastrophe, involving a vast reorganization of the entire body of the planet. Molten drops of iron and associated elements sank to the center of the earth and there formed a molten core that remains largely liquid today.
>
> "As the heavy metal sank to the core the lighter 'slag' floated to the top—the outer layers that are now termed the upper mantle and the crust..."

But if this was the way all the terrestrial planets and the Moon formed, then they also melted to differentiate their materials into layers. Thomas A. Hockey in *The Book of Moon,* (NY 1986), p. 198 asks "Is the Moon differentiated in this way?" He answers, "The geologic and geophysical evidence says 'yes' but not as much as the Earth." Hockey adds, p. 199, "Currently, it is believed that the Moon of today does consist of a crust, a mantle, and possibly a core like that of the Earth." Thus, whatever ancient craters had existed would have melted and disappeared. It is thought the Moon went through this same process.

SAGAN'S BITTER PILL

Albert Einstein, prior to the publication of *Earth in Upheaval,* felt Velikovsky's evidence important enough that he read and reread the manuscript and added margin notes. Einstein's serious and thoughtful attitude contrasts quite strongly with Sagan's unreasoning view that the book, with its evidence, does not even exist and displays quite well the narrow and unscholarly nature of Carl Sagan's approach.

Sagan's refusal to admit that the evidence in *Earth in Upheaval* exists is in great measure like the professors of Galileo's time who refused to look through the newly invented telescope at phenomena that contradicted their own scientific concepts. Giorgio De Santillana in *The Crime of Galileo,* (Chicago 1955), p. 11, writes,

> "Certain doctors, who at least had the courage of their convictions, did actually and steadfastly refuse to look through the telescope, as has been recounted many times. Some did look and professed to see nothing; *most of them, however, gave it the silent treatment,* or said they had never gotten around to looking through it but that they knew already that it would show nothing of philosophical value." [emphasis added]

Sagan, like the learned doctors of yore, has also given Velikovsky's geological, archaeological and paleontological evidence the silent treatment. But he has outdone them in one significant way: while knowing Velikovsky has produced a book which deals with the evidence, Sagan asks where is this evidence to be found. Thus, he asks for evidence to be produced, but when it is produced and documented, Sagan refuses to acknowledge its existence. Sagan has put Velikovsky into a *double bind.* Putting a person into the position of "damned if he does and damned if he doesn't" speaks most eloquently of Sagan's attitude and method.

De Santillana, on p. 97 cites Galileo's answer to those who would condemn his work. "...I am proposing not that this book be not condemned, but *that it be not condemned, as they would, without understanding it, without hearing it, without ever having seen it.*" [emphasis in original] Sagan condemned Velikovsky's work without attempting to understand it, by being unwilling to hear or let others hear its evidence, even without having seen it. Sagan's whole approach is to keep his mind shut to this evidence because it contradicts his fundamentalist uniformitarian scientific philosophy.

Sagan aptly captures the feelings of establishment scientists in his co-authored book *Comet,* wherein he states, "...the idea of periodic (or at least episodic) mass extinctions, visited on the Earth from space, [is] more palatable; but for many scientists and others, it remains a bitter pill to swallow."[87] It is bitter medicine because Sagan and his colleagues have heaped denunciation on Velikovsky for stating this in *Worlds in Collision* and then acting as if his evidence was non-existent. Stephen Jay Gould of Harvard University, in a paper delivered on the topic of catastrophism states, "The *zeitgeist* [of gradualism] is now in for an overhaul... I believe that much of this current advocacy of punctuational [catastrophic] change represents nature reasserting herself against the blinders of our previous gradualistic prejudice...we should reject gradualism as a restrictive dogma."[88] Among the "literature cited" to support evidence of "punctuational change" are, on p. 31, Velikovsky, I, 1950, *Worlds in Collision*...(and) *Earth in Upheaval.* Three words or phrases by Professor Gould apply to the treatment by Sagan of Velikovsky's *Earth in Upheaval*. The first states science's dogma of gradualism has acted as a "blinder" to evidence. Sagan's strong aversion to such evidence caused him to deny its very existence. Gradualism has been a "restrictive dogma"; Sagan and his colleagues have restricted themselves from thinking along catastrophic lines, but have restricted others, not just Velikovsky, by denunciation if they challenged gradualism; and in this sense it has been not only restrictive, it has been "suppressive." Lastly, Gould says, gradualism is in for an "overhaul." Of course, he does not envisage Velikovsky's brand of catastrophism winning the race, but a fair and honest appraisal of the evidence is most certainly in order. Sagan's whole discussion of this material reflects a glib and cavalier attitude. One unfortunately can expect the denunciations to continue.

Sagan stated, "On these questions, Velikovsky seems to have ignored the critical evidence," while the fact is that Sagan totally ignored all the evidence presented by Velikovsky. WHY?

NOTES & REFERENCES

[1]Macbeth, Norman, *Darwin Retried*, op. cit., pp. 110-111.

[2]SCV, pp. 67-68; B.B., pp. 103-104.

[3]Meyerhoff, A.A. & H.A.; "The New Global Tectonics: Major Inconsistencies", *American Associates of Petroleum Geologists Bulletin*, Vol. 56, (1972), p. 270.

[4]*Darwin Retried*, op. cit., p. 116.

[5]*Earth in Upheaval*, op. cit., p. 143.

[6]*Darwin Retried*, op. cit., p. 110.

[7]SCV, p. 66; B.B., p. 102.

[8]Hapgood, Charles; *Path of the Pole*, (Philadelphia 1977), pp. 130-131.

[9]Ibid., p. 136.

[10]*Pensée*, Vol. 3, No. 1 (Winter 1973), Mackie, Euan W.; "A Quantitative Test for Catastrophic Theory," p. 9.

[11]Ibid.

[12]*Earth in Upheaval*, op. cit., p. 144.

[13]SCV, p. 66; B.B., p. 102.

[14]*Earth in Upheaval*, op. cit., pp. 133-134.

[15]McCreedy, Thomas, "Krupp vs. Velikovsky", *Kronos* VI, 3, p. 45.

[16]Schlee, Susan; *The Edge of an Unfamiliar World, A History of Oceanography* (NY 1973) p. 317.

[17]B.B. p. 103.

[18]*Earth in Upheaval*, op. cit., pp. 67-80.

[19]*Darwin Retried*, op. cit., p. 113.

[20]SCV, pp. 66-67; B.B., p. 103.

[21]SCV, p. 67; B.B., p. 103.

[22]Taylor, Ian T, *In the Minds of Men*, (Toronto 1984), pp. 303-304.

[23]Sagan and Druyan, *Comet*, op. cit., pp. 253-254.

[24]*In the Minds of Men*, op. cit., pp. 328-329.

[25]Sagan and Druyan, *Comet*, op. cit., p. 254.

[26]*New Scientist*, (April 23, 1987), p. 24.

[27]*Time*, (Jan. 24, 1972), p. 67.

[28]Ibid.

[29]SCV, p. 67; B.B., p. 103.

[30]*Earth in Upheaval*, op. cit., pp. 98-99.

[31]Black, R.K, Barksdale, W.L.; *Journal of Geology*, Vol. 57, (1949), pp. 105-118.

[32]McQuillin, R.; Fannin, N.; *New Scientist*, Vol. 83, (1979), pp. 90-92.

[33]Gregory, Janet, *FSR*, Vol. 17, (Sept/Oct 1971), pp. 29-30.

[34]SCV, p. 67; B.B., p. 103.

[35]*Kronos*, V, 2, p. 90.

[36]SCV, p. 67; B.B., p. 103.

[37]Gould, Stephen J., "Toward the Vindication of Punctuational Change", *Catastrophes and Earth History: The New Uniformitarianism*, ed. W.A. Berggren and J.A. Van Couvering, pp. 17-18.

[38]Hunt, W.C.; *Bulletin of Canadian Petroleum Geology*, (1977), Vol. 25, pp. 456- 467.

[39]Ibid.

[40]*Earth in Upheaval*, op. cit., pp. 11-12.

[41]Ibid., pp. 164-166.

[42]SCV, p. 67; B.B., p. 103.

[43]SCV, p. 67; B.B., p. 103.

[44]*Science News Letter*, Vol. 14, (1928), p. 81: see Unknown Earth, op. cit., pp. 636-637.

[45]Dort, W,; *Antarctic Journal of the United States*, (Washington 1971), Vol. 6, p. 210.

[46]Huber, B.; *The Physiology of Forest Trees*; ed. K.V. Thimann, (NY 1958), p. 194.

[47]Keith, M., Anderson, B., in *Science*. (Aug. 16, 1983), p. 634.

[48]*In the Minds of Men*, op. cit., p. 316.

[49]Ogden, J.; *Annals of the New York Academy of Science*, Vol. 288, (1977), p. 173.

[50]*Earth in Upheaval*, op. cit., p. 2.

[51]Ibid., pp. 2-3.

[52]Irving, W.N, Harington, C.R., *Science*, Vol. 179 (Jan. 26, 1973), p. 335.

[53]*Earth in Upheaval*, op. cit., pp. 6-7.

[54]*In the Minds of Men*, op. cit., p. 99.

[55]*Earth in Upheaval*, op. cit., p. 7.

[56]Hoyle, Fred; *Ice*, (NY 1981), p. 66.

[57]W in C, pp. 327-328.

[58]*National Geographic on Indians of the Americas*, (1955), p. 167.

[59]*Earth in Upheaval*, op. cit., (Supplement), pp. 251-252.

[60]*Path of the Pole*, p. 102.

[61]*Earth in Upheaval*, op. cit., pp. 153-154.

[62]Ibid.

[63]Ibid., pp. 153-154.

[64]Ibid., p. 155.

[65]Ibid., pp. 59-61.

[66]Ibid., p. 156.

[67]Ibid., pp. 151-153.

[68]Ibid., pp. 61-62.

[69]Ibid., p. 186.

[70]Path of the Pole, op. cit., pp. 286-287.

[71]Ibid., p. 290.

[72]Ibid., p. 292.

[73]d'Orbigny, A.; Voyage dan l'Amérique méridionale, 3 Vols., (Paris 1842), pp. 3, 82, 85, 86.

[74]Pursuit , (Oct. 1969), Vol. 2, pp. 68-69.

[75]Earth in Upheaval, op. cit., p. 207.

[76]Hibben, Frank C.; The Lost Americans, (NY 1946), pp. 168-170.

[77]Earth in Upheaval, op. cit., pp. 42-43.

[78]SCV, p. 67; B.B., p. 103.

[79]W in C, pp. 29-30.

[80]Earth in Upheaval, op. cit., pp. 177-179.

[81]SCV, pp. 67-68; B.B., p. 104.

[82]W in C, p. 362.

[83]Zeilik, Michael; Astronomy, The Evolving Universe, 4 ed. (NY 1985), p. 143.

[84]Hammond, Allen L.; Science, Vol. 175, (1972), pp. 868-870.

[85]Ibid.

[86]Ibid.

[87]Sagan and Druyan, Comet, op. cit., p. 293.

[88]Catastrophes and Earth History, op. cit., pp. 29-31.

SAGAN'S FIFTH PROBLEM

CHEMISTRY & BIOLOGY OF
THE TERRESTRIAL PLANETS

THE ORIGIN OF ATMOSPHERIC OXYGEN

Sagan states, "Velikovsky's thesis has some peculiar biological and chemical consequences, which are compounded by some straight-forward confusions of simple matters. He seems not to know (p. 16) that oxygen is produced by green-plant photosynthesis on the Earth."[1] Let us, therefore, examine p. 16 of *Worlds in Collision* to see just what Velikovsky did say.

> "If, in the beginning, the planet was a hot conglomerate of elements as the nebular as well as the tidal theories assume, then the iron of the globe should have been oxidized and combined with all the available oxygen. But for some unknown reason this did not take place; thus the presence of oxygen in the terrestrial atmosphere is unexplained."

Velikovsky's statement is straightforward and the chemical conse-quences are very simple to understand. Iron, which is a very abundant element in the Earth, would naturally unite with oxygen in the early period of Earth formation. Because iron is a heavy element, thereafter, it would also naturally sink deep into the early molten Earth to form the Earth's iron-nickel core. Thus, there would be no oxygen avail-able to form in the atmosphere. Sagan tells us Velikovsky is confused, but I think it is Sagan who has the problem. The thesis Velikovsky proposed on page 16 respecting atmospheric oxygen is neither idiosyncratic nor unknown to science. Robert H. Dott, Jr. and Roger L. Batten in their book, *Evolution of the Earth*, discuss this very same problem wherein they say,

> "The greatest single problem facing any hypothesis of atmospheric development is to explain the abundance of free oxygen... It has long been assumed that the early atmosphere did not have free oxygen..."[2]

Sagan has informed us that oxygen originated in the early atmosphere from green plants. There are great problems with this idea. One of the problems is the green plants themselves; they are made up of a few simple elements, one of which is "oxygen." If there was no oxygen in the atmosphere, how could plants form—let alone produce oxygen? Theo Loebsack in *Our Atmosphere*, (translated from German by E.C., and D. Rewald), (NY 1959), p. 19 explains. "Other scientists [as

236

Sagan] believe that plants gave rise to the greater part of the oxygen found in the air to-day. This theory has one drawback. How could plants have existed without oxygen, for although they give off oxygen, they also need it to live? How could plants evolve if there were no free oxygen on Earth?" For green plants to grow and produce oxygen they must live in an oxygen environment. But since there was no oxygen for them to carry on metabolism, live plants could not exist. Jeremy Rifkin in his book *Algeny* says this about oxygen in the early atmosphere of the Earth,

> "To begin with, most scientists would agree that life could not have formed in an oxygen atmosphere. If the chemicals of life are subjected to an oxidizing atmosphere, they will decompose into carbon dioxide, water and nitrogen. For this reason, it has long been assumed that the first precursors of life must have evolved in a reducing [oxygen free] atmosphere, since an oxidizing atmosphere would have been lethal... But, in order to posit the theory that life evolved from nonlife, it is essential to assume a reducing atmosphere, because an oxygen atmosphere would destroy the chemicals of life before they could be fashioned into organic compounds by oxidizing or decomposing them back into carbon dioxide, water, nitrogen, and oxygen...
>
> "[A] reducing atmosphere overcomes this first giant hurdle, [but we are] immediately faced with a second hurdle, which is insurmountable. Without oxygen there would be no ozone shield to screen out most of the ultraviolet rays, life could not exist, even on the most primitive level."[3]

Even Sagan is aware of this; in *Broca's Brain* he states that, "If the thin protective ozone layer in our atmosphere, made by sunlight from oxygen, did not exist, we would rapidly be fried by ultraviolet light from the Sun."[4]

Rifkin continues, citing R.L. Wysong who sums up the obvious Catch-22,

> "If oxygen were in the primitive atmosphere, life could not have risen because the chemical precursors would have been destroyed through oxidation; if oxygen were not in the primitive atmosphere, then neither would have been ozone, and if ozone were not present to shield the chemical precursors of life from ultraviolet light, life would not have arisen."[5]

To solve this dilemma, the biologists move the precursors of life under water to protect the molecules from the deadly ultraviolet radiation. Does this solve the problem? Robert Shapiro in his book *Origins* informs us that,

> "Water happily attacks large biological molecules. It pries nucleotides apart from each other, breaks sugar-to-phosphate bonds, and severs bases

from sugars. These reactions are taking place in our cells at this very moment. Fortunately, after billions of years of evolution, our bodies are well equipped to deal with these events. We have developed elaborate mechanisms to repair the damage to our molecules caused by continual attack by water.

"On the early earth, such defenses did not exist. Water continually opposed the assembly of large biomolecules and attacked those that had successfully formed."[6]

Carl Sagan in his co-authored book, *Shadows of Forgotten Ancestors,* (NY 1992), p. 26 agrees:

"Oxygen generated today by green plants, must have been in short supply before the Earth was covered by vegetation. But ozone is generated from oxygen. No oxygen, no ozone. If there's no ozone the searing ultraviolet...from the sun will penetrate to the ground. The intensity of UV at the surface of the Earth in those days may have reached lethal levels for unprotected microbes, as it has on Mars today."

Sagan then performs a feat of magic and pulls out of nothing a fully developed plant that lived on the early ocean shores called a Stromatolite which came into existence in an environment of lethal ultraviolet light. In essence, Sagan has created a form of plant life different than the forms we commonly know today that could live, thrive and reproduce in the presence of ultraviolet light. The entire concept is in reality nothing but legerdemain created with smoke and mirrors.

True, something is peculiar and confused, but it is not Velikovsky's work.

There is a further problem with Sagan's view that the green plants produced the first oxygen of the atmosphere. Sagan informs us in his book *Cosmos,* p. 32, that,

"By one billion years ago, plants working cooperatively, had made a stunning change in the environment of the Earth. Green plants generate molecular oxygen. Since the oceans were now filled with simple green plants, oxygen was becoming a major constituent of the Earth's atmo-sphere, altering it irreversibly from its original hydrogen rich character and ending the epoch of Earth history when the stuff of life was made by nonbiological processes."

Sagan does not explain how the plants could form and survive in an atmosphere that permitted ultraviolet waves to penetrate and fry these first plants. If Sagan is correct, then about one billion years ago oxygen became plentiful and there should be evidence of this in the Earth. The problem for Sagan is that there were great amounts of oxygen produced two billion years ago, or a billion years before the green plants began to change the character of the atmosphere. According to

Fred Hoyle in his book *The Intelligent Universe,* (NY 1983), p. 64, oxygen united with iron to form what are termed "red beds" two billion years ago. Hoyle states, "This rusting has been occurring for a very long time; indeed the oldest red beds were formed about 2,000 million [two billion] years ago. However, in spite of extensive world-wide geological surveys, none much earlier than this have been discovered." The evidence from the two billion year old red beds implies that although plants existed about a billion years ago, they somehow produced great amounts of oxygen about two billion years ago. This is no small problem for Sagan and other scientists. They have concrete evidence that plants changed the atmosphere one billion years ago, yet they have concrete evidence that these same green plants produced oxygen from a period perhaps a billion years earlier. Thus, Sagan's argument respecting how green plants generated the oxygen in the atmosphere is beset by problems which present day science has not solved completely. For a discussion of this see Ronald B. Parker's *Inscrutable Earth,* (NY 1984), Chapter 4.

HYDROCARBONS—CARBOHYDRATES

Sagan states,

> "Velikovsky holds that the manna that fell from the skies in the Sinai peninsula was of cometary origin and therefore that there are carbohydrates on both Jupiter and Venus. On the other hand, he quotes copious sources for fire and naphtha falling from the skies, which he interprets as celestial petroleum ignited in the Earth's oxidizing atmosphere (pages 53 through 58). Because Velikovsky believes in the reality...of both...events, his book displays a sustained confusion of carbohydrates and hydrocarbons, and at some points he seems to imagine that the Israelites were eating motor oil rather than divine nutriment during their forty years' of wandering in the desert."[7]

Sagan has made so many mistakes in this paragraph that it is intriguing to choose just where to begin a response. Firstly, Sagan confused the petroleum liquids that fell to the Earth with the hydrocarbon gases that remained in the Earth's atmosphere. Thus, Sagan argues that liquids and gases fall at the same rate which cannot be. Water vapor is a gas and forms clouds, but when water condenses to form a liquid, it falls as rain. Lastly, Sagan thinks that having fallen to Earth by some new laws of atmospheric dynamics, both at the same rate of descent, the petroleum liquids and hydrocarbon gases were so tempting and that the Israelites were so hungry that they must have eaten the motor oil, or perhaps he thinks they ate the hydrocarbon gases. Sagan's whole paragraph is gas. The confusion I suspect is that when Sagan sees hydrocarbon, he reads carbohydrate; when he reads carbohydrate, he sees hydrocarbon; when he sees liquid, he reads gas, etc. However, the major confusion seems to be that Sagan does not appear to know the difference between the chapter "Naphtha" and the chapter "Ambrosia" of *Worlds in Collision*.

The chapter "Naphtha" deals with the *earlier* period of the near collision between the Earth and an immense comet. There Velikovsky cites ancient sources from around the globe that the Earth burned in many places where liquid petroleum fell from the comet. The chapter "Ambrosia" deals with the later period, after the near collision when hydrocarbon gases left in the atmosphere were converted into carbohydrates. There also, Velikovsky cites legendary evidence of manna forming with the morning dew again from ancient sources around the globe.

JUPITER'S HYDROCARBONS

If the comet was indeed expelled by some fission process from Jupiter and it had hydrocarbons, then Jupiter should also possess these materials. Sagan tells us that,

> "He [Velikovsky] makes no note of the fact that Jupiter is composed primarily of hydrogen and helium, while the atmosphere of Venus, which he supposes to have arisen inside of Jupiter, is composed almost entirely of carbon dioxide. These matters are central to his ideas and pose them very grave difficulties."[8]

The questions are:

1) Are hydrocarbons found on Jupiter?

2) Do comets contain hydrocarbons and liquids?

3) Are there conversion processes to convert hydrocarbon gases into carbohydrates?

4) Does Venus possess hydrocarbons? This last question will be taken up in Sagan's seventh problem, "The Clouds of Venus".

Sagan states that because "Jupiter is composed primarily of hydrogen and helium" it lacks hydrocarbons. Therefore, let us see what a well-known astronomer has to say on this matter which poses "grave difficulties" for Velikovsky. The astronomer just happens to be Carl Sagan, who stated the following at a NASA symposium:

> "...Jupiter is a kind of remnant of the chemistry which was around in the early history of the solar system. It preserves for us the circumstances which were about on the earth in the early history of our planet, and, in particular, the atmosphere of Jupiter is composed primarily of hydrogen and helium, and smaller quantities of methane, ammonia and almost certainly water, although water hasn't been directly detected yet, because it freezes out at the temperatures which we can see in the atmosphere of Jupiter. Now, that same mixture of gases is thought to be the primary

mixture of gases in the early atmosphere of the earth at the time of its origin of life. So it is possible to do experiments in which we mix together those gases, primarily hydrogen, methane, ammonia and water—we can also add, for example, hydrogen sulfide—and supply some energy and see what molecules are made... One such experiment done several years ago in our laboratory at Cornell, in which we had a glass reaction vessel, into which is inserted an ultraviolet source surrounded by a quartz jacket, and in this reaction vessel at this moment are the gases I've just mentioned. They are colorless; the only color...is from the irradiation. The gases are irradiated, and circulated out of the light and then back into it.

"...After the same reaction vessel...and you can see that we have produced some material that is brownish colored, in very high yield. This material we have spent a lot of time trying to analyze. It is a very complex mixture of organic molecules, that is, carbon-based molecules, some of very high complexity. *Most of them of the kind called straight chain hydrocarbons.*"[9] [emphasis added]

The same information can also be found in the *Encyclopedia Britannica* for 1972 which states, "...independent lines of research have led to the suggestion that the upper atmosphere of Jupiter is a giant organic factory producing complex organic molecules that include many of biological importance."[10] Tom Gold, in *New Scientist,* (June 26, 1986), p. 42 states, "Jupiter...contain[s] enormous amounts of methane and other hydrocarbons in [its] atmosphere." Thus are hydrocarbons found on Jupiter? According to Gold and Sagan, apparently they are.

What then of the possibility of life in the clouds of Jupiter? William J. Kaufmann III in *Planets and Moons,* (San Francisco 1979), pp. 147-148, informs us five years after Sagan presented his criticism of Velikovsky that scientists also suggest life may exist in the clouds of Jupiter. After discussing Harold Urey, and Stanley Miller's experiments which produced some types of amino acids he states,

"This experiment has been performed and studied many times, most recently by Cyril Ponnamperuma at the Laboratory of Chemical Evolution at the University of Maryland. Although we are far from being able to create life directly from inorganic chemicals, it is clear that many of the molecular building blocks of living matter are produced in these experiments. Remarkably, the murky brew that is manufactured by these experiments has a distinctive reddish-brown color—exactly the same hues and tints seen in the belts of Jupiter.

"It seems entirely reasonable to suppose that the same chemical reactions are occurring in Jupiter's turbulent atmosphere... The chemicals are there; the electrical sparks are there. So why no amino acids and nucleotides? And since this chemical processing has been occurring for billions of years (rather than a few days, as in the laboratory experiments), perhaps enough time has elapsed for the development of living organisms. At a depth of only 100 kilometers below the Jovian cloud-tops, temperatures and pressures are hospitable—even by terrestrial standards. Simple one-celled creatures could easily survive floating among the clouds at these depths."

These one-celled organisms are therefore considered probable by the scientists just as Velikovsky hypothesized. Kaufmann III, goes on,

"...the Jovian clouds seem to be the next obvious target in the search for extraterrestrial life...

"In 1977, Carl Wose of the University of Illinois announced his identification of extraordinary previously unidentified form of life here on Earth. Unlike bacteria or plant and animal life, this newly discovered life form exists only in oxygen-free environments such as the deep hot springs of Yellowstone National Park and on the ocean floor. These simple organisms superficially resemble bacteria, but they take in carbon dioxide, water and hydrogen and give off methane... These ancient organisms are ideally suited to Earth's primordial atmosphere. And, in addition, they would thrive in Jupiter's clouds. Indeed, it is tantalizing to speculate that some of the methane...we detect in the Jovian atmosphere might actually be caused by biological processes in the belts and zones."

Thus not only are organics known to exist in Jupiter's atmosphere, but the scientific community also speculates that living organisms exist there just as Velikovsky speculated. The scientists suggest these organisms may be responsible for the production of methane, a hydrocarbon gas, just as Velikovsky did, who wrote in *Worlds in Collision*, p. 369, "...Jupiter must possess an organic source of petroleum." One is left to wonder why Velikovsky is so savagely attacked for conceiving of a concept many present-day scientists put forth?

COMETARY HYDROCARBONS

The second question is, Do comets contain hydrocarbons or organic molecules? Again we turn to our well-known astronomer Carl Sagan for information. In his book *Comet,* he states,

"Astronomers tend to be nervous about the word organic—concerned that it might be misunderstood as a token of life on another world. But 'organic' only refers to molecules based on carbon. And organic chemicals would be produced and destroyed even if there were no life anywhere in the universe... Cometary silicates are probably intimately mixed with—perhaps coated by—complex organic compounds... One possible explanation is a large proportion of complex organic molecules (or just plain carbon) in the cometary nucleus which are involatile or which do not produce accessible spectral features. If this interpretation is correct, comets might be as much as 10 percent organic."[11]

In *Earthlike Planets,* Sagan writes on the back cover, "This book is an important one and provides the first serious glimpse in any book that I know of the emerging science of comparative planetology." On page 10 of this book we read, "Such 'trans-jovian' objects have long been inferred to be the source of *comets, which are known to contain large amounts of organic matter and water.*"[12] [emphasis added]

But is this organic matter hydrocarbons? In *Astrophysics* of 1951 we learn that, it was reported by Bobrovnikoff of the Perkins Observatory that he had found from spectrographic analysis the presence of hydrocarbons in the tail of comets.[13]

In *Sky and Telescope,* for March 1987, p. 250, we learn that,

"The Vega infrared team made an important discovery bearing on the carbon in…[Halley's] comet. The spectral features at the position marked 'C-H' on the chart at the upper right signal the presence of carbon and hydrogen joined by a chemical bond. These bonds are ubiquitous in carbon rich materials, such as hydrocarbons and other organic molecules."

Therefore, if an immense comet derived from Jupiter, which Sagan informs us contains hydrocarbons, and like immensely smaller comets, which Sagan informs us possesses organic molecules, or as Bobrov-nikoff relates possesses hydrocarbons, what would happen if its materials entered the Earth's atmosphere? Those that were in liquid form would fall to Earth; those that were in a gaseous state would circulate in the atmosphere.

HYDROCARBONS INTO CARBOHYDRATES

The third question is: are there conversion processes by which hydro-carbon gases can be converted into carbohydrates in the atmosphere of the Earth? Yes, indeed there are. In fact, there are six such natural occurring processes by which the conversion can be accomplished. W.K. Kuong informs us that if:

"…hydrocarbons shrouded the Earth, part of it would mix with hydrogen of the hydrogen layer and another would be oxidized by the oxygen of the oxygen layer. The main products of combustion are carbon dioxide, carbon monoxide and water vapor through cosmic irradiation, as laboratory experiments suggest. The action of cosmic radiation on the carbon dioxide/hydrogen/carbon monoxide/water vapour mixture would generate formaldehyde. Once formaldehyde is formed, various types of sugars and starches would be generated by the process of polymerization and aldol condensation. The general formula for the process is,

$$nCH_2O = (CH_2O)n$$

"In the above equation n is any integer. In n=5, the product is pentose; if n=6, the product (isomers also exist) is a hexose, etc.

"Formaldehyde should be formed during the day when the gaseous mixture is bombarded by particles from the solar furnace. The product would polymerize in the cool of the night, particularly on dust particles, and rain down in the early morning. If one refers to *Exodus* and *Numbers*, one finds that manna was deposited early in the morning with the dew.

"Dr. A.J. Swallow, n his text *Radiation Chemistry of Organic Compounds*, writes, 'The synthesis of organic compounds through the agency of high-energy radiation had been amply demonstrated in the laboratory, an elementary example being the alpha-induced reaction between carbon dioxide and hydrogen to give formaldehyde, which then reacts further. Carbon monoxide can be reduced similarly. The main final product of irradiation *in both cases appears to be a white solid composition* $(CH_2O)n$, which is presumably produced by polymerization of formaldehyde.'"[14] [emphasis added]

In fact, it was pointed out in the *Biblical Archaeology Review* for May-June 1988 by C.L. Ellenberger, L.M. Greenberg, and Dr. Shane Mage, the process described in the article cited above produces "edible carbohydrate or protein-like substances. The product of a similar, if not identical, process of conversion of petroleum into concentrated nutrition [which] is today sold in food stores everywhere as 'primary growth torula yeast.'"

It is quite interesting to note that the description of manna from around the world describes it as *milky* or honey like just as the experiment by Dr. Swallow produced.

MANNA FROM HEAVEN

If this analysis is correct, every once in a great while a very tiny comet, laden with hydrocarbons should enter the Earth's atmosphere and if it does not completely burn up, should leave a small cloud of such material. Therefore, if the hydrocarbons were indeed converted to carbohydrates, they would fall as rain, sleet, hail, snow, or possibly other forms. Charles Fort who reported on materials that fell from the sky, and presented several examples of falls of manna, writes,

"The subject of reported falls from the sky of an edible substance, in Asia Minor is confused, because reports have been upon two kinds of substances... In July 1927, the Hebrew University of Jerusalem sent an expedition to the Sinai Peninsula to investigate showers of 'manna.' See *The New York Times*, Dec. 4, 1927. Members of the expedition found what they called 'manna' upon leaves of tamarisk trees and on the ground underneath, and explained that it was secreted by insects. But the observations of this expedition have nothing to do with data, or stories of falls from the sky of fibrous convoluted lumps of a substance that can be

ground into edible flour. A dozen times since early in the 19th century—
and I have no definitely dated data upon still earlier occurrences—have
been reported showers of 'manna' in Asia Minor... The substance that
occasionally falls from the sky in Asia Minor, comes from far away. The
occurrences are far apart, in time, and always the substance is unknown
where it falls, and its edibleness is sometimes found out by the sight of
sheep eating it. Then it is gathered and sold in the markets.

"We are told that it has been identified as a terrestrial product. We are
told that these showers are aggregations of *Leconora esculenta,* a lichen that
grows plentifully in Algeria. We are told that whirlwinds catch up these
lichens lying loose, or easily detachable, on the ground. But note this:

"There have been no such reported showers in Algeria.

"There have been no such reported showers in places between Algeria
and Asia Minor.

"The nearest similarity that I can think of is of tumble weeds, in the
western [United] states, though tumble weeds are much larger. Well
then, new growths of them when they are not much larger. But I have
never heard of a shower of tumble weeds. Probably the things are often
carried far by whirlwinds, but only scoot along the ground. A story that
would be similar to stories of lichens, from Algeria, falling in Asia Minor,
would be of tumble weeds, never falling in showers on western states,
but repeatedly showering in Ontario, Canada, having been carried there
by whirlwinds.

"Out of a dozen records, I mention that in *Nature,* Vol. 43, page 225,
and in *La Nature,* Vol. 36, page 82, are accounts of one of the showers,
in Asia Minor. The Director of the Central Dispensary of Baghdad had
sent to France specimens of an edible substance that had fallen from the
sky, at Meriden and at Diarbekis (Turkey in Asia) in a heavy rain, the last
of May 1890. They were convoluted lumps, yellow outside and white
inside. They were ground into flour from which excellent bread was
made." [15]

Elsewhere in his book, Fort describes manna falls in Milwaukee, Wis -
consin (1881), and Montgomery, Alabama (1898). Of *La Nature* for
1883 Fort states,

"A correspondent writes that he sent a sample of a substance said to have
fallen at Montussan (Gironde), October 16, 1883. According to a
witness, quoted by a correspondent, a thick cloud accompanied by rain
and a violent wind had appeared. This cloud was composed of a woolly
substance in lumps the size of a fist, which fell to the ground. The Editor
(Tissandier) says of this substance that it was white, but was something
that had been burned. It was fibrous. M. Tissandier astonishes us by
saying that he cannot identify this substance." [16]

SKY OIL

Furthermore, if the hydrocarbons were not completely converted by
the processes described by Kuong above, then they would still

preserve their oily composition. Thus, if this is true of the past, should it not also be true of the present? In this regard Fort informs us:

> "That [in] March 1832, there fell, in the fields of Kourianof, Russia, a combustible yellowish substance, covering at least two inches thick an area 600 or 700 square feet. It was resinous and yellowish: so one inclines to the conventional explanation that it was pollen from pine trees—but when torn, it had the tenacity of cotton. When placed in water, it had the consistency of resin. 'This resin,' [the *Annual Register* states], 'had the color of amber, was elastic, like Indian rubber, and smelled like prepared *oil* mixed with wax.'"[17] [emphasis added]

On the next page, [64], Fort states:

> "April 11, 1832—about one month after the fall of the substance of Kourianof—fell a substance that was wine-yellow, transparent, soft and smelling like rancid oil. M. Herman, a chemist who examined it named it '*sky oil*.'"[18] [emphasis added]

Thus, it is quite clear that both hydrocarbons and carbohydrates can, and do fall to Earth from the sky probably from tiny comets.

THE ORIGIN OF PETROLEUM

This brings us to Sagan's view that "petroleum arises from decaying vegetation, of the Carboniferous and other early geological epochs, and not from comets."[19] One of Sagan's arguments against Velikovsky's hypothesis is that the cometary origin of petroleum is invalid because "It is also very difficult to understand on his [Velikovsky's] hypothesis how it is, if oil fell from the skies in 1500 B.C., that petroleum deposits are intimately mixed with chemical and biological fossils of tens of hundreds of millions of years ago."[20] This statement clearly shows that Sagan does not understand Velikovsky's hypothesis. His misconception lies in his belief that Velikovsky claims the Earth only experienced catastrophes in historical times. Because Sagan refuses to acknowledge the book *Earth in Upheaval,* he could not bring himself to read,

> "As important as the 'world catastrophes' conclusion is, it grows in significance for almost every branch of science when, to the ensuing question, 'Of old or of recent time?' the answer is given, 'Of old and of recent.' There were global catastrophes in prehuman times, in prehistoric times and in historical times."[21]

That is why "petroleum deposits are intimately mixed with chemical and biological fossils of tens to hundreds of millions of years." Sagan raises this point.

> "The amount of downward seepage of petroleum in 2,700 years would not be very great. The difficulty in extracting petroleum from the Earth,

which is the cause of certain practical problems today, would be greatly ameliorated if Velikovsky's hypothesis were true."[22]

Unfortunately, these problems exist today based strictly on Velikovsky's hypothesis. Firstly, Sagan does not seem to understand or remember that the celestial petroleum deposits Velikovsky described were laid down, not as Sagan stated 2,700 years ago, but 3,500 years ago. Secondly, Sagan's inability to understand Velikovsky on Velikovsky's terms, had led him to believe that oil moves downward only by slow seepage or that these events did not occur in earlier periods.

In *Worlds in Collision,* chapter 2, is a subchapter titled "Naphtha" and in the same chapter is a subchapter titled "Earthquake". Earthquakes of the magnitude that Velikovsky describes produce great rents in the Earth's strata. Oil at the surface would run down these cracks deep into the Earth and thus be difficult to extract. Also, Sagan has forgotten that comets would have brought petroleum millions of years ago. Sagan's inability to see outside his gradualist blinders will continue to lead him to evaluate catastrophic evidence by applying gradualistic analyses.

However, let us take Sagan on his own gradualistic terms regarding the age of oil. Sagan apparently accepts the view that oil takes a million years or more to form. He also accepts radiocarbon age dating analysis. Radiocarbon analysis supposedly can age date a substance that is no older than 50,000 or possibly 100,000 years old. Since oil is at least a million or more years old, according to Sagan, it should not give a radiocarbon date. P.V. Smith in a paper in *Science,* titled "The Occurrence of Hydrocarbons in Recent Sediments from the Gulf of Mexico" reported on the radiocarbon date of the oil. He held that the oil radiocarbon dated to be no older than 9,000 years.[23] How can oil take a million years to form and yet be only 9,000 years old by being radiocarbon dated?

If, as Sagan informs us, oil is produced by the decay of green vegetation, we run into still another impossible situation. Grover E. Murray wrote an article in the *American Association of Petroleum Geologists Bulletin,* that methane with minor amounts of propane were found in Precambrian rock in central Australia.[24] Precambrian rock is rock that solidified in the Earth before vegetative life existed. Such materials in the Earth are supposed to be byproducts of petroleum production. How can byproducts of petroleum production be formed in rocks where there were no green plants available to be processed into petroleum? While *Newsweek,* July 8, 1985 p. 56 reports oil from billion year old rock beds in Siberia. Tom Gold, writing in *New Scientist* says,

"There were dissenting voices [regarding the explanation for the origin of petroleum from biological debris in sediments]... The most prominent was Dimitri Mendeleyev, the great Russian chemist. He wrote a treatise on the origin of petroleum which concluded that it had come from the depths of the Earth. He gave many arguments for this viewpoint most of which are still valid today... The geographical distribution of petroleum seemed to point to much larger scale features than the scale of individual sedimentary deposits. The quantities of oil and gas that geologists eventually found turned out to be hundreds of times larger than initial estimates, based on the assumption of biological origin.

"The viewpoint that hydrocarbons could not arise without biology became quite untenable when astronomers discovered that hydrocarbons are the most common form of carbon in the solar system. Jupiter, Saturn, Uranus and Neptune contain enormous amounts of methane and other hydrocarbons in their atmospheres... The comets and the asteroids contain various types of hydrocarbons. Recent investigations into the core of Halley's Comet showed that the core has a surface as *'black as pitch'—most probably because it is of pitch or a similar hydrocarbon.* It is most unlikely that biology created any of this material. [25] [emphasis added]

Professor Gold states that,

"...the greatest problem for the theories of biological origin was the association with helium. Many of the areas of the Earth that bear petroleum and methane are also rich in helium. Natural gas is the source of all commercial helium. Very few areas of the Earth have high concentrations of helium in underground gases without methane being the dominant partner. It is impossible to explain this relationship if the methane originated from biological materials buried in the sediments. Biology can have no part in the process of concentrating helium a chemically inert gas."[26]

Gold then goes on to discuss the problems of petroleum distribution in the Earth based on biological formation.

"The way in which hydrocarbons occur globally also provides problems for the theories of biological origin. Why is the Middle East so rich in hydrocarbons? The mountains of southeastern Turkey, the valley of the Tigris, the folded mountains of Persia, the Persian Gulf and the flat plains of Saudi Arabia have little in common with each other, except that they form one connected region that is enormously well supplied with oil and gas. No one has discovered a unifying feature for the region as a whole. The oil fields span different geological ages, have different rocks, called cap rock, holding down the contents of the reservoirs. Attempts to find sediments rich in biological debris have generally failed, and there is no consensus as to the source that produced such a wealth of oil and gas.

"In many other parts of the world, one can also recognize geographically, patterns in areas where oil and gas is found, spanning quite different geological settings."[27]

Gold is not alone among the scientists who oppose Sagan's view that oil is a product of decay of vegetation. V.B. Porfir'ev in the *American Association of Petroleum Geologists Bulletin* concludes that,

> "the organic theory of the origin of petroleum does not correspond to the modern state of knowledge in the fields of geology, geochemistry, geophysics, thermodynamics, astrophysics and other sciences; in fact, it is an outdated concept. Instead, the general inorganic theory of petroleum meets the requirement of the new knowledge completely..." Porfir'ev goes on to state that the "blind acceptance of organic theory (has) been a waste and a failure. This is witnessed in the summary work of Hedberg; full of disappointments at the indefinite results of many years of scientific work on this problem, he noted that not a single question is solved, nothing is understood and all contradictory hypotheses enjoy equal rights."[28]

Based on Sagan's statement that oil originates from dead plants of the Carbonaceous period, it would be impossible for deposits of oil to develop beneath the ocean far from land. The Atlantic Ocean, according to the currently accepted theory of plate tectonics was formed gradually beginning about sixty five million years ago, long after the Carbonaceous period. Furthermore, it was always an ocean and trees and great amounts of other vegetation do not grow at great depths on the ocean floor. The North Sea, a branch of the Atlantic Ocean, would therefore be a very improbable area from which one might expect to extract oil. It is covered by basalt rock and not sedimentary rock from which oil is ordinarily derived. Yet Thomas Gold in *Power From The Earth,* (London 1987), p. 128 tells us,

> "The early predictions were that the North Sea was a hopeless location for petroleum, and it is said one advisor [probably a scientist] told the British Government that he would drink every cup of oil that was obtained out of the North Sea... [However this region] proved most fruitful..."

How can oil develop under the sea many hundreds of miles from land if it never had immense vegetation growing?

Respecting oil, P.A. Dickey, et. al. in *Science,* Vol. 160 for May 10, 1968 discussed a problem. Taylor summarized it, thus,

> "When drilling for oil and gas, the drill passes through solid rock for thousands of feet, and well drillers have been accustomed to increasing pressure with depth at the rate of about a half pound per square inch per foot depth, so that at 10,000 feet the pressure is 5,000 [units] per square inch. Fairly massive equipment is required to handle these pressures, but occasionally a zone is encountered where the pressure more than doubles, causing difficulty and some danger in the drilling operation. In these circumstances, the drill passes from a zone of high pressure to an

adjacent zone of exceptionally high pressure, and structural geologists have wondered how it is possible for such great pressure differences to have existed side by side for [as Dickey writes in *Science*] 'scores of millions of years'... After all, the reported ages of the oil and gas by the C14 method were only a few thousand years, yet these results are usually dismissed, not for actual technical reasons, but because they do not meet the expectation of a much greater length of time... The high pressures in oil and gas wells are, rather evidence for a youthful age, indicating perhaps thousands of years—rather than millions of years—for rock units." [I.T. Taylor, *In the Minds of Men*, (Toronto 1984), pp. 336-337]

This implies that the Earth was subjected to a catastrophe in recent times and that the pressures in the Earth are still adjusting to the changes caused by this recent event.

Apparently Sagan has all the answers to problems experts in the field have found lead to dead ends based on the organic concept of oil. Velikovsky in 1950 claimed the Earth's petroleum is derived from hydrocarbons in comets. In 1962 A.T. Wilson in *Nature* claimed that all the oil of the Earth was of extraterrestrial origin.[29] In Sept. 16, 1966 J. Oro and J. Han in *Science* described how petroleum can be formed through the interactions of comets with planets. And in 1987 Sagan's Cornell University colleague, Thomas Gold, came to the very same conclusion as Velikovsky. A letter to the editor in *The New York Times* by Robert R. Gallow informs us that,

"It appears that Prof. Thomas Gold of Cornell University has discovered natural gas and oil beneath a meteorite crater in Sweden (news article, March 22). If this finding is confirmed, then vast amounts of hydro-carbons lie deeply hidden in the Earth's crust. This finding would have far reaching implications for energy related industries.

"According to Professor Gold's hypothesis, once the planets were forming, they generated enough gravity to alter the orbits of comets and asteroids. Many of these objects rich in hydrocarbons and other organic compounds struck the Earth. Therefore, natural gas and petroleum were derived from substances that fell from the sky. The conventional view of most scientists is that natural gas and petroleum originated from fossil remains of living organisms. However, the extraterrestrial source of hydrocarbons was suggested much earlier by Immanuel Velikovsky in 1950 in his book *Worlds in Collision*. Velikovsky argued that the Earth's petroleum deposits came from comets. The idea that petroleum came from space was ridiculed at the time. Now it is put forward by others in perfect seriousness..."[30]

Is Sagan unaware of this? In his book *Comet*, he writes, "if the Earth never outgassed at all, comets may still have brought an atmosphere, an ocean and *huge quantities of organic matter.*"[31] [emphasis added] to the Earth. Thus, according to Sagan, "petroleum arises from decaying vegetation," but "comets may still have brought…huge quantities of organic matter." This is double talk. In an article in *Science,* Vol. 249, p. 366 titled, "Commentary Delivery of Organic Molecules to the Earth", a few scientists attempted to assess how much of the Earth's organic materials were brought by comets. One of the scientists who co-authored this article is none other than Carl Sagan. Zdnek Kopal in *The Moon,* (NY 1960), p. 100 explains,

> "the solar system is known to be fairly infested with hydrocarbons-the atmospheres of the major [giant] planets contain tremendous amounts of them—and although oil on the Earth has so far been found under sedimentary deposits, it is perhaps because these rocks are easy to drill and because nobody has looked for it under igneous rock. At any rate should further astronomical work strengthen the conjecture that Lunar carbon may be due to the composition of hydrocarbons, the current theory of the origin of oil on Earth might be due for a drastic revision."

Thus the oil under the basalt rock cover of the North Sea is the *coup de grace* to any suggestion that oil only originates from vegetation.

MARS' POLAR CAPS

Sagan next turns to Mars and states,

> "Reading the text is made still more difficult by the apparent conclusion (page 366) of Martian polar caps made of manna, which are described ambiguously as 'probably in the nature of carbon.' Carbohydrates have a strong 3.5 micron infrared absorption feature, due to the stretching vibration of the carbon-hydrogen bond. No trace of this feature was observed in infrared spectra of the Martian polar caps taken by the *Mariner 6* and 7 spacecraft in 1969."[32]

There are two aspects of the Martian ice caps that exist—materials in it that sublimate (change from a solid state to a gaseous state without becoming liquid) and therefore, disappear in the summer. The other material or materials do not disappear because they require higher temperatures to either melt or sublimate. Sagan has already informed us that,

> "Astronomers tend to be nervous about the word organic—concerned that it might be misunderstood as a token of life on another world. But 'organic' only refers to molecules based on carbon. And organic chemical would be produced and destroyed even if there were no life anywhere in the universe."

Thus, in *Earthlike Planets,* we find "Capping all the units (in the polar regions) are the seasonal accumulations of frozen carbon dioxide and the permanent ice caps."[33] The author concludes that the permanent ice caps are also made up of "carbon dioxide." But, if carbon dioxide sublimates during the Martian summer, how is it that all of it does not sublimate; some form of white material remains. Sagan has assured us it cannot be hydrocarbons and definitely not carbohydrates. However, ten years after Sagan's attack, Harold Masursky of the U.S. Geological Survey in Flagstaff, Arizona, who was involved in analysis of data sent to Earth by spacecraft from Mars, has this to say in *Science Digest,* "The ice caps of Mars have seasonal ice, but then there is the permanent ice... We do not know if it is all water ice, or if there are other fancy constituents that are mixed up with it."[34] Thus, there is carbon, oxygen and hydrogen in the permanent ice caps—carbon from carbon dioxide, oxygen and hydrogen from water. Needless to say, these elements are the bases of both hydrocarbons and carbohydrates. We also know that ultraviolet rays reach the Martian surface and Dr. A.J. Swallow claims that hydrocarbons can be formed through the agency of high energy radiation which is what ultraviolet rays represent. Thus, hydrocarbons can certainly be formed. And isn't this the energy Sagan used in the glass reaction vessel to produce hydrocarbons. But hasn't Sagan informed us that from these constituents "organic molecules could be produced." Thus, the white material Velikovsky describes "probably in the nature of carbon" could most certainly be what Sagan calls "organic" which "refers to molecules based on carbon." According to V.A. Firsoff in *Life Beyond the Earth,* (NY 1961), p. 48 the constituents found in Mars polar ice caps—carbon dioxide and water—can be converted into carbohydrates. Firsoff states,

"Formaldehyde can be formed from carbon dioxide and water vapor without the intervention of chlorophyll or other organic molecules if the mixture is exposed to short ultraviolet rays, as it would be in an atmosphere devoid of oxygen, which in the triatomic form of ozone absorbs these radiations. Ultraviolet rays (in much the same way as chlorophyll) break up the water molecule into its constituent oxygen and hydrogen, and nascent hydrogen thus produced reacts with carbon dioxide to form formaldehyde, the simplest of carbohydrates. Formaldehyde polymerizes readily to sugars..."

Therefore, all the ingredients for the production of "manna" are available and there is a well known process employing ultraviolet radiation, which is known to penetrate the Martian atmosphere and reach the surface, capable of manufacturing it.

Apparently Sagan also tends to be nervous about the word "organic." In his book *The Cosmic Connection,* (NY 1973), p. 118, Sagan tells us this regarding the Martian polar ice caps, "...at the present time its composition is unsettled." The impression that Sagan seemed to give, was that the issue was indeed settled which it most certainly is not.

LIFE IN SPACE—ANYONE'S GUESS

In discussing the catastrophe, Velikovsky states on pp. 184-185 of *Worlds in Collision,*

> "When Venus sprang out of Jupiter as a comet and flew close to the earth, it became entangled in the embrace of the earth. The internal heat developed by the earth and the scorching gases of the comet were in themselves sufficient to make the vermin of the earth propagate at a very feverish rate, like the plague of the frogs ('the land brought forth frogs') or of the locusts must be ascribed to such causes. Anyone who has experience a khamsin, (sirocco), an electrically charged wind blowing from the desert, knows how during the few days that the wind blows, the ground around the villages begins to teem with vermin." On page 187 Velikovsky discusses extraterrestrial life. "Modern biologists toy with the idea that microorganisms arrive on the earth from interstellar spaces, carried by the pressure of light. Hence, the idea of the arrival of living organisms from interplanetary space is not new. *Whether there is truth in this supposition of laval contamination of the earth is anyone's guess.* [emphasis added]

These statements by Velikovsky make it quite clear that he believes that the various plagues described in *Exodus,* and in texts of other cultures, are produced by heat and electrical winds that cause vermin that originate here on Earth to propagate. Nor does Velikovsky know if life can travel through space to Earth on extraterrestrial bodies. Thus, to accuse him of believing that such events occurred, would be dishonest. Sagan states,

> "Even stranger are Velikovsky's views on extraterrestrial life. He believes that much of the 'vermin', and particularly the flies referred to in *Exodus,* really fell from his comet—although he hedges on the extraterrestrial origin of frogs while approvingly quoting the Iranian text, the *Bundahis* (page 183), which seems to admit a rain of cosmic frogs. Let us consider flies only. Shall we expect houseflies or *Drosophila melanogaster* in forth-coming explorations of the clouds Venus and Jupiter? He is quite explicit: 'Venus—and therefore, also Jupiter—is populated by vermin.' (page 369) Will Velikovsky's hypothesis fall if no flies are found?"[35]

Thus, although Velikovsky states directly that one cannot know if life came from outer space, Sagan emphatically attacks Velikovsky and omits the statement that damns his entire conclusion respecting this topic. Velikovsky states explicitly in *Worlds in Collision,* page 369,

> "Venus and Jupiter most possess an organic source of petroleum. On preceding pages, it was shown that there are some historical indications that Venus—and therefore, also Jupiter—is populated by vermin; this organic life can be the source of petroleum."

Nowhere in the literature of petroleum production is there evidence offered to suggest that insects can generate petroleum. However, it is well-known that bacteria can be involved with sources of petroleum. Thus, it seems clear that Velikovsky believes these life forms are bacteria and not insects.

Sir Fred Hoyle, in his book *The Intelligent Universe,* (NY 1983) p. 88 writes that,

> "Particles of bacterial size have been detected in the atmosphere of Venus, Jupiter and Saturn, observations which are not so easy to pass off as coincidental. The particles in the atmosphere of Venus have the same refractive index as biological spores, and those in the atmosphere of Jupiter have the refractive index of rod shaped bacteria, the agreements in both cases again being rather precise—within an accuracy of about half a percent, a figure that speaks against coincidence."

Thus, there is spectroscopic evidence that supports Velikovsky's contention that bacterial life forms may exist in Venus and Jupiter's atmospheres.

Sagan discusses his own conjectures of life on Jupiter in his co-authored book, *Intelligent Life in the Universe,* which says,

> "It is much more difficult to say anything about the possibility of the origin and the present existence of life on Jupiter. For example, we can imagine organisms in the form of ballast gas bags, floating from level to level in the Jovian atmosphere, and incorporating pre-formed organic matter much like the plankton-eating whales of the terrestrial oceans."[36]

Thus, it seems that Sagan believes life can possibly exist in the atmosphere of Jupiter and eat hydrocarbons for energy. How inter-esting, since he earlier implied Jupiter lacks hydrocarbons. Only Sagan conceives of large organisms while Velikovsky and Hoyle conceive of microscopic bacteria.

What is Sagan's view of life in the atmosphere of Venus? At the International Astronomical Union Symposium 40 held October 26-31, 1969, Sagan presented a paper titled "The Trouble With Venus", published in *Planetary Atmospheres,* by the I.A.U., edited by Carl Sagan, T.C. Owen and H.J. Smith (NY 1971), on pp. 125-126 Sagan states the following,

> ...[life], particularly "life based on familiar chemistries, seems implausible on the Venus surface... This leaves the clouds. Especially if the clouds are composed of condensed water—but even if they are not—life in the clouds is not by any means out of the question. There is water vapor, there is carbon dioxide, there is sunlight, and very likely there are small quantities of minerals stirred up from the surface. These are all the prerequisites necessary for photoautotrophs in the clouds. In addition, the conditions are approximately S.T.P. The only serious problem that

immediately comes to mind is the possibility that downdrafts will carry our hypothetical organisms down to the hot, deeper atmosphere and fry them faster than they reproduce. To circumvent this difficulty, and to show that organisms might exist in the Venus clouds based purely on terrestrial biochemical principles, Harold Morowitz and I (1967) devised a hypothetical Venus organism in the form of an isophycnic balloon, which filled itself with photosynthetic hydrogen and maintained at constant pressure level to avoid downdrafts. We calculated that, if the organism had a wall thickness comparable to the unit membrane thickness of terrestrial organisms, its minimum diameter would be a few centimeters... While it is not out of the questions that life exists in the Venus clouds it seems unlikely to have arisen there."

Thus, Sagan has big wind bags in the atmosphere of Jupiter and inch sized wind bags in the atmosphere of Venus. The thought occurs that if life could not have arisen in Venus' clouds, then the little wind bags of Venus are somehow descendants of the big wind bags of Jupiter. Instead of the possibility of vermin, Sagan offers us wind bags.

Could bacteria survive in comets? In his book *Comet,* Sagan discuses the work of Sir Fred Hoyle and N.C. Wickramasinghe who "hold that bacteria and viruses are sprinkled throughout interstellar space; indeed, they boldly propose that interstellar grains—which have about the same size and atomic composition as bacteria—are in fact, bacteria. If true, bacteria would have been incorporated into comets..."[37] Although Sagan shows numerous problems with Hoyle and Wichramasinghe's hypothesis, he tell us, "Just possibly, at some time in the future we will find the right sort of comet, and drill down to the still liquid ocean" to see if bacteria exist there. But, he concludes, "It does not seem likely, though, that anything would be alive down there."[38] Although Velikovsky's idea is still ridiculed by Sagan, it is also held by respected members of the scientific establishment. But they are not ridiculed for their views.

Could Venus have bacteria in its atmosphere? Again Sagan tell us in *The Comic Connection* that, "there are organic molecules—for example, some with a complex ring structure—that would be quite stable under the conditions of Venus."[39] Hence, if bacteria exist in the clouds of Jupiter, and survived in Venus when it was a comet, they may yet exist in the atmosphere of Venus based on Velikovsky's thesis and Sagan's analysis.

This becomes more interesting when we investigate Sagan's other views on life. Bruce Murray in *Journey Into Space,* (NY 1989), pp. 59-61, informs us,

"Sagan had become the chief spokesman for the possibility of extraterrestrial life everywhere, and especially on Mars. Carl's devotion to search for life sprang from his belief that it is the fundamental underlying question in the solar system exploration, not because it was necessarily likely. In public and in private, he elaborated tirelessly on speculative possibilities not *absolutely* excluded by...*Mariners 4, 6* and 7. Inevitably, Carl and I were drawn into scientific conflict. To me, the extraordinarily hostile environment revealed by the *Mariner* flybys made life there so unlikely that public expectations should not be raised. In one private moment of sharp debate, Carl charged, 'Bruce, you at Caltech live on the side of pessimism.' I thought to myself, 'And, Carl, you at Cornell live on the side of optimism.'"

Sagan it seems believes life once existed on Mars as Velikovsky had speculated. Murray tells us on page 65,

> "Perhaps during Mars' brief, early aqueous phase, the chemical means for life to develop did appear. Perhaps microbial life crept forth at that point and then progressively adapted to the much more hostile conditions of the present. Perhaps microbes still lived on Mars. At least this is what Sagan speculated as NASA prepared for the *Viking* mission of 1976."

Although the tests carried out on the Martian soil seemed to give no evidence of life, Murray tells us, "Carl [Sagan] and a handful of others still cling to the hope that life now exists on the surface of Mars de-spite *Viking's* negative results. Virtually all other space scientists dismiss such hopes as unrealistic..."

Sagan has ridiculed Velikovsky's speculations regarding life in the solar system. His own speculations, he seems to believe, are justified even if the face of evidence is negative to it. He believes that life can exist on the Martian surface where ultraviolet radiation would "fry" it rather quickly. He believes life can exist at temperatures below 100 degrees Fahrenheit, where there is extremely little water. And he seems to believe this today. Nearly all scientists dismiss his viewpoint and yet he maintains the possibility of it. Thus, the logic Sagan employs is that the speculations of Velikovsky about extraterrestrial life are unsupported although Velikovsky made it quite clear he does not know if the answer to his speculations are correct. Sagan, however, dogmatically maintains Mars must have life. The contradiction is so glaring that it speaks for itself. Interestingly, if, as Sagan seems to believe, life forms exist on Mars, what do they eat? If there are no organic substances there, they cannot exist. But if they exist, wouldn't these organic materials somehow get into the ice caps?

This contradiction was picked up by Michael Rowan-Robinson in a brief comment titled, "Rebel Without a Cause", in *Nature,* Vol. 276 (Nov. 9, 1978), pp. 150-151, who states,

> "Why should Sagan be allowed to speculate about life on Mars, extraterrestrial civilizations and 'galumping [sic] beasts' out there, and

then ridicule Velikovsky's notion of fly larvae in the tails of comets? More generally, why is it okay for Hoyle and Wickramasinghe to speculate about world-wide epidemics brought by cometary dust, and for Sagan to speculate about the evolution of human intelligence, but not okay for Velikovsky to speculate about astronomy?"

FINE TUNING THE EVIDENCE

Next, Sagan states that in *Exodus,* chapter 9,

"It is said the cattle of Egypt all died, but of the cattle of the Children of Israel there 'died not one.' In the same chapter, we find a plague that affects flax and barley but not wheat and rye. This fine tuned host-parasite specificity is very strange for cometary vermin."[40]

Sagan cites the *Bible* and then implies that this is what is written in Velikovsky's book. Although he gives a chapter as citation from the *Bible,* he strangely gives none from *Worlds in Collision.* Apparently Sagan believes it proper to cite *Worlds in Collision* and the *Bible* as if they were the same book. This is a straight-forward distortion of the evidence, though perhaps not for Sagan. Nevertheless, Velikovsky never claimed that the cattle or the flax and barley of the Hebrews was unaffected by the plagues. Sagan does not cite any evidence for this in *Worlds in Collision* because there is no such statement that he can cite. This is merely more of Sagan's disinformation.

Sagan, having misstated Velikovsky's view respecting life in space, then asks about "fly ablation"; that is, if Venus was incandescent it would surely have fried Velikovsky's flies. However, since Velikovsky clearly stated that whether there is life on bodies in the solar system is anyone's guess, Sagan's assertion is nothing more nor less than another of his balloons full of hot air. This is so because Sagan postulated the concept of tiny balloons surviving between the extremes of the hot surface and the freezing cold of space. Venus, on a cometary orbit, even if it were incandescent, would have an expanded atmosphere in which the extremes of temperature would still produce a region somewhere between these extremes which would be comfortable for life. If Sagan's hot air balloons can survive in just such a region, why couldn't Velikovsky's bacteria?

EVIDENCE FROM MARS

Sagan ends this section with: "Finally, there is a curious reference to intelligent extraterrestrial life in *Worlds in Collision.* On page 364."[41] By intelligent extraterrestrial life, Sagan implies advanced forms such as man that would have left traces of civilization on Mars. This is not what Velikovsky wrote. He wrote, p. 364,

"The contacts of Mars with other planets larger than itself and more powerful make it improbable that any *higher forms of life, if they previously existed there,* survived on Mars. It is rather a dead planet; every higher form of life, of whatever kind it might have been, most probably had its Last Day." [emphasis added]

The dinosaur is an example of a *higher* form of life, but one would hardly refer to it as *intelligent* in the sense Sagan implies. Sagan continues,

"But when we examine Mars as seen by *Mariner 9,* and *Viking 1* and *2* we find that a bit more than one-third of the planet has a modified cratered terrain somewhat reminiscent of the Moon and that it shows no sign of spectacular catastrophes other than ancient impacts. The other one-half to two-thirds of the planet shows fewer signs of such impacts, but instead displays dramatic signs of major tectonic activity, lava flows and volcanism several hundred million years ago. The small, but detectable amount of impact cratering on this terrain shows that it was made much longer than several thousand years ago. There is no way to reconcile this picture with a view of a planet recently so devastated by impact catastrophism."[42]

Clark R. Chapman in *The Inner Planets,* (NY 1977), p. 16 tells us that "In 1965, scientists and laymen alike were shocked when *Mariner 4* revealed a moonlike landscape on the planet Mars rather than the mountains and valleys to which we are accustomed."

Somehow Sagan seems to have forgotten the view his fellow scientists had of Mars prior to *Mariner 4*'s fly-by of the planet. Patrick Moore in his book *Armchair Astronomy,* (NY 1984), pp. 73–74 tells us that,

"In the summer of 1964, I was invited to give a lecture at Cambridge University. My subject was 'Mars.' It was well attended and I gave a general summary of what we knew, or thought we knew about the planet. I made a series of twelve profound statements, each of which was backed up by the best available scientific evidence—and everyone of which turned out to be wrong. "...Before *Mariner 4, we had been confident that Mars had a flattish or at most a gently undulating surface.*" [emphasis added]

Kenneth F. Weaver's "Voyage to the Planets", in *National Geographic,* for Aug. 1970, p. 169–173 describes Mars as Sagan's colleagues expected to find it. Professor Robert Leighton of the California Institute of Technology stated,

"When men first land on Mars—as they may actually do before the end of this century—they will find rather uninteresting terrain for the most part... Everything in the *Mariner* pictures indicates very gentle slopes on Mars. There are no mountain ranges, no great faults, no extensive volcanic fields, in fact, no evidence of volcanic activity. You could stand in a crater on Mars and never know it—even one that appears sharp and clear in the pictures."

Carl Sagan in an article titled, "Mars A New World To Explore", in *National Geographic*, for Dec. 1967 p. 828 also claims that Mars will appear to have a surface of softly undulating hills without sharp variations in elevation. He states "...when men arrive [on Mars], after an eight-month voyage through space, they will wander over a gently sloping landscape and by enormous numbers of eroded flat-bottomed craters." However, the landscape that the scientists saw was totally different than the scientist expected.Clark R. Chapman in *The Inner Planets*, (NY 1977), p. 147 states that the pictures made by *Mariner 4* were "shocking" to the scientists.

The expected description is in no way reflective of the Martian landscape. Mars shows all the signs of a recent violent history. P. Moore in *The New Guide to the Moon*, (NY 1976), p. 193 states, "The Martian scene proved to be utterly unlike what most people had expected. Instead of gentle, rolling plains, there were mountains, valleys, craters and volcanoes."

Velikovsky's view of Mars *Worlds in Collision*, written before 1950 said (pp. 364–365) that Mars, "...is rather a dead planet... The 'canals' on Mars appear to be a result of the play of geological forces that answered with rifts and cracks..." Tidal forces that tore its surface would leave Mars very much with its present appearance. This can be seen in all of the following.

Alan B. Binder and Donald W. McCarthy, Jr. in an article titled, "Mars: The Lineament Systems" in *Science*, Vol. 176 (1972), p. 279, which informs us that,

> "The photographs from the *Mariner 4, 6* and *7* flights were analyzed for linear features, such as polygonal crater walls, linear rilles, linear ridges, linear albedo boundaries and linear scarps. When these features are plotted, they demonstrate the existence of a well developed, planet-wide system of lineaments. This system of fractures might be the consequences of changes in the planets rate of rotation, polar wandering, or similar stresses."

What seems clear is that Mars was subjected to such sudden, strong orbital changes that the planet's surface fractured creating a system of lineaments. This is well in accord with Velikovsky's views of Mars, but not with Sagan's view.

Since the Earth and the Moon also had near collisions and changes in their orbital parameters, especially the Earth with respect to rota-tion, should there not also be lineaments running along longitudinal directions. According to Binder and McCarthy, the Earth and Moon exhibit the same preferential trends in their lineaments as does Mars.

Binder suggests these lineaments were caused by sudden changes in the rotation of these bodies. Since neither Mars' tiny moons nor the Sun could be responsible for Mars' lineaments, some other objects are required.

There is further proof that Mars' crust has been literally violently moved by some immense force. It is generally accepted by planetary scientists that Mars has not experienced a plate tectonic development similar to the Earth which supposedly permitted its surface to wander over the planet. Michael H. Carr in *Solar System,* (Scientific American Book 1975), p. 88 specifically states, "Martian tectonism differs in a very important aspect from tectonism on the Earth. The Earth's crust is divided into a series of plates that are slowly moving with respect to one another and constantly rearranging the Earth's geography. On Mars there is no sign of horizontal crustal motions. The lack of plate motion on Mars may explain why the planet's shield volcanoes are so gigantic." But in *Science News,* Vol. 119, (1981), p. 216, an article titled "The Poles of Old Mars" reports that, "Three spots on the surface of Mars, all of them within 15 degrees of the equator, show signs of once having been at the poles, according to Peter H. Schultz and A.B. Lutz–Garihan of the Lunar and Planetary Institute in Houston... *Viking* spacecraft photos from orbit, he (Schultz) says, shows the region to have carved valleys, like those in the present polar caps, 'pedestal craters' whose shapes suggest that they formed in now vanished ice and signs of laminated terrain reminiscent of the present caps familiar layering, which could indicate cyclic climatic changes." What Schultz and Lutz–Garihan suggest is that the whole Martian crust was redistributed by internal forces and that its poles moved. What is clearly indicated is that Mars' present equatorial region was once near the poles, but was then catastrophically reoriented either by huge meteoric impacts or a near encounter with a large body. But this evidence is clearly in full accord with Velikovsky's analysis.

In fact, to explain the catastrophic appearance of the Martian terrain, one geologist, Bill Beatty, believes Mars must have experienced a Velikovskian episode. Randolph R. Pozos in *The Face of Mars,* (Chicago 1986), p. 60 informs us that Beatty thinks "a Velikovsky type scenario such as a passing close encounter event [occurred to Mars] in the early days." Michael Carr, p. 69, "believes a sudden disaster occured on Mars..."

Nevertheless, is there evidence that shows that the features of Mars were produced by violent forces quite recently?

EROSION ON MARS

And, in fact, it is very easy to demonstrate that Mars has recently been devastated by a planet wide catastrophe. Let us once again assume that the scars and markings on Mars are ancient as Sagan has informed us. According to the scientific estimates, the craters in the southerly hemisphere of Mars are at least 3 billion years old and perhaps older. However, Mars has a small atmosphere which during it northern hemispheric winter season of 6 months often produces dust storms for 3 to 6 of those months over the entire planet. In fact, during one of those planet wide dust storms, *Mariner* was unable to photograph the surface of Mars because the dust obscured it. Dust storms weather (break down) rock and wind erodes (carried small debris) away from where it was weathered. Thus, one would wish to know what the erosion rates are on Mars. In *Aviation Week and Space Technology,* is an article titled "Mariner 9 Data Stir New Questions". We are told that,

> "Using *Mariner 9* wind data, Dr. Carl Sagan of Cornell University calculated erosion rates, assuming a dust storm peak wind of 100 mph blowing 10 percent of the time. This would mean erosion of 10 km (6.2 miles) of surface in 100 million years."[43]

Therefore, on the basis of Sagan's own figures, Mars currently should not have a single small surface feature surviving and few large ones and the entire planet should be a flat, sandy desert. The author adds, "there is no way to reconcile this picture with a view of a planet" having been eroded by dust storms with its present appearance. Many of the craters supposedly in the southerly hemisphere of Mars show sharp, crisp edges. This completely contradicts Sagan's erosional view of Mars. Using Sagan's erosion rate, in 300,000 years, some 100 feet of material would be eroded. This corresponds to the height of a 10 story building removed from the entire Martian surface.

Furthermore, Mars possesses what appears to be river beds. These are not eroded away even though they are often not very deep. But on the other hand, Mars has some of the largest, most spectacular volcanoes in the solar system. Using Sagan's calculations, they should have been eroded away in about 100 million years or so. In *Science,* Allen Hammond writes,

> "...According to Hal Mazursky of the U.S. Geological Survey Laboratory in Flagstaff, Arizona, eolian (wind) erosion is a dominant feature of Mars and is apparently so intense in some areas as to have completely eroded away pre-existing volcanoes. The edge of the largest volcano on Mars, *Nix Olympus,* is apparently being rapidly eaten away exposing a 1 to 2 kilometer cliff around its base."[44]

The absurdity of Sagan's argument is expressed in *Aviation Week and Space Technology* wherein we find,

"Crater counts in old cratered volcanic terrain indicate ages of 3 billion years. Mazursky said, so if this rate of erosion [calculated by Sagan] were extrapolated over an assumed Mars age, the same as the 4-4.5 billion years of existence of the Earth and Moon, Mars would now be about the size of Phobos, one of its two [tiny] moons, he said."[45] [Phobos is only a few miles in diameter.]

Even Sagan is thoroughly aware of this problem of erosion. In his co-authored book, *Intelligent Life in the Universe,* Sagan states,

"Since Mars is much closer to the asteroid belt than is the moon, it should be subject to many more impacts—perhaps 25 times as many. Yet the number of craters of a given size in a given area...must mean that processes exist on Mars which efficiently erode even large impact craters...we see a very large crater, over 100 kilometers across whose ramparts have been seriously breached. Major erosion of its walls has occurred."[46]

Sagan concludes, "Because of the efficiency of Martian crater erosion—regardless of the mechanism—the surface we see is not that of a very ancient Mars."[47] Thus, the "3 billion year old" craters should not exist or be extremely young, nor should river valleys exist, nor small, crisp features exist; but they do.

The problem is that no one seems to know how if water flowed as rivers on Mars it then disappeared. In the journal *Space World* for Feb. 1986, Ray Spangenberg and Diane Moser state on page 14 that, "like the Mona Lisa, the Red Planet is tantalizing us with mysteries. Perhaps the biggest—the story of water—keeps taunting experts with a jumble of apparent contradictions." There is also the question of when the water on Mars disappeared. On page 15 Spangenberg and Moser report:

"While just about everyone agrees that water-a lot of it-must once have flowed on the surface of Mars, controversy has raged long and furiously over the question of how recently it last flowed.

"Planetary geologist Hal Masursky of the U S Geological survey at Flagstaff, Arizona now believes he has evidence that effectively 'drives the nail in the coffin' of the idea that all Mars water channels are very ancient. He said so in a presentation at a recent meeting of the Geological Society of America in Orlando, Florida.

"For the past three years, Masursky has worked on detailed studies of high-resolution *Viking* photos—mapping selected areas [of Mars]...

"In just the past six months, Masursky says excitedly, 'we've found a whole different kind of geology that we hadn't seen before.' At high resolution he and his team were able to distinguish what Masursky calls 'geologic sandwiches'—channels cut by water and overlain by volcanic deposits. 'In other words,' he explains, 'the channel come out of the hills and flows out onto the plain; sitting on top of the channel and burying it are younger geologic-volcanic units.'

"But that's not all: 'We found channels that had formed, and then the lava flow that buried it and then another channel episode that cut a little tiny channel on top of that flow, and then another lava flow on top of that channel.'

"Masursky put the most recent of these in the last 200 million years-only yesterday in terms of planetary geology. 'We have a very interesting complex geologic story of lava flows, of faulting, of stream channel formation. And it's not ancient. It's not 2.4 billion years old. It's within a couple of hundred million years, and it may be younger than that.'"

Since Mars had water flowing on its surface for a few billion years, rainfall had to also exist to supply the ground water for runoff to feed the rivers and their dendritic systems that are observed on Mars. Water erosion on Mars, like that on Earth, would have removed nearly all the ancient craters long ago as it did on the Earth. Furthermore, because the Martian water and atmosphere were stable for billions of years, Mars could not have lost its water by currently understood uniformitarian processes. Only a recent catastrophe could be expected to dispose of Mar's water and atmosphere. With respect to the recentness of the catastrophe the evidence of erosion strongly suggests that Mars lost its water only a few thousand years ago.

Thus, how do the scientists deal with the fact the Mars' surface cannot be even millions of years old? Patrick Moore in *The Unfolding Universe*, (NY 1982), p. 78, tells us that, "...there is a paradox here. [In the channel-river systems] There is no sign of marked erosion even though the Martian atmosphere is dusty, and dust—even fine dust—is highly abrasive. The channels and the craters do not look as though they have been filled up, so that they can hardly be very ancient; tens of thousands of years perhaps, but not millions." What to do with such evidence? Apparently ignore it or say the erosion rate contrary to all evidence is minute.

Now it is quite apparent that with such a powerful erosion rate, the amount of sand created by the dust storms would be enormous. The sand would not, of course, have blown away into space, but have settled back on the surface of Mars. Based on Sagan's analysis, Mars should be today a sandy Sahara with nothing but sand dunes and a few volcanoes rising above the sandy desert floor as its only features. But this strangely is not the case. Why? There is very little sand on Mars compared to what one could expect based on Sagan's calculation alone. There is a huge sand ring region circling the northern hemisphere of Mars, but this is still insufficient for Sagan's analysis.

Are the rock materials of Mars harder than any known type of rock? If this were so, we would have an answer. However, Henry J. Moore, et. al. in the *Journal of Geophysical Research* analyzed this is an articled titled "Surface Materials of the Viking Landing Site" stating in the "Concluding Remarks" that, "Bulk densities of surface materials of Mars cover the range found in natural terrestrial materials" and "It is clear, however, that the surface materials are erodible and can be transported." [48] Furthermore, Michael H. Carr in *The Structure of Mars* informs us that it is near to impossible to reconcile the features of Mars with the efficient erosion by Martian dust storms. He states,

"That wind erodes the Martian surface...cannot be doubted. We have observed storms that stir up so much dust that most of the surface of the planet becomes hidden from view. We have monitored changes in surface markings from storms... The uncertainty regarding wind is not whether it has modified the surface but where and to what extent. Given the violence and frequent occurrence of dust storms, it is somewhat surprising the eolian [wind] effects are not more pervasive. An ancient cratered topography, probably over 3 billion years old, survives over much of the planet. Yet, it has been exposed to wind action for billions of years, and perhaps to hundreds of millions of dust storms comparable to the ones we observed in 1971 and 1977. Many of the plains, such as *Chryse Planitia* may also be billions of years old, but at a scale of 100m [meters] they do not look very different from the lunar maria, having crisp wrinkled ridges and well defined craters. Wind erosion appears to have been highly selective, being very obvious in some areas and negligible in others."[49]

To illustrate the highly improbable nature of Carr's statement, let us assume that the craters are ancient impact creations. Since the southerly hemisphere of Mars has most of the craters we would expect that here the erosion selectively decided not to wear away the evidence of cratering. The largest crater on Mars' southern hemisphere is Hellas and it is expected that the broad plane inside the crater would, over 4 billion years, be impacted many times by smaller meteors. Martin Caiden, in *Destination Mars,* (NY 1972), p. 119, informs us that,

"The portion of Hellas studied by *Mariner 6* and 7 covers at least 1,200 miles of surface, and to that resolution of 1,000 feet, not a single feature could be observed on the desert floor. Scientists were astonished by this fact for another reason—there is no way in which Hellas could have been sheltered from impact by meteorites. The only acceptable conclusion to explain the featureless surface, of course, is that some process is at work on Mars to erase fairly rapidly the effects of meteoroid impacts in the area of Hellas. Scientists admit that higher resolution cameras might well show small craters on the floor of Hellas, but that isn't the point. Large craters should be in evidence and they aren't. So the scientists are left with the most likely conclusion which is that the material making up the floor of Hellas, responds more rapidly than other Martian materials to whatever process of erosion exists on Mars."

Thus, the scientists have little or no erosion in the southerly hemisphere of Mars to save the supposedly ancient cratered terrain there. But inside Hellas in the southern hemisphere either high erosion is at work or the material there is super erodible. This is turning away from logic and evidence and substituting sophistry for them.

MARTIAN ASYMMETRY AND MASCONS

The problem for uniformitarian scientists is how to explain the fact that Mars is asymmetrically cratered according to Michael Zeilik's *Astronomy: The Evolving Universe,* (4th ed.) (NY 1985), p. 169 "extensive photographic survey [of Mars] showed that the two Martian hemispheres have different topographic characteristics: the southern hemisphere is relatively flat, older and heavily cratered: the northern hemisphere is younger, with extensive lava flows, collapsed depressions and huge volcanoes." The boundary between these two hemispheres, however, does not run parallel to the equator, but is inclined to it roughly by about 50 degrees.

It is proposed again and again that geological processes only affect the northern hemisphere but not the southern. What uniformitarian scientists ask be accepted is that some geological process acts exclusively on one hemisphere—the northern—dividing the planet into two distinctly different regions, one with a great many craters, the other with very few at all. What is the unique geological process that can account for this dichotomy? One suggestion offers the idea that volcanism removes craters in the northern hemisphere. The major problem with this concept is that the northern hemisphere is some 3 to 5 kilometers lower than the southern. Volcanism is, geologically speaking, one of the constructional forces of geology in that it builds up continents by adding material over the surface. J. Guest, et. al., in *Planetary Geology,* (NY 1979), p. 145 states that, "Why the northern of the two 'geologic hemispheres' should have apparently been depressed with respect to the other hemisphere is one of the greatest of the planet's mysteries."

This unique planetary asymmetry, which is in some respects quite like the bulge on the Moon, is precisely what one would expect to find according to Velikovsky's thesis. One side of the Moon exhibits several geological formations rarely seen on the other. So too on Mars this dichotomy is evident.

Interestingly Mars also has mass concentrations—mascons—like the Moon below its surface. Mark Washburn in *Mars at Last,* (NY 1977), p. 138 informs us that, "Mascons had been discovered on the moon, but they hadn't been anticipated on Mars." On page 151 he shows that "*Mariner 9* discovered the mascons that were responsible for Mars' lumpy gravitational field..."

Although it is difficult to explain the Moon's mascons it is even harder to explain those of Mars. Mars has volcanoes found on various points of its surface near the mascons, which means magma—molten rock—is found near them below the Martian surface. Victor R. Baker's *The Channels of Mars,* (Austin, TX 1982), p. 22, states that on Mars, "...the volcanoes certainly were active within the last few hundred million years, and it is possible that Olympus Mons is still active today. Crater counts on the volcanoes flanks and margins [are found which] indicates that volcanoes were active over an extremely long time span, certainly for hundreds of millions of years." These molten materials would destroy mascons rather quickly. Since Mars does possess mascons in spite of these hot subsurface phenomena it requires that the mascons are quite young. Again this evidence supports Velikovsky's hypothesis respecting Mars.

Michael H. Carr in *The Planets,* (1987), p. 96, tells us that around many craters is what can only be called "mud". "Particularly striking is the appearance of the material that was thrown out of the craters by the impacts. It appears to have a mudlike consistency and to have flowed across the surface after it fell back to the ground following ejection. The excavated material appears to have contained significant amounts of water." We are once again expected to accept the notion that mud was thrown out and over the Martian surface by the projectiles some 4 billion years ago, but was also not eroded away by dust storms. Thus, we have an intriguing kind of selectivity of erosion on Mars. Whole volcanoes were selected by the wind and dust and were eaten away, but material with a muddy consistency was not.

This brings us to a further contradiction of the evidence. That there has been erosion of Mars cannot be doubted. This is of special interest respecting "Pedestal Craters". These craters stand on elevated mounds that appear to be highly circular. The mounds rise to elevations of several hundred yards above the surrounding ground, but the craters themselves are a great many times too small to have ejected all the material that forms the pedestal mound. These formations are quite unique. Michael H. Carr, "The Surface of Mars: A Post Viking View", *Mercury,* Vol. 12, (Jan/Feb 1983), p. 2) states,

> "Modification of Martian craters also produces patterns not found on
> other planets, especially at high latitudes. Of particular interest are the so-

called pedestal craters. These craters occur at the center of a roughly circular platform that stands several tens of meters above the surroundings. In most cases the platform cannot be formed of ejecta from the crater since the volume of the platform far exceeds that of the craters bowl. Most planetary geologists now view the pedestal craters as indicating that thick blankets of debris used to lie on the surface. In the area between craters, the old debris has been removed, but the debris is retained around craters because of armory of the surface ejecta. The craters are thus left standing at the center of what remains of the old debris blanket. The pedestal craters are found mostly at high latitudes, both north and south, and are regarded as evidence of complex erosion and deposit histories at these latitudes…"

Here then is the further contradiction. Both in the high north and south latitudes and even 15 degrees from the equator several hundred yards of surface has been removed by erosion since the times of last cratering, but somehow mud thrown out by cratering impacts has not been removed. Finally, one is left to ask: Since pedestal craters imply an enormous amount of erosion of the terrain where is all of the eroded material deposited? These prodigious amounts of material had to be deposited somewhere, but its location is undiscovered. One wonders if it is not the scientists who are being selective and not the wind and dust.

A good place to deposit some of this material is the great rift valley, the *Valles Marineris,* a system of canyons extended over 3,000 miles along the Martian equatorial belt. In places the canyon complex is 300 miles wide and 4 miles deep. Its walls are precipitously steep with well defined edges. While dust and debris should easily accumulate in the chasm over billions of years, it too is subject to erosion, and land slides are observed on its flanks. In the *McGraw-Hill Encyclopedia of Science and Technology,* (1987), Vol. 10, "Mars" p. 477, we are told, "While erosion continues to eat away at the canyon walls, that is an unlikely origin for *Valles Marineris* and similar complexes. These enormous chasms have no outlets, and the only obvious way to remove debris is by wind; yet the material to be transported out of the canyons is so great as to cast serious doubt on the effectiveness of this mechanism operating by itself." Thus, there is no explanation of why the canyons are not filled. Yet, over such enormous periods of time, they ought to have been filled. Again, perhaps the dust was highly selective as to where it should settle.

A more honest analysis would state that space scientists do not know how to reconcile powerful erosion with well defined crisp features. The concept that the dust storms were selective about what features to erode will not wash. One cannot have it both ways—dust storms that cover an entire planetary surface, but only erode selectively are unimaginable.

MARTIAN BOULDERS

Finally, dust or sand abrasion of rock produces a smooth often shiny surface on the rock. In the geology text, *The Earth's Dynamic Systems*, by W. Kenneth Hamblin is a description of wind abrasion of desert rock which states,

> "The effects of wind abrasion can be seen on the surface of bedrock in most desert regions, and in some areas where soft, poorly consolidated rock is exposed, wind erosion can be both spectacular and distinctive. The process of wind abrasion is essentially the same as that used in artificial sand blasting to clean building stone. Some pebbles called ventifacts are shaped and polished by the wind... Such pebbles are distinctive in that they have two or more flat faces that meet at sharp ridges. Generally, they are well polished and can have surface irregularities and grooving aligned with the wind direction."[50]

However, when we look at pictures of the rocks on the surface of Mars taken by *Viking 2*, we do not see wind abraded rock. Michael Zeilik in his book *Astronomy: The Evolving Universe*, states,

> "*Viking* landers touched down on Mars in 1969. Seen close up, the Martian surface is bleak and dry. Large rock boulders are basaltic. Some contain small holes...from which gas has apparently escaped; the holes make the rock look spongy."[51]

The rocks shown by the pictures do not look anything like rocks that have been eroded and polished by wind abrasion over billions of years. Mark Washburn in *Mars at Last*, (NY 1977), p. 237, states that pictures of the boulders on Mars, "Contrary to expectations, [show] that there were very few rocks that had been ventifacted to any significant degree. There were wind tails of dust behind many of the rocks, but it did not look like an area where massive sand blasting was occurring."

In *Sky and Telescope* is an analysis in which we find laboratory simulation of Martian conditions. "R. Greeley and his colleagues have used a wind tunnel at NASA's Ames Research Center to simulate erosive conditions on Mars. Atmospheric pressure on Mars is only a few thousandths of that at sea level on Earth, so Martian winds must possess very high velocities to transport dust particles. These high wind speeds should produce a highly erosive environment. Based on these factors, and on wind patterns derived from *Viking* lander data, the researchers calculated that Mars should be eroded at rates of up to 2 centimeters [0.8 inch] per century. But if this were the case, they note the craters visible at the *Viking* sites (which are hundreds of millions of years old) should have been worn away long ago." The researcher's erosion rate was 0.8 inches per century. In five centuries that equals 4 inches; in 1,500 years, a foot; in 15,000 years, ten feet; in 150,000 years, one hundred feet of erosion of Mars' surface, twice the rate calculated by Sagan. One hundred feet is equal in height to a ten story building.

If this much surface material were eroded on Mars, the planet would possess no crisp features; they would all be eroded away or rounded off. All small features, such as small craters and shallow river valleys with their tear drop islands would disappear. These islands are made up of soil and sand, and such materials are easily eroded. All rocks would be abraded down to smooth surfaces. The planet would be a sandy desert with only large features rising above the desert sand dunes. The actual amount of erosion on Mars is about a few feet; thus, its features can only be a few thousand years old.

In fact, there is evidence for extreme erosional effects of the Martian surface. In *Discover,* for November 1987, p. 10 and 14, is a discussion of tornadoes on Mars which states,

"If astronauts ever make it to Mars, they may want to avoid summer in the southern hemisphere. To begin with, that's the season when huge storms tend to blanket the Red Planet with a nasty yellowish dust. And in the late summer, when the dust storms die down, conditions don't improve much; according to a new study, that's when tornadoes scour the Martian surface.

"Tornadoes themselves have not been seen on Mars, but planetary scientists John Grant and Peter Schultz of Brown University think they have seen their tracks: dark lines, less than six tenths of a mile wide, but as long as 46 miles, that cut across hills, faults, and craters. The lines are visible on images made by the *Viking* orbiters between 1976 and 1980. Other investigators had noticed them before, but decided they had to be either sand dunes or thin cracks in the Martian crust. Grant and Schultz think both explanations are unlikely, in part because the lines are so indifferent to topography, but also because they appear in late summer and then disappear.

"Dust devils, less intense whirlwinds that are a common sight here on Earth, have been seen on Mars, but they are not likely in late summer. The conditions then, say Grant and Schultz, are just right for tornadoes: There is a deep layer of warm air above the sun-baked surface, and cold fronts much like the ones on Earth, pass at regular intervals. A wedge shaped front nudges the warm air upward, which is where it wants to go anyway; the strong frontal winds set the updraft a-spinning; and a first class twister ensues.

"So far the tracks have been seen chiefly in a few southern areas, but the two investigators think that's only because the light, sandy surface in those areas makes the dark lines easy to spot; tornadoes are probably just as common in the north. Actually, "common" is an understatement: Grant and Schultz counted between 55 and 60 tracks, formed in a single season, in a typical 400 square mile area—an area smaller than Oklahoma City.

"Visitors to Mars in tornado season won't have to worry about being swept up by dramatic black funnel clouds—there's almost no water vapor in the Martian atmosphere, and so tornadoes would be dry. But they might well encounter winds of 100 to 200 miles an hour. And, says Grant, they'd get very intensely sandblasted."

This type of erosion over billions of years would have hastened the removal of all the ancient features long, long ago.

A MARTIAN FLOOD

Thus, one may properly ask, is there other empirical, or observed evidence to explain the features of Mars? Michael H. Carr's article titled "The Red Planet" in the book *The Planets,* discusses river channels that extend northward (in the southern hemisphere) for over 1,000 kilometers.

"In places the channels are 150 kilometers across and contain numerous teardrop-shaped islands. Geologists have puzzled over the origin of these channels since they were discovered in 1972. Originally there was considerable resistance to accepting the idea that they were formed by water erosion because the Martian surface is so dry; but now most geologists believe that the valleys were formed by floods of water...

"In fact, the channels most closely resemble large terrestrial flood features than typical river valleys. One of the largest floods known on Earth is one crossing the eastern half of the state of Washington [believed to have been produced] 10,000 or 20,000 years ago. The flood is thought to have resulted from rapid emptying of a lake that covered much of western Montana. The lake was dammed by ice, and the flood followed when the ice collapsed. The released water swept over eastern Washington, eroding deep channels, scouring wide swaths of ground, and carving teardrop-shaped islands like those on Mars. Discharge of about one hundred times that of the Amazon River persisted for several days. *The resemblance between the large Martian channels and the eastern Washington flood is so close that few geologists doubt that both were formed in a similar manner.* One difference is that the Martian floods were larger."[52] [emphasis added]

The scablands of Washington (discussed in Sagan's fourth problem) and the Martian flood features resemble each other very closely. But why do both, given the differing erosion rates on the Earth and Mars, still look so very much alike? The teardrop shaped islands on Mars are islands surrounded by sand and soil; these materials are very easily eroded compared to rock. Because it seems quite apparent that the erosion rate on Mars is greater than that of the Earth, all these features would have been completely decayed or removed by wind erosion before the crisp edges of the Martian craters, but they are not (all) highly decayed or removed. Two such similar sets of geological features can only resemble each other in areas where erosion rates differ so markedly only if the one area (Mars) where erosion is greater, is younger than the other area (Washington) where erosion is smaller. It is almost certain that for the Martian flood regions and Washington scabland features to resemble each other so closely, based on the geological evidence, the Martian features must be somewhat younger than those of Washington. Since the flooding event that produced the Washington scablands may have happened about 10,000 years ago (or even earlier), it is evident that the Martian event occurred less than 10,000 years ago or even earlier.

However, for water to flow on the Martian surface, an atmosphere heavy enough to hold the water at the surface must exist. Michael Zeilik tells us "Since Mars does not have liquid surface water now, conditions for it must have occurred in the past and would have required a warmer climate and a denser atmosphere."[53] G. O. Abell informs us that "Geophysicists have pointed out that liquid water can simply not survive in the low pressure atmosphere Mars has today. Various experts estimate that the atmospheric pressure would have to be from 5 to 50 times as great as at present to allow liquid water to flow over the surface of the planet and not immediately evaporate."[54] Thus, it follows from all the evidence that less than 10,000 years ago Mars had a much denser atmosphere. This brings us then to the question, how can an atmosphere and all the water disappear in less than 10,000 years? The answer suggested itself is that Mars had a very near collision with a large celestial body recently and that it flooded and burned away most of the atmosphere and water. Thus, the lunar-like cratered surface in the southern hemisphere of Mars was produced by tidal action and impact. The great lava planes over the rest of the planet were produced by cracks in the crust that poured out copious layers of lava.

TARSIS

The great Tarsis bulge—a bump on Mars 6 1/2 miles high rising above the surface and 3,000 miles in diameter—was probably produced by a collision of a massive meteorite of asteroid size striking almost directly opposite, on the opposing hemisphere where the *Hellas Planitia* crater region (a crater nearly 1,200 miles in diameter) is located. This upward thrust of the Tarsis bulge, which is an anomalous planetary feature according to Carr, opened a rift in the area adjacent forming the 4,000 kilometer long crack in the surface called the *Valles*

Marineris. In short, these features of Mars are of a recent catastrophic origin because they would have been filled in if they were ancient or if they opened gradually.

In fact, Sagan, in his book, *The Cosmic Connection,* (NY 1973), p. 127, describes the Tarsis bulge as mapped by *Mariner 9* as giving the appearance of a huge recent catastrophe. He states,

> "The resulting geological maps reveal an enormous array of linear ridges and grooves that surround the Tarsis Plateau—as if a third or a quarter of the whole surface of Mars were cracked in some colossal recent event that lifted Tarsis. The most spectacular of these quasi-linear features is an enormous rift valley... It [the crack] runs 80 degrees of Martian longitude..."

Running outward from the Tarsis bulge are several great cracks in the Martian surface. The appearance of Tarsis and these great fissures in Mars' surface should also have disappeared long ago by erosion and filling in of weathered material. Thus, the recent aspect to which Sagan alludes implies a very recent catastrophe.

Furthermore, from where did these enormous cracks in the Martian crust come? Jerry Pournelle in *Closeup: New Worlds,* (NY 1977), p. 51 informs us that, "There are dozens of such troughs [cracks] in the equatorial region. They range up to 200 kilometers wide and several hundred kilometers long...[but] there is no generally accepted theory of what caused these astonishing features." On page 52 Bruce Murray, Professor of the Planetary Science Department at the California Institute of Technology, informs us that, "there is *no* theory that adequately explains what is known about them." [i.e., the great Martian cracks] Nevertheless, a planet that recently experienced an enormous celestial catastrophe would be expected as Velikovsky stated, "The canals on Mars appear to be the result of the play of geological forces that answered with rifts and cracks." which Mars undoubtedly possesses. And his theory fully explains not only these unique features but all the enigmatic features of Mars. On page 65, Murray states he is certain of only one thing,

> "we do not understand the geological history of Mars. ...There are huge volumes of material that we can't account for... This isn't just disagreement among competing theories. There are no theories, none at all, that can explain all the existing facts. Something important has happened here that we just don't understand."

HOW TO DEAL WITH MARTIAN EROSION

Sagan has told us that the "cratered terrain [of Mars] is somewhat reminiscent of the Moon and displays no sign of spectacular catastrophes other than *ancient impacts.*'[55] [emphasis added] This, in *Scientists*

Confront Velikovsky and *Broca's Brain*. But, in *Intelligent Life in the Universe*, (pp. 287–289) he states, "the surface we see is not that of a very ancient Mars." Thus, employing Sagan's brand of logic, "ancient impacts" are "not that of a very ancient Mars." In *The Solar System*, Sagan writes,

> "Observations made from *Mariner 9* imply that the winds on Mars, at least occasionally exceed half the local speed of sound. Are the winds ever much stronger? And if they are, what is the nature of a transonic meteorology?...
>
> "The rate of sand blasting by wind-transported grains on Mars is perhaps 10,000 times greater than the rate on the earth because of the greater speeds necessary to move particles in the thinner Martian atmosphere." [56]

Using the same logic that ancient impacts are not very ancient, Sagan has an erosion rate that requires that Mars resemble the sandy Sahara Desert, but is blind to the actual appearance of the planet itself. For Sagan and the other gradualist scientists to reconcile this contradiction to gradualism, they ignore the contradiction altogether. The idea of catastrophism is so bitter a concept that it cannot be faced. In *Mars and the Mind of Man* (NY 1973), p. 97 Sagan states, "so far as I know, none of the previous accounts of Mars...has had a word to say about the pervasive [erosion] aspect of the Martian environment." Thus, like the proverbial ostrich, Sagan deals with the evidence by burying his head in the sands—of Mars.

TIDAL FORCES ON SMALL BODIES

Sagan's last statement on this problem is, "It is also by no means clear why, if all life on Mars were to be exterminated in such an encounter, all life on Earth was not similarly exterminated." [57] Sagan had the answer in part when he wrote, "On page 364, Velikovsky argues that the near collisions of Mars with Earth and Venus 'make it highly improbable that any higher forms of life, if they previously existed there, survived on Mars.'" [58] What Sagan strangely omitted from his citation of *Worlds in Collision*, p. 364 is the part of the sentence that reads, "The contact of Mars with other *planets larger than itself* make it highly improbable." [emphasis added]

Isaac Asimov in his book *The Collapsing Universe*, explains tidal theory, thus,

> "The strength of the gravitational attraction between two particular objects of given mass depends on the distance between their centers. For instance, when you are standing on Earth's surface, the strength of Earth's gravitational pull on you depends on your distance from Earth's center.

"Not all of you, however is at the same distance from Earth's center. Your feet are nearly two meters closer to the Earth's center than your head is. That means that your feet are more strongly attracted to the Earth than your head is because gravitational attraction increases as distance decreases. This difference in the gravitational attraction between two ends of an object is the tidal effect...

"The Moon pulls on the Earth, and it pulls more strongly on the side of the Earth nearest itself than on the part farthest from itself... *The Moon is a smaller body than the Earth is and produces a smaller gravitational pull altogether, and that makes for a decrease in the tidal effect.*"[59] [emphasis added]

Therefore, according to Asimov, the tidal forces of a large planet on a smaller one would devastate the smaller one much more severely than the more massive one during a near collision. Mars is about 11 percent the mass of the Earth and about 13 percent the mass of Venus. Thus, during near collision, Mars should be much more devastated than the Earth or Venus. The same applies to Venus' tidal destruction of the Moon. As an astronomer Sagan should be well aware of the nature of tidal effects. Sagan stated in *Broca's Brain,* p. 78 and *Scientists Confront Velikovsky,* p. 113 that Velikovsky, "believes that Mars being a relatively small planet, was more severely affected in its encounters with more massive Venus and Earth..." If Isaac Asimov believes small bodies are more severely affected by tidal forces from larger bodies, why shouldn't Velikovsky?

PHILOSOPHY AS EVIDENCE

C.R. Chapman and K.L. Jones in *The Annual Review of Earth and Planetary Science,* Vol. 5, (Palo Alto, CA. 1977), p. 529, state, "There is a tendency for planetary geologists to avoid 'catastrophist' interpretations of planetary histories..." This is not very difficult to understand.

Although philosophy is not science, it is certainly used as though it is. We wish to show how uniformitarian scientists use this philosophy as scientific proof. In this respect, the way the ages of the craters were determined gives ample evidence of how the uniformitarian scientists have operated. Clark R. Chapman's, *The Inner Planets,* (NY 1977), pp. 37-39 illustrates this.

"The creation of scientific hypotheses involves a complex interplay of the data with scientists' intelligence, special expertise, and assumptions. When syntheses involving different fields are required—say celestial mechanics and geology—then scientists depend upon mutual communication and faith in each other's expertise. Philosophical assumptions and faith in the judgment of others often become pre-eminent when the questions to be resolved are complex, data are recent and only superficially examined, and implications of the answers are fundamental.

"A case in point is the recent development of an approximate 'planetary cratering chronology' from the hypothesized cataclysm to the present. In the absence of datable rocks from Mars and Mercury there is only one way to establish an 'interplanetary correlation of geologic time' that connects the relative sequence of geology on these planets to absolute dates known only for the Earth and moon: to understand how the quantities and orbits of the asteroids, comets, and other impactors have evolved since the hypothesized cataclysm. If we were to know, for instance, that there have been many more asteroids crashing onto Mars than onto the moon, then the fact that Mars has similar numbers of craters would imply it is a younger, geologically active planet. Unfortunately, we don't know the cratering rates from planet to planet to within a factor of 10; that is, Mars may be struck by ten times as many bodies as Earth, or perhaps just the same number [or less].

"Still, Murray, Soderblom and Wetherill have tried to fashion a rough chronology for Mercury and Mars in this way: first, Larry Soderblom compared the crater densities on the moon to those on Mars and wrote, 'Because the oldest, post accretional [flat], surfaces on Mars and the moon display about the same crater density, it now appears that the impact fluxes at Mars and the moon have been roughly the same over the last 4 billion years.'

"And, if the impact rates have been the same, then any province on Mars cratered similarly to a province of known age on the moon has a similar age. This analogy seems simple and straightforward yet, it *arbitrarily excludes the possibility of differing fluxes and differing chronologies.* [emphasis added] [To resolve this dilemma]

"Bruce Murray carried the uniformitarian analogy over to Mercury. He observed from *Mariner 10* pictures that 'the light cratering on the flooded plains of Mercury is similar to that on the maria [plains] of the moon.' He drew an analogy like Soderblom's supported by the fact that 'similar flux histories for Mars and the moon were independently hypothesized by Soderblom.' But Bruce Murray, who has been chief of more planetary-imagery spacecraft experiments than anyone else, realized he further needed to know the relative number of impacts on different planets. So he turned to an article by George Wetherill, a man who developed his reputation in geochemistry and geochronology but who has evolved into an expert in celestial dynamics. Murray noted that 'relatively uniform impact flux histories throughout the inner solar system for the last 3 to 4 billion years were inferred recently by Wetherill, who concluded that the impacting objects probably originated in [highly elongated] orbits' that would cross the orbits of all the planets. Thus, Murray concluded that a 'straightforward interpretation' supports a similarity in Martian, lunar and Mercurian impact fluxes.' [This is circular reasoning.]

"Murray's conclusion is perhaps less secure than he seemed to realize. First, Wetherill studied the orbits of many different populations of bodies that could have cratered the planets, none of which had identical impact rates on each of the terrestrial planets. But Wetherill had heard that 'recent observations of Mars and Mercury...have suggested to several workers [Soderblom and Murray] the hypothesis that...all of the terrestrial planets have had essentially the same bombardment history.' Wetherill particularly wanted to emphasize that, among all the kinds of bodies he studied, 'the only bodies which [produce] a near equality of flux on the moon and all the terrestrial planets are those derived from the vicinity of Uranus and Neptune.' He went on to conclude that they were responsible for many of the craters, thus, supporting the 'independent' interpretations of Murray and Soderblom. Thus, a plausible, but uncertain hypothesis seems to be independently confirmed, when *in fact the scientists are relying on each other's 'proofs' of equality* in planetary cratering. In the end, uniformitarianism reigns" [emphasis added].

We cite the above to show that not only is there "a tendency for planetary geologists to avoid 'catastrophist' interpretations of planetary histories," but to illustrate that their analyses are only straightforward interpretations of a scientific philosophy which is as dogmatic as any religion. The process of erosion on Mars flies in the face of the concept that the craters on Mars are billions of years old. This erosion phenomenon has been amply discussed in the scientific literature and denies all the crater analysis of the planetary scientists. While the scientists rely on each other's proofs of equality, they completely ignore evidence that invalidates their findings. Yes, uniformitarianism reigns, but only by a dogmatic allegiance to a theory and not to evidence. It is evidence that is fundamental to any science. Why should such a state of affairs exists? The scientists cannot deal with the facts that go against their religion and support a differing view, which to them can only be inaccurate science, or pseudoscience. Another viewpoint on the evidence is so unthinkable that the scientists act as religious zealots acted in the Middle Ages and deny or avoid the damned evidence. In discussing how scientists adhere to uniformitarian philosophy, Chapman states, p. 29,

> "After all, the whole super structure of their science was built on uniformitarian assumptions; to ask a geologist to question them would be like asking a priest to be skeptical of God..."

It appears that uniformitarianism is the new god of the scientific establishment to which Sagan belongs.

NOTES & REFERENCES

[1]SCV, p. 68; B.B. p. 104.
[2]Dott, Robert H., Batten R.L.; *Evolution of the Earth*, op. cit., p. 98.
[3]Rifkin, Jeremy; *Algeny* (NY 1983), pp. 149-150.
[4]B.B., Chap. 12, "Life in the Solar System", pp. 178-179.
[5]*Algeny* op. cit., p. 150.

[6] Schapiro, Robert, *Origins*, (NY 1986), pp. 173-174.

[7] SCV, p. 68; B.B., p. 104.

[8] SCV, p. 68; B.B., p. 104.

[9] *Pensée*, Vol. IV, No.1, "Yet Another Chapter", p. 57.

[10] *Encyclopedia Britannica*, Vol. 13, (1972), p. 142.

[11] Sagan and Druyan, *Comet*, op. cit., p. 153.

[12] Murray B., Malin, M.C., Greeley, R.; *Earthlike Planets* (San Francisco 1981), p. 10.

[13] *Astrophysics*, ed. Hynek, (1951), p. 342.

[14] Kuong, W.K.; "The Synthesis of Manna", *Pensée*, Vol. III, No. 1, (Winter 1973), p. 45ff.

[15] Fort, Charles; *Complete Works of Charles Fort*, (NY 1974), "Lo", pp. 553-555.

[16] Ibid., pp. 62-63.

[17] Ibid., p. 63.

[18] Ibid., p. 64.

[19] SCV, p. 69; B.B., p. 105.

[20] Ibid.

[21] *Earth in Upheaval*, op. cit., p. 240.

[22] SCV, p. 69; B.B., p. 105.

[23] Smith, P.V., *Science*, (Oct. 24, 1952), pp. 437-439.

[24] Murray, G.E.; *American Association of Petroleum Geologists Bulletin.* 49 (1965): 3-24.

[25] Gold, Thomas; *New Scientist*, (June 26, 1986), p. 42.

[26] Ibid., p. 45.

[27] Ibid.

[28] Porfirev, V.B.; *American Association of Petroleum Geologists Bulletin* 58 (1974): 3-33.

[29] Wilson, A.T.; *Nature*, (Oct 6, 1962), pp. 11-13.

[30] Gallo, Robert; *The New York Times*, (March 16, 1987), "Letters to the Editor".

[31] Sagan and Druyan, *Comet*, op. cit., p. 319.

[32] SCV, p. 68; B.B., p. 104.

[33] *Earthlike Planets*, op. cit., p. 110.

[34] Masursky, H.; "Taking a New Tour of Mars", *Science Digest*, (Feb. 1984), p. 87.

[35] SCV, p. 69; B.B., p. 105.

[36] *Intelligent Life in the Universe*, op. cit., p. 329.

[37] Sagan and Druyan, *Comet*, op. cit., p. 311.

[38] Ibid., p. 316.

[39] Sagan, C.; *The Cosmic Connection*, (NY), p. 92.

[40] SCV, p. 70; B.B., p. 106.

[41] SCV, p. 71; B.B., p. 106.

[42] SCV, p. 71; B.B., p. 106.

[43] *Aviation Week and Space Technology*, (Jan. 29, 1973), p. 61.

44 Hammond, Allen; "New Mars Vulcanism, Water and a Debate its History", *Science*, (Feb. 2, 1973), p. 463.

45 *Aviation Week and Space Technology*, (Jan. 29, 1973), p. 61ff.

46 *Intelligent Life in the Universe*, op. cit., pp. 287–289.

47 Ibid.

48 Moore, Henry J.; "Surface Materials of Viking Landing Sites", *Journal of Geophysical Research*, Vol. 82, No. 28, (Sept. 30, 1977), pp. 4521ff.

49 Carr, Michael H.; *The Surface of Mars*, (New Haven 1981), Chap. 11, "Wind".

50 Hamblin, Kenneth W.; *The Earth's Dynamic Systems*, 2 ed., (Minneapolis 1975), p. 268.

51 Zeilik, M.; *Astronomy*, 4 ed., op. cit., p. 170.

52 Carr, Michael H.; "The Red Planet" in *The Planets*, Byron Preiss ed., (NY 1985), pp. 100–101.

53 Zeilik, M.; *Astronomy*, 4 ed., op. cit., pp. 171–172.

54 Abell, George O.; *Realm of the Universe*, 3 ed., pp. 181–182.

55 SCV, p. 71; B.B., p. 107.

56 Sagan, Carl, in *The Solar System*, (Scientic American Books) (San Francisco 1975), p. 9.

57 SCV, p. 71; B.B., p. 106.

58 SCV, p. 71; B.B., p. 106.

59 Asimov, Isaac; *The Collapsing Universe*, (NY 1977), pp. 167–169.

SAGAN'S SIXTH PROBLEM

MANNA

One may ask, why deal with manna since we have done so in the last chapter? Sagan has other fish to fry here and we shall make sure they are properly cooked.

A TASTE OF CYANIDE

Sagan admits that there are hydrocarbons in the "tails of comets," but that "it is now known that comets contain large quantities of simple nitriles—in particular, hydrogen cyanide and methyl cyanide. These are poisons, and it is not immediately obvious that comets are good to eat." [1] Cyanogen is indeed poisonous. Even a grain of its potassium salt touched to the tongue can bring death quickly. However, on the other side, "Hydrogen cyanide is a critical block for the more complex molecules of life." [2] How poisonous are comets?

Sagan deals with cyanide in comets in his book *Comet* stating,

"The molecular fragment CN, a carbon atom attached to a nitrogen atom had...been detected in the comas and tails of many comets, and then confirmed in Halley's Comet as well. It was called cyanogen. But when chemically combined in a salt it has another name—cyanide. Since only a grain of potassium cyanide touched to the tongue is sufficient to kill an adult human, the idea of the Earth flying through a cloud of cyanide generated a certain apprehension. [regarding Halley's Comet's 1910 passage by the Earth] People imagined themselves choking, gasping, and dying, millions asphyxiated by the poison gas.

'The global pandemonium about poison gas in the tail of Halley's Comet was sadly fueled by a few astronomers who should have known better. Camille Flammarion, a widely known popularizer of astronomy, raised the possibility that 'the cyanogen gas would impregnate the atmosphere [of the Earth] and possibly snuff out all life on the planet.' ...it is not surprising that statements like Flammarion's helped to produce a worldwide comet frenzy.

"In fact, it was not even clear that the Earth would pass within the tail of Halley's Comet. In any case, the tails of comets are extraordinarily thin, a wisp of smoke in a vacuum. *The cyanogen is in turn a minor constituent in the tails of comets.* Even if the Earth *had* passed through the tail in 1910 and the molecules in the tail had been thoroughly mixed down to the surface of the Earth, there would have been only one molecule of cyanogen in every trillion molecules of air—a good deal less than the pollution caused even far from cities by industrial and automobile exhaust (and much less than what would happen in the burning of cities in a nuclear war)..."[3] [emphasis added]

Therefore, even if we increased the amount of material in the tail of a passing comet by 100 fold, there would only be one molecule of cyanide in 10 billion molecules; even if we increased the amount of material by 1,000,000 fold there would be one molecule in 1 million molecules. Although comets may not be good to eat, they are hardly poisonous. You can take Sagan's word for it. Nevertheless, there seems to be a distinct odor about this evidence.

Cyanide, as Sagan ought to know, has a smell of bitter almonds. It is immediately obvious that this does not square well with the accounts of manna at all. The reports of people around the worlds said that manna fell with the morning dew and that it had an "aromatic scent."

DO COMETS HAVE FORMALDEHYDE?

Sagan tells us that comets contain "hydrocarbons—but no aldehydes— the building blocks of carbohydrates," but adds, "They may nevertheless be present in comets."[4] Sagan, in his book *Comet*, p. 314, shows how in the Earth's atmosphere, water vapor, methane and ammonia (all constituents of comets: see *Comet*, pp. 149–150) "are broken into pieces...by ultraviolet light from the Sun or electrical discharges. These molecules recombine to form, among other molecules...formaldehyde." The basic chemistry for this has been known for a long time. Apparently Sagan's knowledge of chemistry is limited since he does not seem to know the simple chemical fact that "formaldehyde" is an "aldehyde". An analysis of Halley's comet caused Roger Knacke to write in *Sky and Telescope*, for March 1987, p. 250 that, "The C-H feature at 3.4 microns has smaller bumps superimposed on it, probably because more than one C-H compound is present. Formaldehyde (CH_2O) is an interesting possibility for some of the features. Evidently there is some complex organic chemistry going on..." The odor of formaldehyde, like the odor of Sagan's evidence, seems to get worse.

Formaldehyde, when it polymerizes, becomes carbohydrates. But Sagan still seems to think manna is a hydrocarbon where he states, "In *Exodus*, Chapter 16, Verse 20, we find that manna left overnight was infested by worms in the morning—an event possible with carbohydrates, but extremely unlikely with hydrocarbons. Moses may have been a better chemist than Velikovsky."[5]

In the *Atlantic Monthly*, Mannfred A. Hollinger of the University of California, Davis, Department of Pharmacology, discussed Sagan's knowledge of basic chemistry and pointed out that Sagan clearly doesn't know the simple difference between a *hormone and a hallucinogen.* Hollinger goes on to state that he finds "it distracting that the general pubic is exposed to [Sagan's] *pseudoscientific writing* that would not pass a credible thesis committee." [6] [emphasis added]

A VERY BIG CALCULATION

Sagan then claims by a simple calculation, "during the forty years of wandering, the Earth must have accumulated several times 10^{18} grams of manna, or enough to cover the entire surface of the planet with manna to a depth of about an inch." [7] In this case, Sagan has certainly outdone himself in chemistry. He seems to believe that carbohydrates are not biodegradable.

Then Sagan produces a further calculation with his indestructible carbohydrate (manna). His calculation is supposed to impress everyone with his conclusion that "the mass of manna distributed to the inner solar system by this event is larger than 10^{28} grams. This is not only more massive by many orders of magnitudes than the most massive comet known; it is already more massive than the planet Venus." [8] One can only conclude that Sagan has produced the largest, most massive calculation of "baloney" a reader is ever likely to encounter.

At last Sagan finds his intellectual niche stating B.B. p. 108, "If this indeed happened, it would certainly be a memorable event, and may even account for the gingerbread house in *Hansel and Gretel*" It is at this level of debate that one begins to grasp Sagan's pseudoscientific, fairy tale chemical knowledge. The only thing one can say is that Sagan doesn't seem to know the difference between gingerbread and baloney.

DOUBLE RATIONS—DOUBLE TALK

Sagan, in his footnote, elaborates the following,

"Actually, *Exodus* states that manna fell each day except on the Sabbath. A double ration, uninfested by worms, fell on Friday. This seems awkward for Velikovsky's hypothesis. How did the comet know? Indeed, this raises a general problem about Velikovsky's historical method. Some quotations from his religious and historical sources are to be taken literally; others are to be dismissed as 'local embellishments.' But what is the standard by which this decision is made? Surely such a standard must involve a criterion independent of our predispositions toward Velikovsky's contentions." [9]

Velikovsky's answer is,

"Throughout his paper, Sagan repeatedly stresses that I accept some parts of ancient myths and legends and not other parts, and he wonders why I do not accept either all or else nothing. He suggests that my procedure here is arbitrary and capricious. But Sagan has not troubled to understand my procedure. He complains that I accept ancient legends about manna, but that I do not accept the scriptural account that manna fell in double portions on Fridays and not at all on Saturdays. But I accept the ancient testimonies about manna of the Hebrews, ambrosia of the Greeks, and honey-dew of other peoples from around the Earth, precisely because there is testimony on this from many peoples from many parts of Earth and because there are physical events (the near collisions between Earth and Venus) that could have led to such results. And I reject the report about manna falling in double portions on Fridays and not at all on Saturdays, precisely because that feature of the story does not have a plausible physical basis, is not testified to by other peoples and is therefore to be regarded as an inaccurate elaboration by one people upon what actually transpired." [10]

This raises a general problem about Sagan's method of analysis. Sagan wishes us to digest not Velikovsky's analysis based on plausible physical evidence and testimonies of peoples from around the world, but his own concoctions of the historical and physical evidence. M. Rowan-Robinson, after reading Sagan's chapter on manna, wrote in *Nature* that he found it to be "*a red herring.*" [11] [emphasis added] A red herring is a journalistic term for "baloney". Now one begins to recognize what stinks. Sagan has all along seemed to say comets cannot produce organic molecules such as carbohydrates. In his co-authored book *Intelligent Life in the Universe,* he informs us that, "When radiation falls upon such an orbiting snowbank [which roughly describes the make-up of a comet], chemical interactions among H_2O, CH_4 and NH_3 [water, methane and ammonia] will produce organic molecules, as has been demonstrated in laboratory experiment on simulated comets." [12] In fact, Duncan Lunan in *New Worlds For Old,* (NY 1979), p. 163 tells us that, "with discoveries of organic compounds of increasing complexity in comets...the analysis of carbonaceous condrites are no longer controversial. It seems clear, however, that the compounds were formed 'naturally,' not by the action of life." Lunan claims organic compounds are well known to be components of comets, and that their existence in comets is well accepted by the scientific community so that the suggestion by Velikovsky that organic compounds are found in comets is not in the least controversial. This is made explicit in *Astronomy,* (Feb. 1992), p. 20 wherein we learn, "Comets may be the bringers of life. Last summer Carl Sagan of Cornell University described what a growing number of scientists believe triggered life on the early Earth, 'No one disputes that comets contain abundant organic molecules.'"

So it seems Sagan's argument is a "red herring" with a peculiar odor.

Fred Ferguson reported in the *Overseas Press Club Bulletin,* (Vol. 43 No. 5), for May 1988, p. 1, that Carl Sagan, one of the invited speakers to the 49th OPC Awards Dinner delivered a talk titled, "Dateline Earth: A Report From the Intergalactic Press Club." In his talk Sagan is reported to have presented his views, "with the unmatched dry humor for which Cornell's renowned astronomy professor is known on the lecture circuit." Ferguson further remarks that, "Sagan said the leaders of Earth's nations were obviously not eager to tell the truth on all issues. 'What they really need,' he said, 'is a baloney detection kit. Fortunately there is a community of truth–tellers. In some cases the truth tends to come out.'" Yes, indeed, in some cases it actually does.

Sagan, of course, was speaking about politicians and "political baloney." But there is no doubt that the same criticism also applies to scientists and "scientific baloney." And for some scientists a truly good baloney detection kit and truth–tellers are also sorely needed. Should Sagan now find the criticism here regarding his discussion of manna unpleasant, perhaps he should have offered real criticism instead of the dry humor for which he is famous. Sagan had no qualms about preparing a feast of rancid baloney on gingerbread sandwiches and cups filled to overflowing with cometary cyanide which he then so graciously and eagerly served to the public as criticism of Velikovsky. He emphasized at the opening of his paper that overly polite criticism helps no one. If now he finds the taste of the antidote unpalatable, he and his supporters are admonished to serve only polite, balanced and healthy criticism instead of *ad hominem* and insensitive ridicule.

NOTES & REFERENCES

[1]SCV, pp. 71-72; B.B., p. 107.
[2]*National Geographic*, Vol. 160, No. 1 (July 1981), p. 18.
[3]Sagan and Druyan; *Comet*, op. cit., pp. 143-144.
[4]SCV, p. 71; B.B., p. 107.
[5]SCV, p. 72; B.B., p. 107.
[6]Hollinger, Mannfred A.; *Atlantic Monthly*, (June 1979), pp. 29-30.
[7]SCV, p. 73; B.B., p. 108.
[8]SCV, p. 73; B.B., p. 108.
[9]SCV, p. 72; B.B., p. 108.
[10]*Kronos* III, 2. pp. 24-25.
[11]Robinson, Rowan M.; "Rebel Without a Cause", *Nature*, (Nov. 9, 1978), pp. 150-151.
[12]*Intelligent Life in the Universe*, op. cit., p. 334.

SAGAN'S SEVENTH PROBLEM

THE CLOUDS OF VENUS

CLOUDS OF WATER VAPOR

Of what are the clouds of Venus composed? The atmosphere of the Earth is composed of nitrogen and oxygen, but the clouds above are composed of water vapor. Are the clouds of Venus composed of water? Velikovsky compiled a short list of spectroscopic analyses of the Venusian clouds to determine if they are made of water vapors in some form. The list reads as follows:

"The absence of any response at the water vapor line at 13.5 mm. wavelength, as recorded by *Mariner II,* confirms what was already known from ground based microwave radiometry. There is no water on Venus worth mentioning." [writes] (E.J. Opik, *The Irish Astronomical Journal,* June 1963)

"V.I. Moroz of the Sternberg Astronomical Institute, writing in the *Astronomichesky Journal,* Vol. 40, says: 'The monochromatic albedo curve contains no features characteristic of reflection from ice crystals. Evidently the clouds consist of neither ice nor water but of dust.' Also: 'The form of the monochromatic albedo curve in the 2–2.5 micron range contradicts the notion of a greenhouse effect due to water vapor.'

"G.V. Rosenberg of the Institute of Physics of the Atmosphere, Academy of Sciences, USSR, writes in its *Doklady,* vol. 148, #2: The extreme weakness of the absorption bands of water in the vapor state and in the condensed state in the light reflected from Venus exclude *a priori* that clouds containing water exist on the planet.'...

"A. Dollfus, making a spectral search from a high altitude observatory, and J. Strong, from a balloon, claimed to have discovered a small quantity of water vapor *above* the cloud envelope of Venus. But Rasool and others pointed out that there is a comparable amount of water vapor in the terrestrial stratosphere that could be responsible for that effect.

"H. Spinrad used various techniques and obtained a negative result for water. L.D. Kaplan showed, in a series of papers, and JPL [Jet Propulsion Laboratory] reports that the clouds could not be of water and this is for very decisive physical reasons. His evaluation of *Mariner II* results served NASA in its appraisal of the contents of the clouds.

"Also G. Kuiper was unable to detect water on Venus. "Venus' spectrum [is] incompatible with the ice band..." (*Kitt Peak National Observatory Contribution,* #24, 1963)."[1]

There is practically no water on Venus. This is the opinion of all responsible scientists today based on careful spectroscopic analysis of the clouds. Therefore, it is interesting to note that Sagan had just the opposite view based on his own brand of spectroscopic analysis. In his co-authored book, *Intelligent Life in the Universe,* he states,

> "...From a variety of observations at visual, infrared, and radio frequencies, it has recently been established that the clouds of Venus are indeed made of water: ice crystals in the colder cloudtops, which are seen in ordinary photographs...and water droplets in the bottom of the clouds, which are 'seen' at long wavelengths."[2]

Sagan defended this view of clouds of water vapor on Venus in *Sky and Telescope,* for July 1960 in an article titled "The Venus Greenhouse". Eight years later he was still defending this view in a paper titled "The Case for Ice Clouds on Venus" published in the *Journal of Geophysical Research,* Vol. 73, No. 18, for Sept. 15, 1968. Sagan's argument for "water droplets in the bottom of the clouds, which are 'seen' at long wavelengths" has been proven to be wrong in *Popular Astronomy* for Jan. 1976, p. 26. His claim of ice crystals and water composition of the clouds has been proven to be wrong based solely on the refractive index by J.E. Hansen and A. Arking in *Science,* Vol. 71, Feb. 19, 1971, pp. 669ff.

SAGAN AND WATER VAPOR CLOUDS

However, Sagan used statements that were so unambiguous regarding his evidence, such as "it has recently been established that the clouds of Venus are indeed made of water" or "ice crystals in the colder cloud tops...are seen in ordinary photographs" or "water droplets in the bottom of the clouds...are seen," the impression is given that Sagan has proven his case for water clouds on Venus.

It seems that Sagan, in spite of all the evidence, still clung to his thesis that the clouds of Venus were water ice. Clark R. Chapman in *The Inner Planets,* (NY 1977), p. 104, informs us that,

> "...Sagan has not always taken a detached view of the Venusian clouds, which have been a major part of his serious research since his days as a graduate student. Long after most of his colleagues agreed that his once-accepted water-ice model for the clouds of Venus was incompatible with radio and polarmetric data, the loquacious Sagan continued to press his case. Sagan watchers were forced to conclude that he actually believed in water ice clouds...as most scientists believe in their own theories."

Hence, it seems that Sagan is again often incapable of taking the advice he so freely gives to others that "reasoned criticism of a prevailing belief is a service to the proponents of that belief; if they are incapable of defending it, they are well advised to abandon it." Therefore, we advise Sagan to heed his own advice once again.

VELIKOVSKY'S HYDROCARBON CLOUDS?

Nevertheless, Sagan's statements are completely contradicted by all the evidence of that time and also by evidence since that time. Sagan's work respecting the Venusian clouds was dogmatic. Thus, one perhaps may understand him lashing out at Velikovsky's view. In *Scientists Confront Velikovsky*, p. 73, he writes, "Velikovsky's prognostication that the clouds of Venus were made of carbohydrates has many times been hailed as an example of a successful scientific prediction." On p. 109 of *Broca's Brain*, the prognostication is that the "clouds of Venus were made of *hydrocarbons or carbohydrates...*" [emphasis added]. Apparently, Sagan isn't sure if, according to Velikovsky's hypothesis, the clouds of Venus are made of carbohydrates or hydrocarbons although one can be very sure they are not made of water. Thus, we turn to Velikovsky's statement in *Worlds in Collision*. "On the basis of this research, I assume that Venus must be rich in petroleum gases... The spectrogram of Venus may disclose the presence of hydro-carbon...gases in its atmosphere... Venus has petroleum gases."[3] All of these statements are that Venus clouds contain hydrocarbons and not carbohydrates. Dr. W.T. Plummer also wrote in the journal *Science*, (March 14, 1969), p. 1191, "Velikovsky [predicted] that Venus should be surrounded by a blanket of petroleum hydrocarbons."

Sagan cannot allow this and therefore, writes, "From Velikovsky's general thesis and the calculation just described above, it is clear that Venus should be saturated with manna," a carbohydrate.[4] If this is so, why didn't Plummer or the editors of *Science* suggest this?

Sagan adds, "The idea of carbohydrates in the clouds of Venus was not original with Velikovsky"[5] and "Wildt, unlike Velikovsky, under-stood well the difference between hydrocarbons and carbohydrates."[6] Sagan argues that Velikovsky, a trained doctor, does not know the dif-ference between hydrocarbons and carbohydrates. This is completely the opposite opinion of Harry H. Hess, the late Chairman of the Geophysics Department of Princeton University and President of American Geological Society. He stated in *Velikovsky Reconsidered*, (NY 1977), p. 46, that, "I suspect the merit [of your work] lies in that you have a good basic background in the natural sciences..." Hess was well acquainted with Velikovsky's work and he would have certainly noticed such a confusion if one actually existed. On the other hand, Sagan was shown to be ignorant of the difference between a hormone and a hallucinogen and whose chemistry was called pseudoscience by a recognized authority, and who seemingly doesn't know the simple fact that formaldehyde is an aldehyde. For Sagan to impugn Velikovsky's knowledge of chemistry is absolutely breathtaking. This is especially true since, in the Winter issue for 1973/74 of *Pensée,* (Vol. 4 No. 1) p. 31 Velikovsky explained the difference between hydrocarbons and carbohydrates. "Hydrocarbons are petroleum products consisting of only two elements—carbon and hydrogen; carbohydrates, besides carbon and hydrogen, contain also oxygen." But Sagan never bothers to mention this statement! Sagan states,

> "I pointed out many years ago (Sagan 1961), the vapor pressure of simple hydrocarbons in the vicinity of the clouds of Venus should make them detectable if they comprise the clouds. They were not detectable then and in the intervening years, despite a wide range of analytic techniques used...hydrocarbons...have not been found. These molecules have been searched for by high-resolution ground-based optical spectroscopy from the Wisconsin Experimental Package of the Orbiting Astronomical Observatory OAO-2, by ground based infrared observations; and by direct entry probes of the Soviet Union and the United States, not one of them has been found."[7]

The first argument is that vapor pressure forbids any possibility of hydrocarbons in Venus' upper atmosphere. In 1964 Sagan argued for the presence of hydrocarbons on Venus in an article titled "The Atmosphere of Venus" in *The Origin and Evolution of Atmospheres and Oceans.* Sagan argued that, "the fact that [at the level of the clouds, low vapor pressure organic compounds like CH_4, C_2H_2, C_2H_4, etc.] have not been successfully identified does not entirely exclude the possible existence of some hydrocarbons in the lower atmosphere..."[8]

The second part of Sagan's argument is that direct entry probes have not found hydrocarbons in the lower atmosphere. However, Lewis M. Greenberg reports that "the April 1979 issue of *Popular Science,* (p. 67) reported the following:

> "First [John H.] Hoffman [Head of the mass spectrometer team for *Pioneer Venus 2*] stunned his colleagues by reporting that the atmosphere [of Venus] contains 300 to 500 times as much argon as Earth. Then he found indications in his data that *the lower atmosphere may be rich in methane.* [emphasis added]
>
> "Methane is, of course, the simplest hydrocarbon and is also a constituent of the Jovian atmosphere.
>
> "Scientists pondered the possibility that the methane might be responsible for an unexplained visible glow in Venus' atmosphere... Dr. Thomas Donahue, of the University of Michigan thought it possible, 'But we're being very skeptical about the methane reading.' He added, 'It would be an artifact [a residue from Earth] or it could be an important finding. It will be quite a while to verify...'

"...hydrocarbon clouds on Venus can no longer be claimed as apocryphal."[9]

In *Science News,* (September 12, 1992), page 172 is a follow up on the work of Thomas H. Donahue.

"Donahue and his collaborators...characterize the finding [of methane] as so surprising that they were loathe to publish them...

"The researchers base their unlikely conclusion [that the methane is of volcanic origin] *on the abundance* and composition of methane detected by a mass spectrometer aboard the *Pioneer-Venus Probe* [14 years ago]. *Scientists had known for years that the spectrometer had recorded a sharp rise in methane, beginning at about 14 kilometers above the surface of Venus, during the probe's descent.* But for nearly a decade, Donahue and his co-workers believed the surge merely reflected methane placed in the spectrometer on Earth in order to calibrate the instrument, not activity on Venus...

"'We concluded that the methane sampled was a primeval methane freshly vented from the planet's interior,' says Donahue...

"Donahue estimates that a volcanic eruption spewing out the amount of methane found by the *Pioneer-Venus* would occur only about once every 100 million years. Moreover, *it appears that the probe passed through the [methane] plume near the top of the atmosphere,* where winds would have stretched the vented methane over a wide area, as well as closer to the surface of the planet...

"'*It is embarrassing to invoke such a wildly unlikely event* as a chance encounter between the entry probe and a rare and geographically confined methane plume, but so far we have eliminated all other plausible explanations,' Donahue added." [Emphasis added]

To explain the large amount of methane found in the Venus atmosphere the scientist said that the methane had to have come from an extremely rare volcanic eruption. The one explanation omitted by Donahue is that Venus has a good deal of methane in its atmosphere, just as Velikovsky predicted. Scientists would rather suggest a wildly improbable concept to explain the methane found than give consideration to Velikovsky's prediction. While scientists like Sagan will call Velikovsky's theory extremely improbable, they will propose that it is probable that *Pioneer-Venus* just happened to come down on Venus to experience a unique event that happens once every hundred million years.

Therefore, there are, at present, indications of hydrocarbons in the atmosphere of Venus, which contradicts Sagan's assertion.

OXYGEN ON VENUS

Let us return to Sagan's spectroscopic analysis of Venus' atmosphere. In the original AAAS symposium, Sagan's paper contained this item respecting the assertion by Velikovsky that there was oxygen in the atmosphere of Venus. Sagan, in another of his "absolutely positive" statements said this regarding oxygen, "There is none as has been clearly shown by ground based spectroscopic observation." Unfortunately, along came *Mariner 10* and proved the spectroscopic analysis touted by Sagan as his proof to be dead wrong.

SULFURIC ACID CLOUDS

This then brings us to Sagan's next pronouncement,

"Moreover the question of the composition of the Venus clouds—a major enigma for centuries—has recently been solved (Young and Young 1973; Sill, 1972; Young 1973; Pollack *et. al.,* 1974). The clouds of Venus are composed of an approximately 75 percent solution of sulfuric acid. This identification is consistent with the chemistry of the Venus atmosphere in which hydrofluoric and hydrochloric acid have also been found; with the real part of the refractive index, deduced from polarimetry which is known to three significant figures (1.44); with the 11.2 micron and 3 micron (and now far-infrared) absorption features; and with the discontinuity in the abundance of water vapor above and below the clouds."[10]

This certainly does sound quite scientific and impressive, but I recall Sagan stating, "It has recently been established that the clouds of Venus are indeed made of water." Also regarding oxygen on Venus he said, "There is none as has been clearly shown." Therefore, one might be skeptical about this last pronouncement that "a major enigma for centuries—has recently been solved." In fact, one of the originators of the sulfuric acid model informs us that he thinks the major enigma is *unsolved.* Andrew T. Young, whom Sagan cited as one of the developers of the sulfuric acid theory, had this to say in a NASA report (NASA-CR-142056, Jan. 31, 1975). Young states,

"none of the currently popular interpretations of cloud phenomenon on Venus is consistent with all the data. Either a considerable fraction of the observational evidence is faulty or has been misinterpreted, or the clouds of Venus are much more complex than the current simplistic models. ...A sound understanding of the clouds appears to be several years in the future."[11]

Young does not appear to be the only scientist with this view. J. Gribbin wrote in *Nature,* Vol. 254, p. 657 (1975), a report about the conference "On April 11, 1975, the Royal Astronomical Society held a joint meeting with the Royal Meteorological Society and the Geological Society to discuss the *Mariner 10* results concerning Venus and Mercury. Although one of the participants (Hunt) discussed the possibility of sulfuric acid in the cloud tops, it was clear that it is definitely not known what is closer to the surface."[12] So the enigma of of centuries that Sagan informs us that has recently been solved, is unsolved. *The Encyclopedia Britannica* for 1982, Vol. 2, p. 327 states, "The composition of...[Venus'] clouds remain uncertain; ice and mercurous chloride have been proposed, but there are difficulties with either hypothesis." In fact, John S. Lewis of M.I.T wrote in "The Atmosphere, Clouds and Surface of Venus" in *The Solar System and its Strange Objects,* ed. B.J. Skinner, (Los Altos, CA 1981), p. 93, that there are several different gases being considered as the major component of Venus' clouds. He states,

> "The clouds of Venus have been a favorite topic of controversy, and here the matter is still in a very uncertain state. Half a dozen species [of gases] are currently favored by different individuals as making up the visible clouds. Among the most widely advertised are water or ice, silicate and carbonate dusts, ammonium chloride, compounds of the volatile elements, mercury, arsenic, etc., carbon suboxide, and its polymers, hydrochloric acid solution or solid hydrates of ChI, ferrous chloride dihydrate, etc. Each species [of gas] has more detractors than supporters."

Does this parade of possible constituents for Venus' clouds sound like the enigma that has puzzled the scientists for centuries has been solved, as Sagan claims?

Sagan argues further that "Observations on the wings of the critical 3.5 micron region show not the slightest trace of the C-H absorption feature common to hydrocarbons...(Pollack, *et. al.* 1974). All other absorption bands in the Venus spectrum from the ultraviolet through the infrared are now understood; none of them is due to hydrocarbons."[13] Ralph E. Juergens was skeptical about this statement by Sagan and actually read (Pollack, et. al. 1974) and had this to say, "However, when one checks the paper cited by Sagan, it turns out that (Pollack, *et. al.* 1974) found a great deal of absorption between 3.05 and 3.6 micron, none of which can be attributed to carbon dioxide"[14] [but perhaps to hydrocarbons].

ANALYSIS BY L. D. KAPLAN

Sagan then relates the following story of how the newspapers came to report the story that hydrocarbon clouds were found on Venus by *Mariner 2.* L.D. Kaplan, one of the experimenters, was asked by the newsmen at a press conference to tell how thick the clouds were, how high they were, and of what they were made. Kaplan said that the experiment wasn't designed to test such questions, nor did it. However, he told what he thought was the case, and his thoughts were reported as fact, that Venus had hydrocarbons. However, Sagan, in telling the story failed to explain why Kaplan, a scientist, arrived at the conclusion that hydrocarbons would explain the data. Firstly, he was seeking a molecule that would fit the greenhouse model. Secondly, he was also seeking a molecule that could not only exist, but even polymerize (change form while keeping the same formula) over a great range of temperature in the atmosphere of Venus, and what he came up with that logically fit the spectroscopic and other atmospheric data was hydrocarbons.[15]

Velikovsky answered Sagan and said, "Are there any hydrocarbons, or other organic molecules still present on Venus?"

"After the first American fly-by probe, *Mariner II* passed its rendezvous point with Venus in December 1962, the results were first made public in February 1963, and it was claimed by NASA spokesman Dr. Homer Newell, that the clouds on Venus are rich in hydrocarbons. I have repeatedly read in polemic surrounding my work that this statement at the press conference was a mistake seized upon by the followers of my concepts. It was not a press conference 'mistake.' The conclusion was based upon very careful consideration of the physical characteristics of the cloud layer that was found homogeneous at the top and at the bottom, at temperatures of ca. -35 degrees F[ahrenheit] on the top and over +200 degrees F[ahrenheit] (400 degrees K) at the bottom. Professor L.D. Kaplan, the researcher on the staff of the Jet Propulsion Laboratory, responsible for the statement, discussed the phenomenon in several papers and memoranda and his conclusion was that only the multiple radical CH (hydrogen and carbon bond) has the same physical characteristics at the two ends of the range of temperature as discovered. It is also untrue that JPL revoked the statement made; contrariwise, in *Mission to Venus* (*Mariner II*) published in 1963 the statement is repeated in this form: 'At their base, the clouds are about 200 degrees F and probably are comprised of condensed hydrocarbons.'"[16]

Did Kaplan have second thoughts regarding hydrocarbons in Venus' atmosphere? Velikovsky continues his discussion stating,

"Since the episode in the sociology of science is of interest, so also is a sentence in a letter by L.D. Kaplan (dated April 1963)—not yet realizing why his findings were engendering a storm of protest—to a friend, a member of the Institute for Advanced Study in Princeton. Kaplan wrote that his having identified hydrocarbons caused a violent reaction among astronomers. The word hydrocarbon [Kaplan writes], 'was used only to avoid the use of 'organic compounds' for obvious reasons. The reaction to even 'hydrocarbons' was much too violent.' In a copy of a published report that he sent to his friend he struck out by pencil the word 'hydrocarbons,' changing it to 'organic compounds.'"[17]

Both Kaplan and the Jet Propulsion Laboratory, months after the original statement was made, maintained by publication and by personal communication that based on a logical and scientific interpretation of the evidence supplied by *Mariner II* hydrocarbons (or, as Kaplan stated "organic compounds") probably make up the gases in the clouds of Venus.

Sagan adds that it is, "ironic...that Professor Kaplan was later a co-author of a paper that established a very low abundance of methane, a 'petroleum gas' in a spectroscopic examination of the Venus atmosphere. (Connes, et. al. 1967)."[18] Since methane is a "petroleum gas" one is tempted to believe Sagan that petroleum gases are constituents of Venus' clouds as Velikovsky claimed. But methane is not the only hydrocarbon; there are a great many forms that carbon-hydrogen bonds can make. And, that methane is observed spectroscopically does not exclude other hydrocarbon gases.

Therefore, does anyone know for certain the exact composition of the Venusian clouds? The answer is that no one really does know. Sagan was certain based on spectroscopic analysis that the clouds were made of water and that the atmosphere contained no oxygen. This was in error. Walter S. Adams of the Mount Wilson Observatory and Palomar Observatories wrote a letter to Velikovsky dated July 25, 1955 stating, "Ionized iron and sulfur could not possibly be present in the atmospheres of Jupiter and Venus, because their spectra are atomic and would require very high temperatures for their production." Velikovsky had claimed 15 years earlier that these elements were present. Then, in *Science* for February 23, 1979, p. 753, in *Aviation Week and Space Technology* for February 19, 1979, p. 47 and *The New York Times* for March 3, 1979, p. 12, said that sulfur and/or iron exist in the atmosphere of Venus.

VENUS' SULFURIC ACID AGE

Andrew and Louis Young, whom Sagan cites for the sulfuric acid make up of the clouds; said based on their analysis, iron and sulfur were constituents of Venus' atmosphere. Young and several others claimed sulfuric acid composed the clouds. But Peter R. Ballinger, a researcher in organic chemistry, also had this to say about the possibility of sulfuric acid in Venus' clouds:

> "'It is likely that sulfuric acid would be gradually decomposed by solar radiation of ultraviolet and shorter wavelength particularly in the presence of iron compounds...to give hydrogen and oxygen. This process would also be expected to result in the preferential retention of deuterium, as discussed in another context...' [Ballinger, who wrote this in Transactions of the Faraday Society, Vol. 61, (1965) page 1681, went on to say], 'Because of this and other chemical reactions *sulfuric acid might well have a relatively short lifetime, consistent with a recent installation of the planet in its present orbit.*'"[19] [emphasis added]

In 1985, Lawrence Colin in an article titled "Venus the Veiled Planet" in *The Planets,* B. Preiss ed., (NY 1985), p. 282 states flatly, "The chemical composition of the air [of Venus] remains the most contro- versial aspect of our knowledge of the Venusian atmosphere."

If the sulfuric acid model of the clouds is accepted, Venus could not be 4.6 billion years old because solar radiation would decompose its sulfuric acid and therefore Venus could only have existed, as Ballinger puts it, since "recent" times. Geologically speaking, recent is within the last 10 thousand years.

VENUS' CARBON DIOXIDE AGE

This brings us to a question Sagan raised earlier about the atmosphere of Venus. He states, "the atmosphere of Venus, which...is composed almost entirely of carbon dioxide" is inconsistent with Velikovsky's hypothesis.[20] Ralph E. Juergens, however, informs us that,

> "something much like this [Venusian atmosphere] could be expected on a planet with a very hot surface and an atmosphere polluted with petroleum gases. *Venera* reports even include terms like 'photochemical smog' in connection with the stratified hazes. Photochemical reactions... with lots of sunshine and lots of automobile exhaust, could be expected to release sulfur dioxide in abundance from heavy hydrocarbons (in addition to that vented at the surface of a hot, youthful planet) and quickly oxidized it to SO_3, which would then bind up all available water to form sulfuric acid solutions. And again, as on Earth, the sulfuric acid would probably end up on top."[21]

Therefore, the character of Venus' atmosphere agrees fairly well with Velikovsky's description of what occurred.

But let us examine the question of Venus' carbon dioxide on Sagan's terms. First of all, carbon dioxide is not stable in the presence of ultraviolet radiation from the Sun. Ultraviolet rays break carbon dioxide down into carbon monoxide, CO and oxygen molecules, O_2. Once these molecules of carbon monoxide and O_2 form, they do not combine again easily. Therefore, one would expect to find a great deal of carbon monoxide in the upper and middle layers of Venus' atmosphere if Venus is billions of years old. U. von Zahn et. al., in a paper titled "The Composition of Venus' Atmosphere" in the book *Venus,* published by the University of Arizona, deals with this problem. They write:

> "*Photochemistry of CO_2.* The central problem of the photochemistry of Venus' middle atmosphere is to account for the exceedingly low abun- dance of CO [carbon monoxide] and O_2 [molecular oxygen] observed at the bottom of the middle atmosphere. In fact, O_2 has not been detected even at 1 ppm [part per million] level. Due to low abundances of O_2 and

O_3 [ozone, which absorbs ultraviolet radiation]...solar ultraviolet of sufficient energy to photolyse [breakdown by light action] CO_2...penetrates down to 65 km [or 39 miles above the surface of Venus]...

"The 3-body [3 elements or compounds] recombination reaction with a rate constant kb [based on temperature] is, however, spin forbidden. Consequently at typical temperatures of the Venus middle atmosphere (200k) this [recombination] reaction has a very small rate ... [But at this temperature] ... oxygen is converted to molecular oxygen...with a rate constant kc which is 5 orders of magnitude higher than kb. Neglecting for a moment the effect of trace gases in Venus' atmosphere, CO_2, CO, and O_2 are nonreactive with each other and we therefore expect a fairly rapid transition (on geologic time scales) of the CO_2 atmosphere to one dominated by CO and O_2. CO_2 would disappear from the upper atmosphere within a few weeks, and from the entire middle atmosphere *in a few thousand years.* [emphasis added]

"Indeed, these arguments describe the situation correctly for the upper atmosphere of Venus, provided we take into account also the various dynamic processes exchanging gas between the upper and middle atmosphere. The above arguments, however, fall short in explaining the observed composition of the middle atmosphere which at least close to its lower boundary is characterized by an extreme dearth of CO_2 photolysis [break down], that is CO and O_2."[22]

There is at present no *observed* and delineated process to save the situation. For the abundance of carbon dioxide to persist in the middle atmosphere of Venus, the planet must be only a "few thousand years" old. We have also learned from Ballinger, above, that sulfuric acid "is consistent with the *recent* installation of the planet [Venus] in its present orbit." Thus, the atmosphere of Venus is consistent with the view that it was born less than 10,000 years ago. If it were billions of years old, its atmosphere would be very rich in carbon monoxide and oxygen, but this is simply not the case.

Clark R. Chapman in *The Inner Planets,* (NY 1977), p. 114, informs us that, "The quantity of carbon contained in carbonate rocks in Earth's crust is roughly the same as that contained in the carbon dioxide in Venus' atmosphere." If Venus is indeed a new planet similar in size to the Earth, should it not possess a comparable amount of carbon dioxide? Furthermore, Venus also has the same quantity of nitrogen in its atmosphere as that of Earth. This too is in accord with the evidence that Venus is a new planet. There are also other constituents of Venus' atmosphere that point to its extreme youth which we will discuss elsewhere.

MARS' CARBON DIOXIDE

The carbon dioxide problem is also applicable to the planet Mars. Over billions of years its very thin atmosphere of carbon dioxide should have been completely converted to carbon monoxide and oxygen by the Sun's ultraviolet rays. But like Venus, it is still in the early process of this conversion.

Charles A. Barth in an article titled "The Atmosphere of Mars" in the *Annual Review of Earth and Planetary Sciences,* Vol. 2, (Palo Alto, CA 1974), edited by F.A. Donath, F.G. Stehli and G.W. Wetherill, p. 356 states, "Photodissociation of carbon dioxide to produce carbon monoxide and atomic oxygen takes place from the top of the atmosphere all the way down to the surface. In the upper atmosphere the known recombination reactions are not rapid enough to balance the photoproduction of atomic oxygen to explain the low abundances" of carbon monoxide and oxygen. A kind of circulation called Eddy diffusion is invoked to transport oxygen to the lower atmosphere where oxygen would mix with other constituents and become reconverted to carbon dioxide. However, Mars' atmosphere, even at the surface, is so thin that these recombination processes will occur slowly. Furthermore, ultraviolet radiation which reaches "all the way down to the surface" will photodissociate the carbon dioxide as it reforms and thus the problem remains.

MARS' NITROGEN-15

If Mars lost its atmosphere recently, then other evidence for this should exist. Charles A. Barth (p. 357) also tells us the amount of water vapor in Mars' atmosphere is photodissociating and hydrogen is escaping into outer space and that there is only 21,000 years of water vapor left in Mars' atmosphere. S. Yanagita and M. Imamura in *Nature,* Vol. 274, (1978), p. 234, show nitrogen-15 which is an unstable isotope and does not have a long lived radioactive parent is present in excess amounts in Mars' atmosphere. It is produced by cosmic radiation interacting with oxygen-16. Therefore, the nitrogen-15 had to be produced quite recently. Cosmic rays do not penetrate a deep atmosphere such as that of the Earth. If Mars' atmosphere was destroyed recently, cosmic rays could then interact with oxygen-16 to produce the abundant nitrogen-15 which has not decayed. The evidence of carbon monoxide, water vapor and nitrogen-15 all indicate Mars' atmosphere experienced a very recent catastrophe.

Velikovsky claimed Venus left part of its atmosphere with Mars. Therefore, carbon dioxide on Mars is in full conformity with Velikovsky's hypothesis.

Thus, on Venus, ultraviolet radiation has changed all the CO_2 in the upper atmosphere to CO, but not in the middle atmosphere. On Mars, the conversion is well on its way, but carbon dioxide is still more abundant than the next two greatest constituent gases found there which are carbon monoxide and oxygen.

VENUS' ARGON-36 AND ARGON-40 AGE

The Argon-36 age of Venus' atmosphere indicates that it was produced very recently, no more than in the last few thousand years.

On the other hand argon-40 is also an indication that Venus' atmosphere is young. Argon-40 is a decay product of radioactive potassium-40. Therefore, over time argon-40 should increase in amount to levels comparable to the argon-40 levels found on the Earth if Venus is as old as the Earth. But, interestingly this is not the case. Billy P. Glass in *Introduction to Planetary Geology,* (NY 1982), p. 314 informs us that,

> "the ratio of the mass of radiogenic 40 Ar [Argon-40] to the mass of Venus is smaller by amount, a factor of 15 than the value for the Earth. Since 40 Ar within a planet increases with time due to radioactive decay of 40 K [potassium-40] the amount of 40 Ar should be higher if the primary degassing took place late in the planet's history."

That Venus has both too much argon-36 and too little argon-40 are clear indications pointing to an extraordinarily young age for Venus. If Venus were as old as the Earth, its argon-36 would have decayed to only a tiny fraction of its present amount. If Venus were an old planet, its argon-40 would have increased in amount to that contained in the Earth.

The problem with Venus' atmosphere is argon-36. Argon-36 is a primordial product from ancient times. "The atmosphere of Venus contains as much argon-36 as you would expect to find in the planet's original atmosphere" (according to M. McElroy, *Pioneer* experimenter in the *Washington Post,* Dec. 11, 1979, p. A6). If Venus were 4.6 billion years old, its Argon-36 would have decayed to a level comparable to that found on the Earth. Venus has hundreds of times as much Argon-36 as the Earth. In fact, it has what appears to be exactly the amount of Argon-36 that Venus would have if it were born in the last few thousand years.

VENUS' OXYGEN AGE

Furthermore, there is the oxygen problem of Venus' atmosphere. Venus is volcanically very active and one of the products of volcanic activity is water. But there is very little water on Venus. This is explained by having some of the water vapor rise to the upper levels of the atmosphere where the Sun's rays dissociate it into hydrogen and oxygen. The hydrogen could and would supposedly have escaped from the atmosphere, but not the oxygen. Eric Burgess in his book *Venus an Errant Twin*, tells us,

"Oxygen is also important to the question of what happened to the water. If water molecules were broken down into hydrogen and oxygen, the disappearance of the oxygen has to be explained, since very little of this gas is present in the atmosphere today. No completely satisfactory explanation is yet available for what happened to the oxygen."[23]

But if Venus is a new born planet, the oxygen problem is solved. The photodissociation of carbon dioxide, sulfuric acid and water require Venus to possess abundant oxygen.

Recall Sagan's remark about green plants producing oxygen. We pointed out that Jeremy Rifkin in *Algeny*, pp. 149-150 stated that scientists have long assumed that life developed in an oxygen free atmosphere. That is, young terrestrial planets would possess little or no oxygen. This is precisely the case for Venus, and it again indicates Venus is a young planet based on scientific historical considerations.

In fact, biologist Carl Woese of the University of Illinois, having become critical of the prebiotic soup paradigm for the origin of life on Earth, has developed a proposal that life may have begun in an environment like that of Venus; more specifically, life could have originated in the clouds. Woese's "An Alternative to the Oparin View of the Primeval Sequence", in *The Origin of Life and Evolution,* (NY 1980) pp. 65-76, states, "The Oparin thesis has long ceased to be a productive paradigm: It no longer generates novel approaches to the problem; and its overall effect now is to stultify and generate disinterest in the problem of life's origin. These symptoms suggest a paradigm whose course is run, one that is no longer a valid model of the true state of affairs."

What Woese then proposes is that life originated in the earliest period of the Earth's history, before the planet had been completely consolidated. Since the materials that would eventually form the crust and mantle had not differentiated, there existed large amounts of iron exposed at the surface. Under those conditions the new metal com‐ bined with the atmosphere in a series of chemical reactions, giving rise to a new atmosphere dominated by carbon dioxide and hydrogen. Then as the Earth outgassed water, the CO_2 and H_2O initiated a "runaway greenhouse effect" that mimicked the present atmosphere of Venus. Cometary and meteoric bombardment added materials to the atmosphere that were carried in its lower levels. There, subject to lightning and other reactions over millions of years, each water droplet was both a laboratory and an oasis in which countless experiments could have been tried. According to Woese, the first life forms that evolved would have been methanogens, which derive energy from combining organic substances with hydrogen and oxygen.

What the methanogens did was remove CO_2 from the early atmosphere by combining it with hydrogen. In so doing, they destroyed the conditions that permitted the runaway greenhouse effect to operate, and the Earth subsequently cooled. With the cooling of the atmosphere, whatever water vapor existed in it condensed and fell as rain to form oceans in which the methanogen evolved.

It is interesting that a scientist involved in research into the origin of life should draw the conclusion that the Earth's *primeval atmosphere* exhibited conditions currently found in the atmosphere of Venus. In other words, without saying so, Woese has reinforced Velikovsky's concept that the atmosphere of Venus is of a *recent* origin.

THE EARTH'S HELIUM
Thus, what is clear is that the atmosphere of Venus is not compatible with a long history of billions of years. And while we are about it, what about the atmosphere of the Earth? Ian T. Taylor tells us that,

> "During the radioactive decay of uranium and thorium in the Earth's crust, alpha particles are given off, and these become helium 4, the most abundant isotope of helium...eight alpha particles are produced as each uranium 238 atom decays to lead 206. Estimates have been made of the total uranium and thorium in the earth's surface, and from this, the rate of production of helium is reckoned to be 3×10^9 grams per year. In addition to this about the same quantity of helium is generated each year in the upper atmosphere by bombardment of cosmic rays. If helium 4 has been released into our atmosphere at this rate for some four billion years, then the total quantity of helium 4 present today should be about 10^{20} grams. In fact, the actual quantity found is a thousand times less than this figure."[24]

Velikovsky, however, in *Worlds in Collision* recorded from ancient writings that "a shower of meteorites flew toward the earth" (p. 51). That there rained from the sky, "a rain of bitumen" (p. 54). That there fell "strange rains and hails and showers inexorably and utterly consumed with fire" (p. 55). That there were exchanges of thunder-bolts between the Earth and the comet. All of this fire and electric lighting bolts would have destroyed most of the available helium.

In this regard, it was reported in the journal *New Scientist* that,

> "Physicists studying the upper atmosphere have to explain how the Earth came to lose nearly all its helium. They have tried to argue that the gas escapes from the Earth's upper atmosphere steadily and continuously, sufficiently fast to reduce its abundance to known levels. A recent laboratory experiment conducted at the Central Radio Propagation

Laboratory for the U.S. National Bureau of Standards has led to the suggestion by Dr. E.E. Ferguson, head of the atmospheric collision processes section there, that instead, '*some catastrophic event may have made most of the helium boil off in geologically recent times.*'[25] [emphasis added]

Recent times is, geologically speaking, within the last 10,000 years.

NOTES & REFERENCES

[1] Velikovsky, I.; "The Weakness of the Greenhouse Theory", *Kronos* IV, 2, pp. 29-30

[2] *Intelligent Life in the Universe*, op. cit., pp. 323-324.

[3] W in C, p. 369.

[4] SCV, p. 73; B.B., p. 109.

[5] SCV, p. 74; B.B., p. 110.

[6] SCV, p. 74; B.B., p. 110.

[7] SCV, pp. 74-75; B.B., p. 110.

[8] Sagan, Carl; "The Atmospheres of Venus" in *The Origin and Evolution of Atmospheres and Oceans*, ed. P.J. Brancazio and A.G.W. Cameron, (1964) p. 280.

[9] Greenberg, Lewis M.; "Velikovsky and Venus", *Kronos* IV, 4, p. 6.

[10] SCV, p. 75; B.B., p. 111.

[11] *Age of Velikovsky*, op. cit., p. 126.

[12] Ibid.

[13] SCV, p. 75; B.B., p. 110.

[14] *Kronos* III, 2, p. 95.

[15] Ibid., p. 96.

[16] Velikovsky, I.; "Venus' Atmosphere", *Pensée*, Vol. 4, No. 1, (Winter 1973), pp. 33-34.

[17] Ibid, p. 34.

[18] SCV, p. 78; B.B., p. 113.

[19] *Kronos* III, 2, p. 45.

[20] SCV, p. 68; B.B., p. 104.

[21] *Kronos* III, 2, p. 95.

[22] Von Zauh, U., et. al., "The Atmosphere of Venus", *Venus*, (Tucson 1983) ed. D.M. Hunten and L. Colin, T.M. Donahue, VI. Moroz, pp. 338-339.

[23] Burgess, Eric; *Venus an Errant Twin*, (NY 1985), p. 133.

[24] *In the Minds of Men*, op. cit., pp. 334-335.

[25] *New Scientist*, Vol. 24, (1964), p. 631.

SAGAN'S EIGHTH PROBLEM

THE TEMPERATURE OF VENUS

THE GREENHOUSE EFFECT

Sagan's entire argument respecting the temperature of Venus is that it is heated by a runaway greenhouse effect. He states that,

"We now know from ground based radio observations and from the remarkably successful direct entry and landing probes of the Soviet Union that the surface temperature of Venus is within a few degrees of 750 degrees K... The surface atmospheric pressure is about ninety times that of the surface of the Earth, and is comprised primarily of carbon dioxide. This large abundance of carbon dioxide, plus the small quantities of water vapor which have been detected on Venus, are adequate to heat the surface to the observed temperature via the greenhouse effect."[1]

Therefore, according to Sagan, Venus' heat is derived from the Sun from an atmospheric phenomenon. Velikovsky's thesis is that Venus is a new planet in the early cool down state and that its heat is derived from its hot core which is producing the high surface temperature.

A greenhouse requires four elements to produce heat above average from sunlight. One of these elements is a glass–enclosed structure through which light can pass and which will stop the heated air inside from escaping to mix with the surrounding atmosphere and dissipating. The second requirement for the process is plenty of light entering the glass-enclosed structure which is converted from short wave radiation to long wave radiation. The other two requirements are water and carbon dioxide in sufficient quantities to trap and hold the long wave radiation so that it does not reradiate away quickly. Water and carbon dioxide do this work extremely well. But can sunlight in a deep, heavy carbon dioxide atmosphere with practically no water raise the temperature of that atmosphere to about 900 degrees Fahrenheit, which is a temperature that will melt lead?

VENUS' WATER AGE

Venus certainly possesses sufficient carbon dioxide. But what about the quantity of water? Andrew Young, cited earlier, and Louise Young have this to say about water in Venus' atmosphere; "...studies at radio wavelengths have established once again that there is no more than 0.1 or 0.2 percent water vapor in the lower atmosphere, and the true value is probably close to 0.01 [1/100 of a] percent. The cloud tops are drier still."[2] Therefore, the requirement of sufficient water vapor to do the job is not established.

This raises an interesting question. If Venus went through the same early evolution as the Earth billions of years ago, it would have, over time, out-gased an ocean of water at least comparable to that of the Earth. Young and Young tell us that,

"If one assumes that Venus once had as much water as the Earth has now, it is necessary to explain how all but one part per million of it was lost. There is a known mechanism by which a planet with abundant water could lose a large portion of it: Water vapor in the upper atmosphere could be dissociated by ultraviolet radiation and the hydrogen could be lost to space, either by thermal escape or through the influence of the solar wind. That effect, however, could not produce an atmosphere so thoroughly desiccated as that of Venus. Of the water Venus has today, very little reaches the upper atmosphere and therefore, it is not dissociated; at the present rate Venus would not have lost a significant amount [of water] in the history of the solar system."[3]

If we accept the present gradualist assumption that the Earth and Venus had similar early histories, both planets would early have had atmospheres and water. The currently accepted notion for the development of oceans is the "outgassing hypothesis" presented by W.W. Rubey in 1951. The hypothesis suggests that gases expelled by volcanoes and hot springs contain steam [water], carbon dioxide, nitrogen and carbon monoxide. This process, it is believed, can explain the atmospheric constituents of the Earth. Thus, the seas of Venus would take at least as long to form as those of the Earth. However, this outgassing process continues all throughout the 4.6 billion year history of planets so that the depth of the oceans steadily increases over the aeons. If this is so, even if Venus had lost its first atmosphere and oceans say 3 or 4 billion years ago, (after the first atmosphere and water of the planet was removed by the solar wind and a new atmosphere of carbon dioxide was baked out of the surface rock), then outgassing during the subsequent 3 to 4 billions years would have produced a new ocean of smaller depth. Protected by the new atmosphere, this ocean would not, or could not, have escaped from the planet.

William K. Hartmann in *Moon and Planets,* 2 ed. (Belmont CA 1983), pp. 430–431, informs us that on Venus,

"The H_2O is too heavy to escape thermally in the lifetime of the solar system. Thermal escape of H [hydrogen], produced by photodissociation of the H_2O was thought to have caused the loss of H_2O from Venus. However, *Pioneer* discovered the 285-K exosphere temperature, and calculations show that the H escape time from such an exosphere is 20

Gy! [20 billion years] So, how could H and hence, H_2O, have been depleted? If the exosphere had once been heated to 1,000 K or so, the H escape time could be brought down to a tiny fraction of the age of the solar system. In any case, *Pioneer* scientist (Stewart A. and others, 1979, "Ultraviolet Spectroscopy of Venus: Initial Results from the Pioneer Venus Orbiter", *Science,* Vol. 203, p. 777) concluded, 'If Venus ever possessed a large amount of water, it cannot have lost it by escape mechanisms known to be operating now.'"

Lawrence Colin in the book *The Planets,* (NY 1985) tells us, " *Overwhelming evidence* suggests that in its past Venus had much more water, perhaps as much as the Earth today—*a whole ocean.* " (p. 282) [emphasis added]

J. Kelly Beatty in an article titled "Venus the Mystery Continues", *Sky and Telescope,* Vol. 63, (1982), p. 134, asks,

"Where has all of Venus' water gone? Theorists have asked this question for years. It doesn't make sense to them that a planet so like the Earth in size and distance from the Sun should have 10,000 to 100,000 times less water. After all, the pair have comparable amounts of carbon dioxide and nitrogen, so the water was probably there at the outset, but has somehow disappeared over geologic time."

Thus, even when we accept the established views of geophysicists, we face a contradiction. Based on the established view, Venus should have enormous amounts of water vapor in its atmosphere, but this is not the case. Why?

The evidence clearly implies that whatever water Venus did possess was burned off in its early history when it was a stupendously hot, brilliant comet. Thus, its lack of water is in full harmony with Velikovsky's thesis.

If, as Velikovsky claims, Venus is a new planet, then it has not had time to outgas sufficient new water vapor into the atmosphere and therefore, it should have very little or practically none, which is the case. The amount of water vapor in Venus' atmosphere contradicts the view that Venus is an ancient planet. If the amount of water is one one-hundredth of one percent as Young and Young inform us (or even less), then Venus could be no older than 10,000 to 20,000 years.

DOES SUNLIGHT REACH VENUS' SURFACE?

The second requirement of the greenhouse is sufficient light to heat up the atmosphere. Sagan states, " *Venera 9* and *10* missions...obtained clear photographs, in sunlight...of surface rocks. Velikovsky is thus, certainly mistaken when he says (page ix) 'light does not penetrate the cloud cover...'"[4]

Lewis M. Greenberg, however, has this to say,

"When four Venus probes plunged toward the planets surface—two in daylight and two in darkness [according to *Popular Science,* April 1, 1979, page 67].

"One instrument carried aboard each probe was a nephelometer designed to detect clouds by monitoring variations in light. Within each nephelometer was a sensitive radiometer that reacted to even small changes in outside light. Those *radiometers showed only gloom around the two probes dropping through darkness*—until the temperature sensors failed." [emphasis added]

"At almost exactly that instant, the radiometers detected a faint glow in the atmosphere. The glow grew brighter and brighter" [as the space probes left the cloud cover above them.][5]

If the clouds were fairly dark, then the light that was observed to get brighter and brighter must have been coming from the surface of Venus and not from the Sun. As one moves nearer to a source of light, the light becomes brighter, thus the light that was observed could not have been sunlight. It is also quite clear that the clouds trap, reflect and occlude most of the sunlight that enters Venus' atmosphere. Sagan, nevertheless, argues that the light observed could only be sunlight that allowed *Venera 9* and *10* to take clear pictures. But did the *Venera 9* and *10* probes carry flood lights? Professor Greenberg informs us that,

"Soviet scientists recently disclosed new information about Venus obtained from the *Venera 9* and *10* landings of Oct. 1975. The findings were presented at a space research meeting (COSPAR) in Philadelphia. V.A. Avduevsky, deputy director of the Soviet Space Flight Control Center, announced that pictures taken of Venus' surface by *Venera 9* revealed a rock-strewn terrain which cast distinctive and unanticipated shadows.

"As reported in the Philadelphia *Inquirer* of June 14, 1976, 'the atmosphere of the surface [of Venus] was much brighter than scientists had expected. The photographs showed very dark shadows *even when flood lights were turned on.*' [emphasis added]

"According to Avduevsky, there should not be any shadows because *sunlight was diffused by the* [15 kilometer thick] *cloud cover.* He and M.Y. Marov of the Institute of Applied Mathematics of the Soviet Academy of Sciences, said that indicated *a direct light source on the surface,* but they could not guess what that was." [emphasis added]

"In reporting on the same meeting, *Science News,* (Vol. 109, June 19, 1976, p. 388) noted that, 'The Venus clouds turn out to be more tenuous than anyone had thought...more like a haze than heavy clouds. As a result, the surface illumination is brighter than anyone expected and photography is much easier there. With the sun at a 30 degree angle from the zenith, the light flux at the surface is about 100 watts per square meter, an illuminacy of about 14,000 lux.' M.V. Keldysh, a former

president of the USSR Academy of Sciences, observed that 'this value
corresponds to the illuminacy at the terrestrial mean latitude in the
daytime with overcast clouds.'

"Although it discussed the surface light on Venus in some detail,
Science News significantly refrained from explicitly ascribing the source of
the light to the Sun. Thus, the surface illumination of Venus still remains
an inexplicable puzzle to conventional thought for it evidently cannot be
attributed to solar light...

"Thus, the data so far retrieved from the latest Soviet space probes go a
long way towards confirming Velikovsky's conclusions about the planet
Venus. At the same time, the new information is a telling blow to Carl
Sagan's claim [SCV, p. 82 BB, p. 117]...that '*Venera 9* and *10*
missions...obtained clear photographs in sunlight [sic] of surface rock.'"[6]

Only a small amount of hazy sunlight reaches the surface of Venus, the
rest is produced by some sources in Venus' atmosphere near or at the
surface. This accords well with Velikovsky's hypothesis that there are
chemical fires on Venus.

A REAL GREENHOUSE

So far, the amount of water needed to trap sunlight is not established,
and sunlight of sufficient quantity is also greatly in question. However,
the greenhouse model for Venus lacks one of the most important
properties. Clark R. Chapman in his book *The Inner Planets* explains
what is basically wrong with the thinking of meteorologists regarding
the greenhouse effect.

"It was recently pointed out to embarrassed meteorologists who have
debated the relevance of their greenhouse calculations, that this effect
may not be important for greenhouses. Outside ground warmed by the
sun heats adjacent air, which then floats upward to where the barometric
pressure is less. The air parcel expands, cools, and settles into
equilibrium. Meanwhile, at the ground the warmed air is replaced by
cooler parcels from above. This process...warms upper regions and keeps
the air near the ground from getting too hot. Air on Earth begins to
convect whenever the temperature begins to drop with altitude more
quickly than about 6 1/2 degrees C per-kilometer [of altitude]. So
except in an inversion, when the upper air is relatively warm [er than the
surface] convection maintains the 6 1/2 degrees C—per kilometer
profile, which is why mountain tops are cool. The reason it is warmer
inside than outside a greenhouse is mainly that the [glass] roof keeps the
warmed-up inside air from floating away by convection...There is no lid
on Venus and the dense carbon dioxide is free to convect."[7]

The super hot air of Venus, therefore, must rise and carry away the
surface heat of the planet to the upper atmosphere where there is no
covering. There the heat will radiate back to space. This upward
motion or convection of gas by heat will allow it to pass right through

a cloud. Hence, Venus must convect and radiate heat into space from its surface long before the surface reaches anything like 750 degrees K. To achieve a relatively high surface temperature, Venus requires a cover encapsulating the entire planet to keep the hot air at the surface from mixing with the freezing cold air of the upper atmosphere. There is no such mechanism that can halt the mixing and convection of the hot surface air with the cold upper atmosphere. This is an immense problem for the greenhouse effect of Venus.

THE SECOND LAW OF THERMODYNAMICS

What is most unusual is that Sagan has it just the opposite way around. He has the clouds of the atmosphere which are cooler convecting downward to the hotter surface to heat it. A basic college physics textbook, *Physical Science* by Verne H. Booth and Mortimer L. Bloom defines the "Second Law of Thermodynamics" thus: "Heat does not of its own accord flow from cold regions to hot regions."[8] Velikovsky long ago pointed out that the heat gradient of Venus shows that heat is flowing from the surface of the planet to the clouds. Sagan's hypothesis has the clouds absorbing most of the Sun's light. It is there in the clouds where the short wave radiation is converted to long wave radiation. The clouds have to be hotter than the surface for the air to move downward. Richard A. Kerr, the editor of *Science* tells us that,

> "Perhaps the most perplexing of the atmospheric problems lingering after *Pioneer* is the 460 degrees C temperature at the bottom of the [Venus] atmosphere. The much ballyhooed greenhouse effect of Venus' carbon dioxide atmosphere *can account for only part of the heating,* and evidence for other heating mechanisms is now in a turmoil. The question concerns how the sun's energy behaves once it penetrates the highest clouds. When *Pioneer Venus'* probes looked at the atmosphere, each one found more energy being radiated up from the lower atmosphere than enters it as sunlight."[9] [emphasis added]

How can cooler clouds convect to a hotter surface? Velikovsky knew this and showed long ago that Sagan's greenhouse theory violated the second law of thermodynamics. R.E. Newell states,

> "I have previously raised objections to the greenhouse theory though most have been rejected for publication. But recently even the greenhouse advocates have begun to note certain problems. Suomi et. al. [in the *Journal of Geophysical Research,* Vol. 85 (1980), pp. 8200-8213] notes that most of the visible radiation is absorbed in the upper atmosphere of Venus so that the heat source [the cloud cover] is at a low temperature while the heat sink [the surface] is at a high temperature, in apparent violation of the second law of thermodynamics."[10]

Thus, why is the surface hotter than the clouds? And how does one explain this phenomenon? Sagan explains it this way in *The Cosmic Connection:*

> "I had earlier proposed a specific theory, in terms of the greenhouse effect, to explain how the surface of Venus could be at such a temperature. But my conclusion against cold-surface models did not depend upon the validity of the greenhouse explanation: *It was just that a hot surface explained the data and a cold surface did not.*" [11] [emphasis added]

Sagan is unable to explain this phenomenon and Velikovsky also pointed out that a few years ago Sagan wrote, "An explanation of how the surface (of Venus) stays as hot as it does is one of the key unsolved problems in understanding the Venus environment." [12]

John S. Lewis' "The Atmosphere, Clouds and Surface of Venus," in *The Solar System and its Strange Objects,* ed. Brian J. Skinner, (Palo Alto CA 1981), p. 93 tells us that," It is not yet known whether the familiar green house effect can provide such high [Venusian] surface temperatures in a rigorously self-consistent model." If the model was truly correct it would do just that.

HOT AIR RISES

Therefore, how does one somehow get cooler air from the clouds down to heat the hotter surface? The answer is that the Hadley Cell does it, or so we are told. The Hadley Cell mechanism works in the following way. Sunlight heats the clouds which move away from the equatorial regions of Venus to the polar regions. Thus, in the polar regions the hot air moves down to the surface and the air there spreads around the planet. So far so good. However, there is one little problem with this scenario. For the Hadley Cell mechanism to work, the polar surface regions must be cooler than both the equatorial surface regions and the clouds. However, Shane Mage has presented reports of the surface temperature that proved "Above the poles [of Venus], temperatures are now shown to be 40 degrees F[ahrenheit] higher than at the equator." [13] According to Billy P. Glass in *Introduction to Planetary Geology,* (NY 1982), p. 312, "The warmest part of the atmosphere (of Venus) visible to the *Pioneer Venus* orbiter is the North Polar region." Therefore, hot clouds from the equator will not travel to the hotter poles unless one wishes again to violate the second law of thermodynamics. The motion of the Hadley Cell must be reversed and the hot air at the poles must travel toward the equator where the surface is also hotter than the clouds. This Catch-22 situation makes it extremely improbable that Venus is heated by a greenhouse effect.

For a greenhouse mechanism to work there must be Hadley cells to circulate the cool gases at the top of the atmosphere down to the surface, but the scientists have not found these operating on Venus. James B. Pollack in *The New Solar System*, 3 ed. (Cambridge, MA 1990), p. 97 states "The east-west winds on Venus blow in the direction of the planet's rotation at all altitudes." But if Hadley cells were operating there would be northward rising winds carrying warm air from the tropical zone of Venus to the polar regions and surface cool winds blowing to the equator. Clark R. Chapman in *Planets of Rock and Ice,* (NY 1982), p. 99 tells us "...theorists began to expect their might be traditional Hadley cell winds...in the Venusian tropics. But after *Pioneer-Venus,* the Hadley model remains a plausible hypothesis in search of confirmation." There is no evidence of greenhouse atmospheric circulation on Venus. It is all a theoretical concept presented on paper but unsupported by the actual movement of the Venus atmosphere.

V.A. FIRSOFF'S OBJECTIONS

Let us examine several other interesting problems with Sagan's greenhouse theory. V.A. Firsoff, the British astronomer, raised the following objection to Sagan's hypothesis:

"Increasing the mass of the atmosphere (Venus has 91 Earth atmospheres) may intensify the greenhouse effect, but it must also reduce the proportion of solar energy reaching the surface, while the total of available energy must be distributed over a larger mass and volume. Indeed, if the atmosphere of Venus amounts to 75 air masses...the amount of solar energy per unit mass of this atmosphere will be about 0.01 of that available to the Earth. Such an atmosphere would be strictly comparable to our seas and remain stone cold, unless the internal heat of Venus were able to keep it at a temperature corresponding to the brightness temperatures...[of Venus'] microwave emission." [14]

According to Firsoff, the greenhouse effect will not heat Venus very much hence, its heat must come from below the surface.

Firsoff, in his book *The Interior Planets,* expands on this.

"The greenhouse effect cannot be magnified *ad lib*. Doubling the thickness of the greenhouse glass may enhance its thermal insulation, so raising its temperature, but it will also cut down the transmitted sunshine, so reducing its heat. In the end, the procedure becomes self-defeating; the loss of sunshine is no longer compensated by increasing insulation, and the temperature of the greenhouse begins to drop. The sea is a perfect 'greenhouse' of this kind—none of the obscure heat from the bottom can escape to space. But it is not boiling hot; in fact, it is not much above freezing point. Sagan's deep atmosphere would behave in exactly the same way..."

"For all the seeming refinements of mathematical treatment, the basic reasoning is very simple. In a convective atmosphere, the temperature decreases upward at a constant rate (adiabatic lapse), so that, however cold such an atmosphere may be at its upper boundary, any arbitrarily high temperature can be reached at its base if the atmosphere is made sufficiently deep. If, though, the atmosphere is heated by the Sun from above, this reasoning is naive: a deep atmosphere will cease to be convective (adiabatic) at a certain level below the tropopause and become isothermal—the temperature will stop rising.

"An adiabatic atmosphere of a mass envisaged by Sagan is possible only if it is heated from below. In other words, the surface of Venus would have to be kept at a high temperature by internal sources."[15]

According to Glass, in Introduction to *Planetary Geology*, (NY 1982), p. 311, "The pressure at the surface (of Venus) is approximately 90 bars, which is equivalent to the pressure in the ocean on Earth at a depth of nearly 1 km below sea level." That is a depth in the ocean of 3,000 feet. Even at the equator at a depth of 3,000 feet, the ocean is only about 40 degrees Fahrenheit or *eight degrees above freezing*. Firsoff claims that Venus totally covered by an ocean of gas equivalent to a shallow ocean would not get as hot as Sagan's greenhouse model requires. Ralph E. Juergens who first cited this material added that,

"An atmosphere of a mass even greater than that 'envisaged by Sagan' has, of course, been shown to exist on Venus. Probes have also established it to be an adiabatic atmosphere, with temperatures decreasing upward from the surface at a fairly constant rate..."[16]

Venus, like the ocean, is supposedly heated by sunlight from above, but the ocean is warmer at the top and cooler toward the bottom; and at a certain depth downward, the temperature is almost uniformly cool. The same should occur on Venus. Instead, Venus' heat falls steadily from the bottom to the top in the opposite direction as the ocean's thermal gradient.

THE RUNAWAY MECHANISM

Shane Mage has also raised some very interesting problems regarding Sagan's greenhouse theory. He writes,

"The crucial part of this [greenhouse] theory is expressed by the term *runaway*: the massive envelope of Venus would have been formed when, after Venus lost its primordial atmosphere, thus exposing its surface to the intense heat of the young Sun, solar radiation boiled the CO_2 out of surface carbide rocks, increasing the massiveness and hence radiation trapping capacity of the envelope until all available CO_2 was released and temperature-equilibrium attained.

"It is this 'runaway' mechanism that has now been invalidated by the argon-36 discovery. According to *Pioneer* experimenter Dr. Michael

McElroy, (in the *Washington Post,* Dec. 11, 1979, p. A6), 'The atmosphere of Venus contains as much argon-36 as you would expect from a planet's original atmosphere.' But if Venus never lost its primordial atmosphere, the 'runaway greenhouse' would never have had the chance to get started because the surface of Venus would always have remained shielded from direct solar radiation."[17]

It follows that the Sun could never have heated the unshielded surface of Venus and carbon dioxide could never have been liberated to create the greenhouse effect.

VENUS' NEON AND ARGON AGE
Anthony Feldman in his book *Space,* (NY 1980), p. 85, informs us in the same general context that,

"A recent discovery about the composition of the Venusian atmosphere has cast doubt on the popular theory accounting for the formation of the solar system. The theory suggests that the Sun and planets formed at the same time.

"The innermost planets—Mercury, Venus, Earth and Mars—are thought to be small and rocky because the Sun drew their lighter constituents away. If this idea is correct, the closer a planet is to the Sun, the less likely there is to be lighter gases in the atmosphere. But in the atmosphere of Venus, the opposite appears to be true. In particular, there seems to be 500 times as much argon gas and 2,700 times as much neon as in the atmosphere of Earth.

"So far scientists cannot explain why these gases were not drawn away from the planet during the birth of the solar system." The author then draws this startling conclusion, "Further discoveries about Venus may soon force a revision of the most basic ideas about how the Sun and the planets were formed."

ASHEN LIGHT
Mage adds one final telling blow to Sagan's greenhouse hypothesis saying,

"The most striking disclosure by Professor Donahue (Dr. Thomas Donahue, University of Michigan physicist who designed the experiments performed in the Venusian atmosphere), however, is his statement that, 'as we approach the surface of the planet from a distance about ten miles above the night-side, a faint glow was detected that got brighter and brighter and brighter until the probes touched down. That glow, I think, is almost literally the surface and some of the gases in the atmosphere on fire. Chemical reactions that produce light, I think, are occurring in that high-temperature environment where many reactive gases were found out by mass spectrometer.'

"This picture of a 'chemical soup' containing 'many reactive gases' and 'almost literally on fire' supports the thesis of an extremely young Venus.

If its atmosphere were many millions of years old, [as Sagan contends] reactive gases would have been consumed aeons ago. Only as a newborn planet could Venus reasonably be expected to have an atmosphere still on fire." [18]

THERMAL BALANCE

This brings us to Sagan's criticism of Velikovsky's views of the temperature of Venus. He states,

"...Velikovsky is trying to say here that his Venus...is giving off more heat than it receives from the Sun, and that the observed temperatures on both the day and night sides are due mostly to the 'candescence' of Venus than to the radiation it now receives from the Sun. But this is a serious error. The bolometric albedo (the fraction on sunlight reflected by an object at all wavelengths) of Venus is about 0.73, entirely consistent with the observed infrared temperature of the clouds of Venus of about 240 degrees K; that is to say, the clouds of Venus are precisely at the temperature expected on the basis of the amount of sunlight that is absorbed there." [19]

Sagan claims that Venus is in thermal balance; the amount of solar radiation absorbed by Venus is equal to the amount of energy emitted by Venus. If Venus had additional heat from its core, then obviously it would emit more heat than it receives from the Sun. Venus would have a thermal imbalance if Velikovsky's theory for its heat production is correct. Professor Greenberg, however, informs us that,

"...we have Sagan...clearly on the record against Velikovsky's thermal prediction for the planet Venus" [his] "position has been adamant and unyielding. It is, therefore, gratifying to discover that there is now tentative evidence indicating that Venus does indeed radiate some 15 percent more energy than it receives from the Sun. This most recent finding was reported in *New Scientist,* 13th November, 1980, under the title 'The mystery of Venus' internal heat' [p. 437]. It is reprinted...with permission of *New Scientist*...[working with data from the *Pioneer Venus* orbiter, F.W. Taylor], '...found that Venus radiates 15 percent more energy than it receives [from the Sun]. To keep the surface temperature constant, Venus must be producing this extra heat from within.'

"All the inner planets, including the Earth, produce internal heat from radioactive elements in their rocks. But Taylor's observations of Venus would mean the planet is producing almost 10,000 times more heat than the Earth—and it is inconceivable, according to present theories of planetary formation, that Venus should have thousands of times more radioactive elements than the Earth does. Taylor's suggestion is met with skepticism—not to say sheer disbelief—from other planetary scientists." [20]

Reginald Newell informs us that,

"The presence of an internal heat source would naturally lead to an energy imbalance at the top of the [Venus] atmosphere and this has, in fact, been observed (albeit there are certain experimental uncertainties) with about 10 w/M2 more energy leaving the planet than arrives from the Sun."[21]

If the greenhouse theory should be found inadequate, then this imbalance may be far greater than 15 percent.

Sagan briefly mentions Jupiter which is in thermal imbalance, that is, it emits more heat than it receives from the Sun. He states,

"In 1949, Kuiper...suggested that Jupiter gives off more heat than it receives, and subsequent observations have proved him right. But of Kuiper's suggestion, *Worlds in Collision* breathes not a word."[22]

What Sagan breathes not a word about is how the thermal imbalance resembles that of Venus. Ralph E. Juergens informs us that in *Nature*, for September 6, 1974, "When *Pioneer 10* scanned the dark side of Jupiter in 1973, its sensors found that infrared emission from that region, as in the case with Venus, is as intense as that from the Jovian sunlight hemisphere. This was hailed as confirmation of earlier indications that Jupiter gives off more heat than it receives from the Sun."[23] But what is ever worse for Sagan is the fact that Shane Mage pointed out that on Venus the "equatorial night side temperatures [are shown to be] 4 degrees F[ahrenheit] above the day side level."[24] Thus, one of Venus' hemispheres during the 58 long day period when it is in the darkness of night time emits four degrees Fahrenheit more temperature than the opposite hemisphere that is in daylight.

THERMAL BALANCE DENIED
By taking the known surface temperature of the Sun and calculating the distance to Venus, Venus' thermal balance or imbalance is determined. At Venus' distance, the solar radiation will be diminished to a particular value. One then calculates the diameter of the planet with its atmosphere to determine how much area intercepts this solar radiation value. What is most important for the rest of the calculations is to determine, as precisely as possible, how much light is reflected by the planet's clouds and how much light or radiant energy is absorbed. This ability of clouds to reflect light is called its *albedo*. The higher the percentage of albedo, the more light is reflected; the lower the percentage of albedo, the more light is absorbed. This albedo figure must be known precisely, before it can be determined, at the cloud top temperature of 240 K, whether or not Venus is in thermal balance. Only a few percentages of difference, say, 0.76 and 0.78, or 2%, at the cloud tops, would have a pronounced effect on Venus and

would indicate about 10% more heat coming from the planet than is supplied by sunlight. What do the measurements actually show?

V. I. Moroz presented some of the earlier measurement of Venus' albedo. G. Muller measured it to be 0.878, or 87.8%, in 1893. Andre Danjon derived a spherical albedo for Venus of 0.815 in 1949, while C. F. Knuckles, M. K. Sinton and W. M. Sinton derived a spherical albedo for Venus, of 0.815, in 1961.[25] What Moroz did was average the readings to get the best possible indication of the actual albedo.

The more acceptable measurements are the most recent and use the latest technical developments, especially those taken from spacecraft orbiting Venus. These are fully discussed by F. W. Taylor, et. al. in an article titled, "The Thermal Balance of the Middle and Upper Atmosphere of Venus." The albedo found does not indicate thermal balance, but thermal imbalance. The reader will learn that these readings and their implications of a *thermal imbalance* were so distressing to the investigators that they rejected all the measurements showing that Venus was emitting more heat than the sunlight was delivering. F. W. Taylor states:

> "Measurements of albedo are more difficult to calibrate than those of thermal flux, because of the problem of obtaining an accurate reference source. Using Earth–based measurements, Irvine (1968) calculated a value for A [albedo] of 0.77 ± 0.07, which was later revised upward to 0.80 ± 0.07 by Travis (1975). The *Pioneer Venus* infrared radiometer had a 0.4 to 4.0 μm channel calibrated by a lamp from which Tomasko et. al. (1980b) obtained a preliminary albedo for Venus of 0.80 ± 0.02.
>
> "Another approach to determining the albedo is simply to assume that the atmosphere is in net radiative balance...[by equation]. In this way, a value of 0.79 + 0.02-0.01 has been obtained from *Venera* radiometry (Ksanfomality, 1977, 1980b) and [a value] of 0.76 ± 0.006 [has been obtained] from *Pioneer Venus* emission measurements (Schofield et. al., 1982).
>
> "*Clearly the Pioneer measurements of emission and reflection are not consistent with each other if net radiative balance applies.* A source inside Venus equal in magnitude to 20% of the solar input (i.e., accounting for the difference between A = 0.76 and A = 0.80) is very unlikely, since Venus is thought to have an Earth–like makeup, which would imply internal heat sources several orders of magnitude less than this. Also, even if such sources were postulated, it is difficult to construct a model in which these fairly large amounts of heat can be transported from the core to the atmosphere via a rocky crust without the latter becoming sufficiently plastic to collapse of the observed surface relief. This could be avoided if the transport was very localized, i.e., via a relatively small number of giant volcanoes. Although large, fresh-looking volcanoes do appear to exist on Venus...and the composition of the atmosphere is consistent with vigorous output from these, a simple comparison with terrestrial

volcanism shows that the volcanic activity on Venus would have to be on an awesome scale to account for the missing 5×10^{15} W [watts], or so, of power. A more acceptable alternative is that the preliminary estimate of 0.80 ± 0.02 for the albedo from the P. V. [*Pioneer Venus*] measurements is too high, since the uncertainty limit is now known from further work to be too conservative. (J. V. Martonchik, personal communication.) A fuller analysis of the P. V. [*Pioneer Venus*] albedo data—still the best, in terms of wavelength, spacial and phase coverage, and radiometric precision, which is likely to be obtained for the foreseeable future—is likely to resolve this puzzle. In conclusion, then, the best thermal measurements of Venus *with the assumption of global energy balance* yield a value for the albedo of 0.76 ± 0.01; this is the most probable value.[26] [Emphasis added]

Let us examine what was assumed as the truth regarding the reasons for Venus being in radiative balance. First, Taylor and his colleagues assumed that each of the albedos measured by all the other investigators was wrong or that the instrumental error range in every other investigator's case was always on the minus side. There were three measurements: one by Irvine of 0.77 ± 0.07, revised upward by Travis to around 0.80 ± 0.07; another by Tomasko et. al., using *Pioneer Venus* instruments to obtain a value of 0.80 ± 0.02; and another, based on the assumption that Venus was in thermal balance, by Ksanfomality, who used *Venera* instrumentation to calculate an albedo of $0.79 + 0.02{-}0.01$.

Thus, three albedo measurements indicated that the reflection of light, in terms of albedo, clustered around 0.80, the last two with tolerance levels plus or minus well above 0.76, accepted as the emission albedo by Taylor and his colleagues. If Venus reflected 80% of the light incident to it, but allowed 4% more light than was found by the emission albedo of 0.76, then the planet was emitting 20% more heat than sunlight delivered. Usually when a group of scientific measurements cluster around a particular value, as the 80% albedo did, scientists assume that the averaged value of these readings is a fairly good indication of the actual figure. By averaging the measurements, scientists achieve the best approximate value. However, since it was unthinkable that Venus could be in radiative imbalance, even based on one measurement that assumed balance but arrived at a figure of A = 79%, J.V. Martonchik reevaluated the *Pioneer Venus* data that gave an albedo of 80% $\pm$ 0.02% and recalculated the possible instrument error to make it large enough so that, on the minus side, it would agree with the lower 76% emission albedo. This was presented in a *private communication* that could not be analyzed by other investigators to determine whether or not Martonchik's correction was itself without error.

Second, Taylor and his colleagues assumed that Venus must possess an internal heat source almost equivalent to that of the Earth. Since the Earth generates an internal heat value much smaller than the value that suggests a 20% imbalance on Venus, they concluded that Venus, itself, could not be responsible for this additional heat. If Venus' emissions were actually this much greater, they said, Venus would have to be enormously volcanic, and—in the next breath—said Venus appears to exhibit a highly volcanic surface. They also admit that the Venusian atmosphere appears to contain the gases consistent with vigorous, volcanic outgassing. But all these correlations mean nothing. Taylor and his colleagues say such agreement between thermal imbal‑ance and Venus' volcanic surface and atmosphere is too incredible to accept, and reject all of this on the *one assumption* that Venus *must be* in thermal balance. Yet their conclusion states that the 0.76 ± 0.01 albedo is the *most probable* value. There is no evidence to suggest this. Nevertheless, if this assumption is correct, it should be corroborated by all the other readings taken by *Venera* and *Pioneer Venus* probes. These readings should exhibit demonstrable evidence of radiative balance throughout the rest of the Venusian atmosphere. This, in fact, is not the case. In fact, *all* the other readings deliver a death blow to any assertion that Venus' atmosphere is in thermal balance.

If the runaway greenhouse effect is correct, not only must the cloud tops exhibit thermal balance between solar input and infrared thermal output, but the lower atmosphere must show the same. If, as Velikovsky claims, the greenhouse effect presents only a minor contribution to Venus' high thermal emission, then as one gets closer to the surface of the planet, the measurements should show an even greater radiative imbalance than the 20% suggested by the cloud top readings. The nature of establishing radiative balance in an atmosphere is explained thus,

> "Radiative balance occurs [on a planet] at every level when the amount of downward, directed solar radiation that is absorbed is equal to the amount of infrared radiation that is emitted upward. When *local temperature* satisfies this balance, the atmosphere temperature is maintained."[27] [emphasis added]

M. G. Tomasko further reinforces this concept:,

> "In a steady state, the algebraic sum of the atmospheric heating and cooling rates due to all physical processes should [equal] zero at each location. At a given location, the heating and cooling rates depend on the state of the atmosphere: its temperature, pressure, composition, radiation and wind fields. A successful steady-state model for the thermal

balance of Venus' lower atmosphere will include the relationship of the heating and cooling rates to the atmospheric state, and show that the atmospheric structure leading to net zero heating or cooling at each location is equal to the observed structure." [28]

What, then, do the *Pioneer Venus* probes that entered the atmosphere show regarding this? According to Tomasko,

"Among the most accurate measurements of the temperature-pressure structure of the lower atmosphere of Venus [we find] those made by the four *Pioneer Venus* (PV) probes... The probe entry locations...vary in latitude from 30° South to 60° North [and measure] Day and Night...temperature profiles..." [29]

According to Richard A. Kerr, the editor of *Science:*

"When *Pioneer Venus* probes looked at the temperature, each one found more energy being radiated up from the lower atmosphere than enters it as sunlight... To further complicate the situation, the size of the apparent upward flow of energy varies from place to place by a factor of [two], which was a disturbing discovery." [30]

It is impossible to believe that in one area the greenhouse effect is twice as efficient as in others. From Velikovsky's theoretical view-point, parts of Venus' surface, known to the scientists as *coronea,* or hot spots, are much hotter than other areas of the surface. This would clearly explain these measurements.

From the National Aeronautics and Space Administration (NASA) publication, *Pioneer Venus,* we have the following:

"The measured, infrared fluxes [upward from Venus] show several anomalies, the origin of which is still being debated. Taken at face value, the anomalies suggest that parts of the atmosphere are transmitting about twice the energy upward than is available from solar radiation at the same level." [31]

Wind motions, pressures and all the other atmospheric conditions in Venus' atmosphere will not create a 50% imbalance. While the cloud top indicated a 20% thermal imbalance farther down in the atmo-sphere, the *Pioneer Venus* probes revealed a 50% radiative imbalance. This is in full agreement with Velikovsky's concept that the extremely hot planet's surface is the major heat source, indicated by greater fluxes of infrared closer to the thermal source. This finding was anathema to the investigators, as was the imbalance measured at the top of the cloud cover; but, What was to be done about these *most accurate readings?* Tomasko tells us exactly what some scientists did and why:

"The thermal flux profiles are surprisingly variable from site to site, in view of the great similarity in temperature profiles measured at these sites. In addition, at both the Night and North probes sites, they are much greater than the globally averaged solar net profile at low altitudes, *implying a substantive radiative imbalance in the lower atmosphere.* In view of the large and variable nature of these flux measurements, the investigators have searched for instrumental problems which could have affected the measurements and have found one *that could have* systematically increased the measured thermal net fluxes. ...The authors believe that they understand the vertical dependence of the flux errors, and by *adjusting the fluxes to reasonable values,* at low altitudes, *they have derived corrected fluxes."* [32] [emphasis added]

Corrected values are not really valid values. Not only did the scientists correct the two highest readings to "reasonable values"—meaning thermal balance values—they corrected every probe reading because every probe showed strong radiative imbalance at all levels of the atmosphere—much as was found for the cloud tops. Based on the investigators' beliefs, they simply made the data fit the theory of a greenhouse in radiative balance. However, one may be quite sure that, if the readings corroborated their beliefs of thermal balance, we would be showered by this direct confirmation of theory. But these measurements did not exhibit any evidence for balance. As Kerr admitted, "The much ballyhooed greenhouse effect of Venus' carbon dioxide atmosphere can account for only part of the heating, and heating, for other mechanisms is now in turmoil." [33]

Essentially, the thermal imbalance at the cloud tops was corroborated by *Pioneer Venus* probes below the clouds, so one can be confident that such is also the case throughout the entire Venusian atmosphere. What must also be pointed out is that the *Pioneer Venus* probes did not measure radiation all the way down to the surface. According to Seiff, "[t]emperature data were not obtained by the *Pioneer* probes below 12 kilometers [7 miles] altitude." [34] Tomasko indicates that "[t]he data from the temperature sensors of all four [*Pioneer Venus*] probes terminated at 13 kilometers altitude." [35] From this fact, one would expect that, based on Velikovsky's concept, the *Venera* probes that *did* approach the surface, measuring the solar fluxes in and infrared thermal fluxes emitted from the surface, would show an even greater radiative imbalance than did the *Pioneer* probes. In fact, that is what they did! Seiff states:

> The *heating rates* needed to warm the atmosphere from the Day probe [*Pioneer Venus*] profile to that of *Venera 9* integrated over altitude... is *45 times the midday solar heat absorbed* at 30° latitude... *This is also true for the Venera 10, 11 and 12 data* relative to the large probe data, for which necessary heating rates integrated over altitude are [somewhat less than] *40 times the mean dayside solar input* for the albedo of 0.71." [36] [emphasis added]

On page 219 of his article in *Venus,* Seiff essentially admits the same thing. The average reading from the *Venera* probes showed solar radiation absorbed by Venus. The surface infrared fluxes emitted were around forty times more than enters Venus' atmosphere as sunlight. The tremendous rise in infrared heat nearer to the surface is similar to the heat emitted by a white-hot block of metal, in that, as one puts one's hand near the block, the heat rises, but, at a certain closer distance to the block, the heat rises immensely.

This makes sense in terms of the greenhouse effect as well. The regions where most of the Sun's radiant energy is absorbed would, at those altitudes, generate the strongest greenhouse effect. Bruce Murray points out that,

> "On Venus...a smaller fraction of the incident solar energy penetrates the atmosphere all the way to the surface. Most of it is scattered back into space to provide the bright image that is seen through the telescope; much of the remainder is absorbed]within the atmosphere. Thus, Venus' atmosphere is heated more at the top and middle than at the bottom, and, in this sense, resembles more the shallow seas on Earth than its atmosphere." [37]

The ocean is an excellent greenhouse, but sunlight is primarily ab-sorbed at the upper 300 feet and this is the warmest region of the oceans. This explains why the solar greenhouse effect is greater at the cloud tops compared to Venus' output—by 80% of infrared heat energy, while the greenhouse effect's energy becomes reduced to 50% well below the clouds, and to 2.44% of infrared emissions given off by the planet at the surface. The regions of greatest solar absorption have the strongest greenhouse effect, compared to that emitted by the planet. The regions of least solar absorption have the weakest green-house effect, compared to that emitted by the planet.

This apparently is true at Venus.

Furthermore, all the readings showed the same clustering of values, of nearly forty times greater than the solar radiation input, which strongly suggests that the readings were basically correct. The scientists found these measurements so repugnant to their theoretical green-house concept that they were dismissed. Seiff tells us that "it is clear that the *Venera 9* day probe differences cannot be induced by solar heating but *must* be ascribed to other processes or to measurement uncertainties." [38] All, and I stress, *all* of the measurements, from the cloud tops to the lower atmosphere to the surface, gave consistent readings of radiative imbalance—contrary to everything greenhouse advocates claim. No single set of probe readings by *Pioneer Venus* or *Venera* suggests, in any way, that scientists have clear measurements supporting their contention of thermal balance. Every reading is another nail in the coffin of their theory. Realizing this, their actions to ignore or deny these findings point to a rigidity that has no place in science. This rigidity of thought is expounded in *Pioneer Venus,* a NASA publication. Let us remember Werner von Braun's admonition that the greenhouse effect is unproved by experiment.

> "One of the primary objectives of the *Pioneer Venus* multiprobe mission was to test... *the belief* that the "greenhouse effect" is responsible for the high surface temperature. [emphasis added] [After describing the measured thermal imbalance described above, they continue.]
>
> "Possible instrumental errors in this difficult measurement may be responsible. A possibility is that two of the probes entered regions that are unusually transparent to thermal radiation, but this is rather unlikely because much of the absorption [of infrared] is due to ubiquitous carbon dioxide, which makes up nearly all the atmospheric gas. The suggestion has been made that heat balance oscillates around an average state and that the anomalous measurements were made during the heating phase [Venus' temperature goes up and down over time with respect to its balance temperature and the probes, just by coincidence, descended on Venus during the heating period]. In spite of these difficulties in interpreting some of these observations, the greenhouse effect, coupled with global dynamics, is now well-established as the basic explanation of the high surface temperature." [39]

At every step of the investigation, every measurement showed radiative imbalance wherever readings were taken in the Venus atmosphere; and, at every step, every measurement that contradicted the runaway greenhouse effect concept of thermal balance was either changed, culled or set side based solely on a theoretical belief that Venus must be in thermal balance. We have observed this irrational behavior, of setting aside evidence negative to greenhouse assump-tions, from the beginning of this investigation right up to the end, based only on the theoretical greenhouse consideration.

The entire concept of the runaway greenhouse is supported by expediently using procedures that accountants and economists call "smoke and mirrors," not to say, pure denial of the facts. Velikovsky's theory regarding the thermal nature of Venus is in complete harmony with the real evidence and the actual measurements.

LIES, DAMNED LIES AND SAGAN'S CHART
This brings us to Sagan's chart which he claims definitely proves that Venus is not cooling down as Velikovsky claims. Sagan had stated earlier that "The steps in reasoned argument must be set out for all to see." And "*Experiments must be reproducible.*" [emphasis added]

Seven years prior to Sagan's first presentation of his criticism, Velikovsky wrote a paper titled "Venus, a Youthful Planet" which was published in the April 1967 issue of *Yale Scientific Magazine*. Velikovsky outlined the experiment by which his hypothesis could be proved correct or incorrect. In the article he stated,

> "I maintain that Venus' temperature is slowly but unmistakably decreas-ing. A measurement from the ground surface of Venus cannot be performed with methods now available with accuracy sufficient to detect the phenomenon in a matter of a few years; but, with a bolometer or thermocouple, a drop, even if in only fractions of a degree, could be detected from the cloud surface of Venus; *such measurements need to be repeated at each successive synodical period of Venus of which there are five in eight terrestrial years. It is understood that only figures obtained by one and the same observatory and from the same surface segments, preferably all taken during a quiet period of the Sun, can be profitably compared. On this new test, I am once more prepared to rest my case.*"[40] [emphasis added]

Sagan then informs us that, "An unbiased presentation of a microwave brightness temperature of Venus...are exhibited in Figure 1 below."[41] If indeed Sagan wishes to offer an "honest experiment" respecting this test, one would expect Sagan to have made sure that his chart does not violate any of the criteria which Velikovsky outlined as the basis for testing the cooling of Venus. In going over the chart, I find that Sagan has, in fact, departed from employing any of these criteria whatsoever. First, the spacing of the measurements show, by their random places along the time line, that these measurements were not "repeated at each successive synodical period of Venus." If these measurements were taken only at "successive synodic periods" they would be uniformly spaced. They are not.

Secondly, if the measurements were taken, as Velikovsky's carefully outlined experiment requires, there would be five measurements every eight years; but Sagan's chart has the first six measurements in a three-year period, the next seven measurements in a five-year period, the next nine measurements in another five-year period and the next four measurements in a year and a half period.

The measurements appear not all from the "same observatory" or of the "same segment" of Venus' cloud surface, nor do we know the behavior of the Sun during these measurements. Ralph E. Juergens remarks,

> "A new feature of Sagan's 'A.A.A.S. paper' is a figure purporting to show that contrary to Velikovsky's expectation that Venus must be cooling off with the passage of time, 'an unbiased' compilation of data shows nothing of the kind... What he (or someone else) has assembled to prepare this figure is the history of microwave brightness temperatures for Venus, as inferred from observations since about 1957. Since these observations were made, in general, at different wavelengths as the years

went by, what they actually show, if anything, is a progression of probings toward the truth of Venus' high temperature. In no way do they indicate the thermal history of Venus; the only history illuminated is that of the technological improvements in radio telescopes."[42]

Apparently Sagan doesn't seem to know the difference between the history of the technological improvements of radio telescopes and the thermal history of Venus. He also seems to think that violating all Velikovsky's criteria for an experiment is reproducing Velikovsky's experiment. Sagan has stated "There is a planetary double standard at work" regarding Velikovsky's hypothesis.[43] However, looking at this evidence of Sagan's, one wonders if there are any standards whatsoever at work regarding his chart.

Juergens above remarked that "this is Sagan at his worst." But perhaps it might be more precise to state that Sagan's chart which disregards all criteria for an experiment is really poor Sagan at his best.

A THERMODYNAMIC ANALYSIS

George R. Talbott also carried out an analysis of whether Venus may or may not be cooling and he sent his fully quantitative analysis to *Science*. Dr. Talbott at the time was

"doctor of science at Indiana Northern Graduate School... He is a scientist and author whose experimental background is divided between research and engineering applications. His scope and competence in mathematical physics and physical chemistry are evident from his two-volume *Philosophy and Unified Science,* published in 1977 by the international firm of Ganesh and Co. in Madras. Dr. David Lee Hilliker's introduction to this work lists some additional publications of Talbott, including a book in thermodynamics (*Electronic Thermodynamics*). Talbott also holds a basic patent, with D.R. Paxton, in the field of thermo-dynamics and is the creator of SPLAT, a program for nuclear radiation shielding. He has been chairman of the Dept. of Mathematical Studies at Pacific State Univ. in Los Angeles and presently earns his living making thermal predictions for complex space applications. Talbott's biography is to be found in *Who's Who in the West,* and in *Who's Who,* published out of Cambridge, England."[44]

Here is what Talbott states about the attack by Sagan and others respecting Velikovsky's theory of Venus' temperature:

'Suffice it is to say here that allegedly 'devastating attacks' on Velikov-sky's thesis of the recent origin of Venus are based upon incorrect thermal reasoning, and this is not a matter of opinion, but of thermody-namics and heat transfer. The calculations given in *Scientists Confront Velikovsky* are not being criticized for being simple and clear—these are virtues. The criticism is based upon their demonstrable irrelevance."[45]

Talbott states that a candescent Venus 3,500 years ago would be at a temperature about 750 degrees K, its present measurement today; this is based on the law and physics of thermodynamics. He states,

"...the relevant point is proven in a *fully quantitative manner* that a massive, molten body—quantitatively, a mass equivalent to Venus and having the Venus surface area and molten at between 1,500 degrees K and 2,000 degrees K—will transfer heat internally by flowing magma and will radiate its heat in such a way that in *exactly 3,500 years* its temperature is expected to be *exactly* 750 degrees K, which by measurement it *is*."[46] [his emphasis]

Talbott adds this about Sagan's greenhouse model,

"It is really impossible to argue intelligently that the present measured temperature of Venus can be *deduced exactly* from a [greenhouse] model which is 'naive' or 'inappropriate.'"[47]

When we compare Talbott's fully quantitative proof with Sagan's "biased" chart, we see the difference between Talbott's science and Sagan's science fiction.

SAGAN VERSUS SAGAN

Sagan states that Velikovsky,

"believes that Mars, being a relatively small planet, was more severely affected in its encounters with the more massive Venus and Earth and therefore, should have a high temperature... In the same section, he baldly states, 'Mars emits more heat than it receives from the Sun' in apparent consistency with his collision hypothesis. This statement is dead wrong."[48]

Velikovsky's answer is as follows:

"Sagan starts this section with the discussion of the temperature of Mars and accuses me of wrongly stating that according to the data known before the publication of *Worlds in Collision* in 1950, 'Mars emits more heat than it receives from the Sun...' Sagan says, 'This statement is however, dead wrong.'

"But, *Sky and Telescope*, March 1961, reported Sagan's opinion: 'It has long been known that the observed surface temperature of Mars is about 30 degrees centigrade higher than would result from the sun shining on an airless planet at its distance.'"

And in the *McGraw-Hill Encyclopedia of Science*, Vol. 8, (1971), an article titled "Mars", Velikovsky also informs us that, "Between the theoretical and observed temperature values of Mars—*reflecting* as a grey body—there is an actual excess of heat given off by Mars."[49]

Sagan has told us in both *Scientists Confront Velikovsky* and *Broca's Brain* that,

322   Charles Ginenthal

"The temperature of Mars has been measured repeatedly...and the temperature of all parts of Mars are just what is calculated" and that "Mars emits more heat than it receives from the Sun...is however dead wrong... What is more, this was well-known in the 1940's." [50]

However, in complete contradiction, Sagan informs us in *Sky and Telescope* for 1961 that, "It has long been known that the observed temperature of Mars is at 30 degrees centigrade higher than would result from the Sun shining on an airless body." Thus, about 20 years after it was known that Mars is not hotter than expected, Sagan tells us it is hotter than expected. Sagan adds,

"It is difficult to understand this set of [Velikovsky's] errors, and the most generous hypothesis I can offer is that Velikovsky confused the visible part of the electromagnetic spectrum...with the infrared part of the spectrum... But...Mars turns out to have exactly the temperature everyone expected it to have, we do not hear of this as a refutation of Velikovsky's views. There is a planetary double standard at work." [51]

After learning of Sagan's contradictory statements respecting the temperature, I am in full agreement with him that there is indeed a "double standard at work."

Sagan remarks that, "The conclusion is clear, Mars, even more than Venus, by Velikovsky's argument, should be a 'hot planet.'" [52] This remark is based on Sagan's belief that Mars should be a "hot planet" in terms of *his* interpretation of Velikovsky's hypothesis of the temperature of Mars. But to be quite accurate, it is really Sagan's misinterpretation, represented as Velikovsky's hypothesis of the temperature of Mars.

SAGAN AND HIS MARTIAN GREENHOUSE EFFECT

Nevertheless, let us examine Sagan's view of why Mars is supposed to be somewhat warmer than believed. Velikovsky (in *Kronos III*; 2, p. 47) states that "Sagan wishes to ascribe the 30 degree difference to a greenhouse effect produced by the very rarefied atmosphere on Mars and for this he assumed the presence of a certain (not proven) quantity of water (vapor? at what temperature?) besides carbon dioxide." Having gone through the greenhouse effect on Venus and having found it wanting we need only cite one source to show Sagan's Martian greenhouse work is ignorant. Martin Caiden in his book, *Destination Mars,* (NY 1972), p. 196, writes:

"As one of the *Mariner* scientists explained, the Martian temperatures have nothing to do with air [atmosphere] of Mars, but with the surface. What we measure with our instruments is the amount of solar radiation, mainly infrared, that strikes and is absorbed by the Martian surface.

"There cannot be air temperature as we understand this term because the Martian atmosphere is so thin it can't retain heat. There simply aren't enough molecules present, so that a thermometer, as we know this instrument, can record variations in temperature.

"The atmosphere of Mars in many ways is like the environment of space. It's so thin it's an atmosphere in scientific terms only."

So much for Sagan's understanding of the temperature of Mars which is like his understanding of Velikovsky's hypothesis of Mars—pure fantasy.

WHAT OTHER SCIENTISTS SAY

Finally, let us see what other scientists have to say about Sagan's greenhouse effect hypothesis. These were reported by Professor Greenberg:

"The University of Wisconsin's Vener Suomi claims that 'the infrared losses from the [atmosphere's] deep layers are too large to be supplied by the sunlight reaching the surface' of Venus. 'Thus, the greenhouse effect is unable to account for the high surface temperatures.' (See *Science News*, Vol. 116, 11/3/79, p. 308)

"Jonathan Eberhart, Space Sciences editor for *Science News* is also well aware of the greenhouse quandary. He writes, 'Venus is very hot (perhaps 200 degrees hotter than airless Mercury, which is barely half as far from the Sun) that the greenhouse must be an amazingly effective one...the scientists are faced with explaining an almost perfect system and there is still disagreement about whether the job has been done... The question is still open.' (*Sciquest*, Vol. 53, No. 1, January 1980, p. 12)...

"...Terence Dickinson—science writer and the present editor of *Star and Sky*—remains especially openminded to the question of the greenhouse effect. [He states], 'The mathematical models for the Venus greenhouse effect have been controversial, at least in their detailed application, for years now. Although the unexpectedly large [sic] amount of water vapor in the lower atmosphere tends to support the greenhouse theory, Velikovsky's contention of internally generated heat cannot yet be ruled out.' (See "Venus—To Hell and Back", *UFO Report*, August 1979, p. 77)

"V.A. Firsoff remains justifiably skeptical where the greenhouse effect is concerned while, in a recent article on Venus for *Science* (1/18/80), Richard A. Kerr failed to endorse that [greenhouse] effect as a foregone conclusion...

"...Moreover, on Dec. 9, 1978, it was found that 'the region of the highest thermal emission on this day was on the nightside of the planet, at roughly 45 degrees N.' This sharply contrasted with earlier observations made on April and May 1977 at an identical solar phase angle, in which the region of intense emission was consistently on the sunlit side of the terminator [the line between day and night]. This striking change of appearance indicates that at least *some of the major thermal features are not solar fixed."* (*Science,* Vol. 203, 23 February 1979, p. 789). [emphasis added][53]

Velikovsky has reported, "Two discoveries made since this paper was written in 1963 need to be recorded. One of them concerns the problem of Venus' temperature. Measurements were made of the 11 cm wavelength emission from Venus, at Green Bank Observatory and reported by K.I. Kellermann in the September 1966 issue of *Icarus,* and strongly suggest that the heat of the ground surface of Venus comes, not from the atmosphere above, but from the *sub-surface."*[54] Professor Greenberg notes, "It should be noted that Carl Sagan was an Associate Editor of *Icarus* at the time. Was he oblivious of Kellermann's findings?"[55]

Werner von Braun in his co-authored book, *New Worlds: Discoveries From Our Solar System,* (NY 1979), p. 129, states,

> "...scientists would like to be sure that the greenhouse effect is indeed the cause of the hot surface and lower atmosphere. So far it is only a theory, not yet proven by experiment." While Richard A. Kerr, in *Science* for Jan. 18, 1980, pp. 292-293, states, "The much ballyhooed greenhouse effect of Venus' carbon dioxide atmosphere can account for only part of the heating."

Reginald E. Newell explains what is wrong with the greenhouse effect theory quite simply with this analogy.

> "...imagine a well insulated domestic oven and inquire what happens when only a 10 watt bulb giving an energy input of about 10 w/M2 [having a 10 watt bulb on every square meter (about a square yard) of Venus' surface]. Rarely does the air temperature inside the oven rise to 750 K! Now imagine the oven is placed on top of the molten lava of an active volcano; it will soon reach the required temperature. The energy for the process is that which has come by conduction from the process keeping the rock molten. It seems that the latter situation, rather than the former, corresponds to Venus..." [See footnote 10]

If the Greenhouse Effect mechanism is correct it would explain the high surface temperature of the planet; it doesn't, and this is admitted. It would explain why the heat source—the clouds—are cooler than the heat sink—the surface; it doesn't, and this is admitted. It would explain the super-rotation of the upper atmosphere around the planet in four to five days; it doesn't. Why doesn't it account for any of these phenomena observed in the Venusian atmosphere which a good theory ought to do?

Sagan has told us that, "not all scientific statements have equal weight." To paraphrase him: The science of thermodynamics and the first and second laws of thermodynamics are on extremely firm footing. Sagan has accused Velikovsky of attempting to rescue religion and astrology. What could be more dogmatic than offering a theory that violates the second law of thermodynamics and not explaining how this can be done with a process that circumvents this law? Since there are no Hadley cells operating in the Venusian atmosphere there is no way to move the cool greenhouse upper atmospheric gases to the surface. Apparently Sagan has a solution for this problem or seems to believe in miracles.

NOTES & REFERENCES

[1] SCV, pp. 81-82; B.B., p. 116.

[2] Young, Andrew and Young, Louise; "Venus" in *The Solar System*, (San Francisco 1975), (Sci Am Books), op. cit., p. 53.

[3] Ibid., p. 56.

[4] SCV, p. 82; B.B., p. 117.

[5] Greenberg, Lewis M, *Kronos, IV*, 4. p. 9.

[6] Greenberg, Lewis M, *Kronos*, II, 1, pp. 104-105.

[7] Chapman, Clark R.; *The Inner Planets*, (NY 1977), pp. 102-103.

[8] Booth, V.E., Bloom, M.L.; *Physical Science*, (NY 1972), p. 219.

[9] Kerr, Richard A.; *Science*, Vol. 207, (1980), p. 289.

[10] Newell, R.E.; *Speculations in Science and Technology*, Vol. 7, No. 1, (1984), pp. 51-57.

[11] Sagan, C.; op. cit., *The Cosmic Connection*, p. 84.

[12] *Kronos*, IV, 2, p. 31.

[13] Mage, Shane; "Back to the Drawing Board", *Kronos IV*, 4, p. 14.

[14] Firsoff, V.A.; *Astronomy and Space*, Vol. 2, No. 3, (1973).

[15] Firsoff, V.A.; *The Interior Planets*, (London 1968), p. 102.

[16] Juergens, Ralph E.; *Kronos IV*, 2, pp. 74-75.

[17] Mage, Shane; *Kronos IV*, 4, p. 14.

[18] Ibid., pp. 14-15.

[19] SCV, pp. 79-80; B.B., pp. 114-115.

[20] Greenberg, Lewis M.; "Venus' Internal Heat", *Kronos VI*, 2, pp. 21-22.

[21] Newell, R.E.; *Speculations*, op. cit., pp. 51-57.

[22] SCV, p. 80; B.B., p. 115.

[23] Juergens, Ralph E.; *Kronos IV*, 2, p. 74.

[24] Mage, Shane; *Kronos IV*, 4, p. 14.

[25] V. I. Moroz, "Magnitude of Albedo Data," *Venus*, ed. Donald Hunten, Lawrence Colin, Thomas M. Donahue, V. I. Moroz (Tucson, Arizona, 1983), p. 30.

[26] F. W. Taylor et al., "The Thermal Balance of the Middle and Upper Atmosphere of Venus," *Venus*, ed. Donald M. Hunton, Lawrence Colin, Thomas M. Donahue and V. I. Moroz (Tucson, Arizona, 1983), pp. 657-658.

[27] *Encyclopaedia Britannica, Macropaedia*, 19 vols. (Chicago, 1991), Vol. 2, p. 167.

[28] M. G. Tomasko, "The Thermal Balance of the Lower Atmosphere of Venus," *Venus*, ed. Donald M. Hunten, Lawrence Colin, Thomas M. Donahue, V. I. Moroz (Tucson, Arizona, 1983), pp. 620-621.

[29] *Ibid.*, p. 606.

[30] Richard A. Kerr, "Venus: Not Simple or Familiar, but Interesting," *Science* 207 (1980): 289.

[31] *Pioneer Venus*, op. cit., p. 127.

[32] Tomasko, op.cit., pp. 612-613.

[33] Ker, loc. cit.

[34] Seiff, op. cit., p. 223.

[35] Tomasko, op. cit., p. 606.

[36] Seiff, op. cit., p. 226.

[37] Bruce Murray, Michael C. Malin and Ronald Greeley, *Earthlike Planets* (San Francisco, 1981), p. 55.

[38] Seiff, loc. cit.

[39] *Pioneer Venus*, loc. cit.

[40] Velikovsky, I.; "Venus a Youthful Planet", *Yale Scientific Magazine,* op. cit.

[41] SCV, pp. 81-82; B.B., p. 117.

[42] Juergens, Ralph E.; *Kronos* III, 4, p. 98.

[43] SCV, p. 79; B.B., p. 114.

[44] *Kronos* IV, 2—back cover.

[45] Ibid., p. 9.

[46] Ibid.

[47] Ibid., p. 10.

[48] SCV, p. 78; B.B., pp. 113-114.

[49] *Kronos* III, 2, p. 47.

[50] SCV, p. 78; B.B., pp. 113-114.

[51] SCV, p. 79; B.B., p. 114.

[52] Ibid.

[53] Greenberg, Lewis M; *Kronos* V, 2. pp. 87-88.

[54] *Kronos* IV, 3, p. 67.

[55] Ibid., footnote.

SAGAN'S NINTH PROBLEM

THE CRATERS OF VENUS

TECTONIC FEATURES OF VENUS

In this section, Sagan turns to Venus' topography to prove that Venus is at least as ancient as the Earth. He states that, "radar observations reveal enormous linear mountain ranges, ringed basins and a great rift valley, with dimensions of hundreds to thousands of kilometers. It is very unlikely that such extensive tectonic or impact features could be stably supported over a liquid interior by such a thin and fragile crust" that Velikovsky's thesis requires. [1] Here, Sagan has told us that the topography of Venus is produced by plate tectonics, as are the topographic features of the Earth are thought to be. He informs us that Venus has "enormous linear mountain ranges"; these would be comparable say to the linear chain of mountains running from the southern tip of South America to Alaska. The "great rift valley with dimensions of hundreds of thousand of kilometers" in length would be comparable to the rift valleys of the Earth. But are these topographic features really the product of plate tectonics on Venus? More to the point, does Venus possess a set of tectonic plates? Or even more to the point, does Venus exhibit tectonic topography? In this regard C. Leroy Ellenberger reports that,

> "G.W. Wetherill and C.L. Drake reported [in *Science,* Vol. 209, (July 4, 1980), p. 103] that 'Preliminary radar data obtained from *Pioneer Venus* suggest that plate tectonics is also absent on that planet' and G.H. Pettengill *et. al.* reported [in *Scientific American,* (August 1980), p. 65, that] 'The tectonic motion of large plates appears not to have played a dominant role in altering the surface [of Venus] that tectonic motion has on the Earth.' While H. Masursky et. al. stated flatly [in the *Journal of Geophysical Research* 85: A13, (Dec. 30, 1980), p. 8232, that] 'An integrated global pattern of subduction troughs or mid-basin ridges, indicative of global tectonism, has not been identified from the altimetry data [for Venus]'" [2]

Regarding tectonism, which requires that Venus' surface be broken into several pieces or plates, Isaac Asimov states from this radar *Pioneer* map of Venus' surface,

> "It would appear that Venus' crust is quite different from that of Earth. Earth's crust is relatively thin and is broken into half a dozen large plates and a number of smaller ones. These plates move very slowly relative to

each other, crushing together or slipping one under another, or pulling apart. These movements, called 'plate tectonics' produce earthquakes and volcanoes at the joints, build mountain ranges where the plates crush together, ocean deeps where one shoves under another, rifts where they pull apart.

"Venus, on the other hand, may have gone through such a period early in its history, but it seems that at present, its crust is more or less one piece. Most of Venus' surface seems to be the kind we associate with continents rather than sea bottom…whereas on Earth, we have a vast sea bottom making up 7/10 of the planet, with the continents placed within it like large islands. Venus has a huge super continent that covers 5/6 of the total surface, with small regions of lowland making up the remaining sixth."[3]

To be precise, Venus' topography is almost exactly the opposite of that of the Earth. Deep ocean basins make up 7/10 of the Earth's topography, while only 1/6 of these basins make up the topography of Venus; and these basins are shallow, not deep. Conversely, 3/10 of the Earth's topography contains continental crust, while on Venus, continental crust make of 5/6 of the surface. While continental rises comprise thirty percent of the Earth's surface, Venus' two tiny continental rises comprise less than five percent of its surface. Thus, the features of Venus appear not to be the products of plate tectonics, notwithstanding Carl Sagan's statement.

VENUS' MANY ANCIENT IMPACT CRATERS

Sagan also tells us that,

"In 1973 an important aspect of the surface of Venus, verified by many later observations, was discovered by Dr. Richard Goldstein and associates… They found from radar that penetrates Venus' clouds and is reflected off its surface, that the planet is mountainous in places and cratered abundantly; perhaps, like parts of the Moon, saturation cratered —i.e., so packed with craters that one crater overlaps the other."[4]

Sagan confidently says that the evidence appears to show that Venus is cratered like the Moon. This is quite remarkable, because Isaac Asimov tells us such pictures are rather fuzzy. He writes,

"Beams of microwaves, striking Venus and being reflected, can yield information about the surface just as reflected light would. Light, of course, would be better because we can analyze the light directly by eye, whereas for radar we must use complicated instruments. Then, too, microwaves have much longer wavelengths than light does, and that gives us a fuzzier picture, one that misses the fine details. It is as if a long legged human being stepped over irregularities without noticing them, while an ant having to clamber over those same irregularities, would be aware of each in detail.

"Nevertheless, microwaves are long enough to go through the clouds on Venus both coming and going, while light waves can't. It is better to have a fuzzy picture than none at all...

"Even so, Earth based radar studies showed the presence of a couple of mountain ranges, a possible large volcano, and a giant canyon."[5]

Asimov does not mention that the Earth based radar was able to see lots of craters on Venus, like those on the Moon. Greenberg reports,

"According to a report in *Science*, (1/18/80, p. 292), Gordon Pettengill of MIT and team leader of the *Pioneer* orbiter's radar mapper experiment feels that 'There are few, very few features that *Pioneer Venus* has seen that I would call impact craters for sure.'"[6]

But Sagan has Venus cratered like the Moon with craters overlapping craters, and craters inside craters. Rick Gore informs us,

"Until the orbiter's cloud penetrating radar began, crudely mapping the Venusian surface, we knew relatively nothing about the planet's terrain. Then a surge of new radar imaging began backing up the emerging portrait of *Venus as a volcanic cauldron*. Remarkable computer enhanced data from radio telescopes at Arecibo, Puerto Rico, and Goldstone, California began revealing features with shapes suspiciously like lava flows across the planet.

"Stunning images from the Soviet Union's *Venera 15* and *16* orbiters revealed not only abundant evidence of volcanism, but also *far fewer ancient meteoric impact craters than on the Moon or Mars*."[7] [emphasis added]

If Sagan is still unconvinced of the fact that Venus is not cratered just like the Moon, I refer him to that well-known astronomer, Carl Sagan, who wrote in his book *Comet* the following:

"...a radar image of the surface of Venus obtained with the largest telescope on Earth...[shows] the smaller crater at the center of Lise Meitner, about 60 kilometers across. There is still debate about whether such craters are of impact of volcanic origin, *but the sparseness of craters on Venus* shows that the surface is continually being modified—probably by volcanism."[8] [emphasis added]

Thus, the logic seems to be that the surface of Venus, which according to Sagan, appears to be cratered like the Moon, is not cratered like the Moon, nor is Venus definitely impact cratered.

VOLCANISM ON VENUS

Sagan has mentioned that volcanism is changing the surface of Venus. Why? In *Science Digest*, the reason is given that, "...volcanism is the only plausible way for Venus to rid itself of internal heat. On Earth, heat is vented through constant movement of crustal plates; that mechanism seems to be lacking on Venus."[9] How then does this volcanism affect the topography of Venus?

The answer is that lava flows that reach the Venusian surface are at great temperature. However, they encounter a surface that is also extremely hot; therefore, they may take a longer time to cool and have more than sufficient time to travel and to reach the deepest elevations of the planet. Furthermore, even after these lava flows are in place, the heat at the surface will maintain them in a kind of fudge-like state so that they continue to deform slowly like fudge over the surface. If this analysis is correct, Venus should exhibit a topography of relative flatness or smoothness with generally insignificant differences in the overall height and depth of its topography. M.Y. Marov informs us in the *Annual Review of Astronomy and Astrophysics* that,

> "Radar measurement of the near equatorial belt of Venus (within + 10 degrees by latitude)...did not reveal significant elevations on the surface. The maximum deviation from flatness on a horizontal scale of 200-400 km are 2-3 km. Thus, Venus is smoother than the Earth... This is confirmed by the new radar mapping of several spots on Venus' surface with much better horizontal resolution up to 12-15 km. The many visible craters are much smoother than those on the Moon, Mercury or Mars. The planetary topography was also surveyed with bistatic radar in a number of regions along the orbit traces of *Venera 9* and *10* orbits... Both flat and mountainous terrain were found with maximum height variations less than 3-5 kilometers... One may thus conclude that the microrelief of the Venusian surface...is in general also smoother than, for example, that of the Moon."[10]

Thus, it appears very probable that volcanism dominates the surface topography of Venus. Lawrence Colin, in *The Planets,* (NY 1985), p. 280, informs us that the three major highland features—Aphrodite Terra, Ishtar Terra and Beta Regio—are almost certainly volcanic creations. He writes, "Ishtar Terra is made up of a giant volcano plus an uplifted plateau. Aphrodite Terra...also displays some features which may be volcanic. Beta Regio appears to consist of two side by side giant volcanoes, both of which may still be active."

If this is so, how can Venus have two small continents about the sizes of the United States and Africa? The answer seems to be that they are recent lava flows that were very great and that have not yet settled. Rick Gore tells us that,

> "Besides sterilizing the planet, Venus' intense heat may be generating an exotic kind of erosion. On Earth, rain wind and waves erode continents. On Venus, 'viscous creep' may be the great leveling agent.
>
> "'As rocks heat,' MIT's Sean Solomon explains, 'they soften and can flow. Mountains may actually creep away, like Silly Putty spreading out. Viscous creep on Venus could, at least be as effective as water erosion on Earth in reducing mountainous relief. *If so, that further implies that all the high features we see on Venus are young.*'"[11] [emphasis added]

If, as Dr. Solomon tells us the high features of Venus are young, then the low features such as the larger craters cannot be 4 to 4.6 billion years old as Sagan seems to believe.[12] If "viscous creep" is anywhere nearly as effective as water erosion on the Earth, then not a single ancient crater, from the early bombardment of Venus 4.6 to 4 billion years ago, would have survived this erosion process.

Sean C. Solomon et. al. analyzed the viscous creep in a paper titled "On Venus' Impact Basin; Viscous Relaxation of Topographic Relief", in the *Journal for Geophysical Research,* Vol. 87, (1982), pp. 7763-7771 and showed that high surface and crustal temperatures on Venus would erase all large impact craters in geologically short times. George E. McGill, et. al. in *Venus,* (Arizona University 1983), p. 95-96, discussed Solomon's work:

> "Creep rates of rocks are strongly temperature dependent. If the high surface temperature of Venus implies a much hotter crust than on earth the entire crust of Venus may be much less resistant to creep than the crust of Earth... Cordell and Solomon et. al. have shown that large features such as impact basins will suffer essentially complete relaxation of their topographic relief in times on the order of 1 Gy [one billion years] even if the [Venusian] crust is dry...

> "This suggests that present elevated regions are young or continuously renewed, and that (unless we illogically assume that the relief of Venus is uncharacteristically high at present) elevated regions would have risen and spread to oblivion many times in the history of Venus."

Therefore, if Venus is an old planet all the high surface features should have relaxed into softly rounded hills, but this is not the case. Venus has some steep topography. All the impact craters, if such do exist, should have disappeared billions of years ago. All the great rifts and valleys should have had their valley walls flow together to cause them to disappear. The very existence of such topography logically demands that all the surface features of Venus are extraordinarily young or they would not exist as they do in their present state.

There are, indeed, craters on the Earth, but none has survived with rims and basins intact since the beginning. There is not a single crater surviving on the Earth that is 4 to 4.6 billion years old. Venus would be no different given an erosion rate "at least as effective as water erosion on Earth."

WEATHERING OF VENUS' ROCKS

Again, if Venus is 4.6 billion years old, there would also be weathering of its rocks at the surface. Eric Burgess informs us that,

> "The rocks of Venus undergo different types of weathering [breakdown than the Earth]. Chemical weathering would be expected to decompose

olivines, pyroxenes, quartz and feldspars into magnesite, tremolite, dolomite and sulfides and sulfates. Mechanical weathering would be expected to disintegrate rocks by spalding and preferential chemical weathering and possibly by wind erosion...

"Although winds on Venus near the surface do not blow at high velocity, they represent the movement of extremely dense air by terrestrial standards, sufficiently dense to move particles up to several millimeter diameter across the surface of Venus."[13]

Asimov explains that Venus',

"...surface winds were recorded that weren't very fast, only a little over 11 kilometers (7 miles) an hour. Since the atmosphere of Venus is so dense, however, such winds would have the energy of earthly winds blowing at 105 kilometers (65 miles) an hour. The 'gentle' wind is just about equivalent to a hurricane on Earth." [14]

These winds should blow debris into basins and form dunes of material similar to sand dunes. However, Burgess tells us that, "The radar data are...inconsistent with Venus being covered by vast areas of windblown debris." [15] This has been confirmed by Magellan data. Venus lacks a planetary soil covering its entire rocky surface.

VENUS' HYDROFLUORIC—HYDROCHLORIC AGE

Sagan has stated that the sulfuric acid cloud model, "is consistent with the chemistry of the Venus atmosphere, in which *hydrofluoric and hydrochloric* acid have also been found."[16] What Sagan has not discussed is that these acids, when they react with rocks, are quickly neutralized. If this is so, over 4.6 billion years of history a great deal of new volcanic rock has been constantly emplaced on Venus' surface, and these gases as they reacted with the new surface rock would all have been neutralized. Young and Young discuss this and relate that,

"Among the more exotic materials proposed for the clouds only one has been detected spectroscopically. It is hydrogen chloride and it was found along with hydrogen fluoride by William S. Benedict of the University of Maryland in the spectra recorded by the Conneses. Both gases are highly corrosive; when they are dissolved in water, they yield hydrochloric acid and hydrofluoric acid. Their abundance is too low for them to be in the clouds, but that they should be present in the atmosphere at all is a surprise." [17]

The amount of water vapor in the Venus atmosphere, though small, is sufficiently large enough to convert hydrogen fluoride and hydrogen chloride quickly into hydrofluoric and hydrochloric acid.

In the chart at the bottom of page 53, Young and Young show hydrochloric and hydrofluoric acid moving to and from the surface of Venus. Thus, these acids interact with the surface rock. They go on to say, "Such strong acids could not survive for long in the Earth's atmosphere; they would react with rocks and other materials and soon be neutralized."[18] Therefore, it is assumed *ad hoc* that the high temperature cooks hydrogen chloride and hydrogen fluoride out of the surface. But this theory assumes that these gases would not be neutralized as they formed acids in the rocks. Young and Young say, "A number of assumptions are implicit in this hypothesis: that the rates of chemical reactions at the surface are high, that the atmosphere and the surface are in chemical equilibrium and that the effects of circula - tion in the atmosphere are small enough to be neglected."[19] With enough "ifs" it seems one can prove anything that fits the gradualist picture of Venus' atmosphere and surface. But if one or more of these *ad hoc* explanations is incorrect (and out of three variables, this is quite probable) then there is no adequate explanation for these gases other than a very youthful Venus has not neutralized them as yet.

Sagan has told us that, "radar observations reveal enormous linear mountain ranges." However, in *The New Solar System,* the book in which Sagan wrote an introduction, we are informed, "Venus high - lands are unlike any topography seen on the smaller terrestrial planets."[20] Lawrence Colin informs us as late as 1985, "...an integrat- ed pattern of ridges and valleys together [associated] with continental plates, has not been identified."[21] Nearly every statement by Sagan is contradicted by the literature. How can an astronomer be so ignorant of the findings of other authorities in his own field? Shouldn't Sagan explain why the literature on Venus contradicts most of his assertions?

WHAT HOLDS UP VENUS' HIGH MOUNTAINS?

For example, Sagan states,

> "In 1967 Velikovsky wrote: 'Obviously, if the planet is billions of years old, it could not have preserved its original heat; also any radioactive process that can produce such heat must be of a very rapid decay [*sic*], and this again would not square with an age of the planet counted in billions of years.' Unfortunately, Velikovsky has failed to understand two classic and basic geophysical results. Thermal conduction is a much slower process than radiation or convection, and, in the case of the Earth, primordial heat makes a detectable contribution to the geothermal temperature gradient and to the heat flux from the Earth's interior. The same applies to Venus. Also, the radionuclides responsible for radioactive heating of the Earth's crust are long lived isotopes of uranium, thorium and potassium—isotopes with half-lives comparable to the age of the planet. Again, the same applies to Venus.
>
> "If, as Velikovsky believes, Venus were completely molten only a few thousand years ago—from planetary collisions or any other cause—no more than a thin outer crust, at most ~100 meters thick, could since have

been produced by conductive cooling. But the radar observations reveal
enormous linear mountain ranges, ringed basins, and a great rift valley,
with dimensions of hundreds to thousands of kilometers. It is very
unlikely that such extensive...features could be stably supported over a
liquid interior by such a thin and fragile crust."[22]

Sagan's attack essentially boils down to the view that if Venus is as hot
in its interior as is assumed, it is not possible for Venus to support its
small continents. Of course, Sagan's assumption that Venus would
only have a 100 meter thick crust is only an assumption. If Venus'
surface temperature is 750 degrees K, to become molten at say about
2,000 degrees Celsius, the depth of the crust might easily be one or
two miles, and thus, the small continents of Venus could be supported.
But this too, like Sagan's analysis, is only conjecture. Therefore, what
is needed is another celestial body to compare with Venus that
astronomers know is extremely hot inside and see if such a body can
support large mountainous masses of material. Such evidence would
not be conjecture, but well-known physical and observational fact that
could help settle this question.

Does such a terrestrial body exist with extreme internal temperature?
The answer is 'yes': it is Jupiter's Galilean moon Io! Because of the
nature of Io's orbit, this moon is stressed by Jupiter's gravity so that
the tidal forces on this small satellite are enormous. Io is deformed so
strongly that its surface is literally being changed rapidly by volcanic
activity. Pictures of the satellite show volcanic explosions of great
intensity. According to Joseph Veverka in *The Planets,* edited by
Byron Preiss (NY 1985) p. 126, between the observations of *Voyager I*
made in March 1979 and those of *Voyager 2* made four months later, a
completely new volcano developed on Io's surface. Veverka states
specifically that, "Io's surface is changing at an astonishingly rapid
pace." The pace, however, is far more rapid than that on Venus.

Therefore, one asks how hot is it inside this satellite? Bradford A.
Smith, et. al., in an article titled, "The Jupiter System Through the
Eyes of Voyager I" in *Science,* Vol. 204, for 1979, p. 951, tell us that,

"Probably the most spectacular discovery of the *Voyager* mission has been
the existence of active volcanoes on Io, erupting materials to heights of
several hundred kilometers above the surface. The first discovery of an
active volcanic eruption is described by Morabito, *et. al.;* it appeared as
an enormous umbrella shaped plume rising 270 km (162 miles) above the
bright limb. Since this discovery, six additional volcanic plumes have
been found; most have been seen several times. In the likely case that the
trajectories are ballistic, the altitudes measured on images taken in the
clear filter, imply eruption velocities of several hundred meters to [about]
1 km/sec... What causes such violent activity?"

Heat from radioactivity has been ruled out because Io is thought to be at least 4.6 billion years old. Tidal flexing of Io would produce a tremendous amount of frictional heat in Io's interior. Thomas Gold has calculated the amount of heat needed to expel materials up to velocities of 1 kilometer per second in *Science,* Vol. 206 (1979), p. 1071. The temperature Gold arrives at, 6,000 degrees K, would have to be generated inside Io to produce these plumes. The surface temperature of the Sun is also 6,000 degrees K. Io, however, is a small body (compared to the Earth which also has a 6,000 degree K core temperature), so its heat affects its surface.

Earlier we described the large volcanic craters on Io which are quite like the craters on Venus.

What is quite apparent is that Io is enormously hot inside like Venus and has a surface that is changing much more rapidly than the surface of Venus. Thus, the question: Does Io possess high mountain ranges? Io is tiny by comparison with Venus; it is somewhat larger than our Moon while Venus is somewhat smaller than the Earth. Therefore, large aggregates of mountains on Io would be comparable to small continents on Venus. In this respect, Joseph Veverka, cited above, in *The Planets,* pp. 224–226, states that on Io, "*There are also high mountains and steep cliffs.*" [emphasis added] However, to save the uniformitarian view, Veverka adds, "The high mountains also indicate that Io's crust must be fairly solid (at least in certain areas), even though the intensity of tidal heating must mean that the satellite is molten at a relatively shallow depth." This is essentially assuming all that is to be proven. High mountains with steep cliffs exist on Io, but why they have not sunk almost immediately back into the liquid mantle is without explanation. The truth is, no one really knows conclusively about the nature of the inside of planets in spite of highly sophisticated techniques developed to analyze planetary interiors. Venus should easily support small continents if Io can support high mountain ranges. Sagan's argument and assumption is thoroughly contradicted by Io's high mountain ranges. Once again, the astronomical evidence he offers is not to be taken seriously.

Not only does Io present a problem for how high mountains can be supported by a body in a highly molten state, but Mars presents a similar problem although it is not in a highly molten condition. The Tarsis bulge on Mars is a continental sized bulge 5,000 kilometers in diameter and 10 kilometers high. Basically, the Tarsis bulge is too heavy to remain at its present level. The *McGraw-Hill Encyclopedia of Science and Technology,* Vol. 10, (1987), p. 477, discussed this problem. "Most of Mars' tectonic features are associated with Tarsis and it affects approximately one-quarter of the entire surface. Its formation appears to have begun some 3 to 4 x10^9 [3 to 4 billion] years ago, long before the volcanoes that now dot its crest and surroundings. A fact puzzling to scientists is that Tarsis is, in effect, top heavy—its bulk should not be sitting so high on the crust, but rather sink in much as an iceberg submerges most of its volume beneath the ocean's surface." Thus, Sagan's argument is reduced to this: On both Io, a highly molten body, and on Mars, a body that is not molten, masses of material are supported high on the surfaces where they should sink into the crust, but they don't, but Venus, which is perhaps midway between the Io and Martian state, should not support a large mass high on its surface. It is extremely difficult to accept Sagan's logic on this.

According to Robert D. Reasonberg of M.I.T., in *Astronomy,* for Sept. 1982, p. 10, the high Venusian elevations described by Sagan cannot even be supported by a crust that is as thick as that of the Earth. After discussing the high elevations of Beta and Aphrodite, he states, "From what we understand about planets, there's simply no way in which these [high] anomalies could be supported by the strength of the interior of the planet. They are probably supported by dynamic mechanisms, upwelling of the mantle." Thus, even if Velikovsky never presented his theory, the scientists cannot explain why the high elevations on Venus do not sink into the planet; no more can they explain this than the high elevations on Io which do not sink into it. If upwellings support this concept, why shouldn't upwellings support Velikovsky's concept?

VENUS' ROTATION

There is further evidence that contradicts Sagan about Venus' birth. For example, Venus rotates on its axis in the opposite direction of the other planets. This should not be if, as Sagan assures us, that all the planets were formed by an accretion process about the same time billions of years ago. Neither science nor Sagan offers a clear and persuasive explanation of Venus' retrograde rotation based on unifor-mitarian theory. Velikovsky, however, claims that the past motions of Venus were so disturbed by planetary near collision which certainly could have changed Venus' rotational direction. This explanation does seem to fit this evidence of gross anomalous rotational behavior quite well. Sagan, in his co-authored book, *Intelligent Life in the Universe,* p. 325, says, "The cause of the retrograde rotation of Venus is unknown." However, in his book *Comet,* p. 204, Sagan speculates about the retrograde rotation of Venus and Uranus.

> "One of the regularities of planetary motion is that the planets tend to rotate on their axes and revolve about the Sun in the same direction as the Sun spins...suggesting that the Sun and planets all condensed out of

the same flattened spinning cloud of gas and dust. The rotation of Venus and Uranus do not follow this regularity, *and must be due to other, perhaps catastrophic later events.*" [emphasis added]

In this case, Sagan appears to agree with Velikovsky that the rotation of Venus was disturbed by an astronomic catastrophe. Although Sagan does not believe the catastrophe is the same one that Velikovsky proposes, he and the scientific establishment have belatedly begun to see some planetary phenomena from the catastrophic viewpoint. The last paragraphs of this section will show just how far the astronomers are willing to go to save uniformitarianism.

VENUS LACKS A MAGNETIC FIELD

Another problem is Venus' magnetic field. Michael Zeilik, discussing Venus' composition states,

> "With the mass and size [of Venus] in hand, we can find Venus' bulk density: 5200 kg/m^3, almost the same as that of the Earth. We guess that the interior of Venus should closely resemble the Earth's interior...a rocky crust (which the Venus landers confirmed), a mantle, and a metallic core. Because Venus has a lower density than the Earth, we imagine that it has a somewhat smaller core.
>
> "But this model has a severe problem. A metallic core, liquid in part, implies by comparison with the Earth, that Venus' should have a planetary magnetic field. Because Venus rotates 243 times more slowly than the Earth, we expect its internal dynamo to be weaker and the magnetic field to be less intense than the Earth's, but still there. If one exists, it must be at least 10,000 times weaker than the Earth's magnetic field! That's much weaker than expected from a simple dynamo model. What a magnetic mess—Mercury has a planetary field and Venus does not! Mars has a barely detectable magnetic field... Perhaps the dynamo model [for magnetic field production] is not the correct one to apply to all planets."[23]

To save the dynamo theory, it is assumed that Venus is "in the middle of a [field] reversal, [when] the magnetic field is essentially zero. Venus may be midway through a reversal and so has a very weak field."[24] This is merely an *ad hoc* assumption to save another uniformitarian concept. But if Venus is immensely hot below its surface, the dynamo theory requires that such a hot interior would not generate a magnetic field; and this, too, supports Velikovsky's hypothesis because iron, above 1500 degrees F. austenitizes and austenitic iron possesses no magnetic properties!

NO EROSION OF ROCK ON VENUS

Since Venus is supposedly as ancient as Earth and had erosional processes, most of its surface rock should be eroded by forces such as heat, wind, pressure and acid. However, the *Venera* pictures of the surface show mostly sharp edged rocks which indicate they are extremely young. C.J. Ransom remarks that,

> "Russian probes recently soft-landed on Venus and took photographs. These photographs reveal sharp-edged rocks which were classified as young-looking. The *Venera 9* and *10* photos show a young-looking surface that inspired [this] speculation, [in] (Aviation Week and Space Technology, Nov. 3, 1975) that Venus is in an, *'early cool down phase of evolution rather than in final stages of suffocation in a thickening atmospheric greenhouse.'* [emphasis added]
>
> "It was suggested that on an evolutionary scale, Venus should be classed with the 'young still living planets.'
>
> "In a *Nature* article, Sagan said that the rocks should look young because there should be very little erosion. [*Nature*, Vol. 261, p. 31, 1976]
>
> "He also provides a possible explanation of erosion for those rocks that might look old, and he describes a source for rocks that might look young and actually be young. It is wise to consider as many cases as possible, but choosing one of the diverse possibilities is not proof that the young looking rocks are not young.
>
> "In 1968 Jastrow and Rasool [in *Introduction to Space Science*, 2 ed. 1968, p. 792] noted that Venus has many theoretical resemblances to the Earth, but in actuality it is a strikingly different planet."[25]

In *Science News,* Vol. 108 (1975), p. 276, an article titled "Grand Unveiling of the Rocks of Venus" describes *Venera 9* photographs of the surface, stating,

> "The initial photo, apparently taken with the camera looking almost straight down (suggesting that mission officials wanted to ensure at least one picture before moving anything), contains a remarkably clear view of some sharp-edged, angular rocks. According to Boris Nepoklonov, one of the mission scientists quoted by the Soviet news agency Tass, 'This seems to knock the bottom out of the existing hypothesis by which the surface was expected to look like a desert, covered with sand dunes because of constant wind and temperature erosion.' In fact, he says, 'even the moon does not have such rocks. We thought there couldn't be rocks on Venus—they would all be annihilated by erosion—but here they are, with edges absolutely not blunted. This picture makes us reconsider all our concepts of Venus.'"

Sagan seems unwilling to reconsider his own concepts of Venus. Imagine surface winds of Venus blowing day in and day out with the force comparable to a hurricane for 3 or 4 billion years, remembering also that the gases rushing over the surface are *highly corrosive;* based on any reasonable, gradualist suppositions, the result would be a global sandy Sahara! The sharp edged rocks indicate Venus is a young planet. G. Benford in *Closeup: New Worlds,* (NY 1977), pp. 96-97 states,

"How these [Venusian] craters could persist despite the rapid erosion by super-heated, dense, dust laden winds, remains a mystery." Thus Venus' craters would have also been eroded away under its present conditions unless they are extremely young.

THE VENUSIAN REGOLITH

The problem for the space scientists is that there is no evidence for the existence of a thick layer of soil which scientists term *regolith* covering Venus. The planet has supposedly been subjected to strong erosional processes for billions of years and therefore these processes should have created a very thick regolith over the Venusian surface. Nevertheless, George E. McGill, et. al., in *Venus* cited above on page 94 state that,

"radar and *Venera* lander observations imply that most of the surface of Venus cannot be covered by unconsolidated wind-blown deposits; bulk densities on near-surface materials are not consistent with aeolian sediments... Thus, present-day wind—blown sediments on Venus cannot form a continuous layer over the entire planet."

On page 70 McGill, et. al. state, "Radar data are not consistent with a thick, widespread porous regolith as on the Moon." However, there is no known process that would cement wind-blown sediments together while they lie on the Venusian surface and this is admitted by the space scientists. Bruce Murray in dealing with the lack of erosional material on Venus states, "...Russian closeups of Venus were surprising. I had presumed its surface was buried under a uniform blanket of soil and dust. Chemical weathering should be intense in such a hot and acid environment... [Apparently] unknown processes of topographic renewal have to outstrip degradation and burial." [Bruce Murray, *Journey Into Space*, (NY 1989), p. 126]. To explain the lack of a Venusian regolith the scientists have invented an *ad hoc* process that has no scientific basis for its action to reconsolidate the detritus or soil on Venus back to rock.

Erosional processes are clearly at work on Venus' surface producing detritus and dust, and there is no known method operating to recon-solidate this surface debris back into rock. Thus the radar data should have given evidence of a thick Venusian regolith but it did not. If Venus is an ancient planet this is precisely what should have been found. A planet recently born and molten only a few thousand years ago would exhibit just such a surface topography. The porous nature of the rock on Venus is most probably the result of bubbling that occurred when the rock was in a molten state and not of recementing of surface materials because there is no known process that can do this.

Furthermore, the age of the rock found at all sites on Venus is extremely young based on radioactive substances found in it. Yu A.

Surkov in "Studies of Venus Rocks by Veneras 8, 9, 10", in *Venus* above p. 158 makes this explicit stating, "the discovery of the Venus surface rock very rich in uranium, thorium and potassium suggests that these elements melted from a primitive material deposited in the interior of the planet." This strongly implies that Venus' rocks which are found to be of igneous origin, [from lava] are volcanic and that Venus is a volcanic cauldron. A volcanic cauldron is, once again, just what would be expected based on Velikovsky's scenario.

VENUS' VOLCANISM LIKE IO'S

If Venus was so recently molten one would expect that it exhibit volcanic outpourings comparable to and quite similar to those of Io, Jupiter's satellite. David Morrison's "The Enigma called Io," in *Sky and Telescope* for March 1985, p. 201 describes Io's style of volcanism:

"Some of Io's volcanic features look a great deal like their terrestrial counterparts: low, shield-shaped constructs with calderas at their peaks and flows of erupted materials on their sides. However, most of Io's calderas are not at the tops of mountains but instead appear to lie scattered amid the plains."

Io exudes its sulfur flows this way because it is tremendously hot internally. If Venus is, as Velikovsky suggests, also tremendously hot internally it too should vent its magma in a similar manner. That is, Venus should also have lava pouring out of circular vents on the plains of its surface, and this it does.

In *New Scientist* for Nov. 4, 1989, p. 34, a brief article titled "Radar shows lava flows on Venus", states,

"The flat plains of Venus consist of lava that has flowed from the planet comparatively recently, according to latest radar results. And *an appreciable amount of the planet's interior heat may escape through these lava flows, rather than through the large volcanoes and rift valleys that geologists have known for some years.* [emphasis added]

"A team of American researchers from Cornell and Brown universities, and the National Astronomy and Ionospheric Center, used the large Arecibo radio telescope in Puerto Rico to map four areas of Venus in great detail (*Science*, Vol. 246, p. 373). Two covered volcanoes; the other two recorded regions covered by flat plains.

"In the plains, the researchers found dozens of small vents, which have oozed lava without forming volcanic cones. The researchers say 'the large number and wide distribution of vents in the lowlands strongly suggest that plains volcanism is an important aspect of surface evolution and contributed to heat loss on Venus.'"

What was the position of the scientists regarding volcanism on Venus? Isaac Asimov in "The Unknown Solar System," in *Discover* for Oct. 1989, p. 40, remarks that, "For years astronomers had believed that

Venus was a geologically dead place. Although quakes, volcanoes, and other activity surely wracked the planet at one time, it seemed certain that Venus was quiet today."

However, when *Magellan* mapped Venus the astronomers saw that the surface of the planet was covered by about one hundred thousand volcanic formations, tens of thousands of volcanic domes, and many collapsed circular formations called coronas with diameters between seventy five to over 1,200 miles. According to Harry Y. McSween Jr., "A Goddess Unveiled," *Natural History*, (November 1993), p. 64, these coronas were caused by plutons or immense chambers of rising lava which after rising then collapsed. McSween asks,

"Why Venusian magmas should collect into such huge blobs remains a mystery... Venus also displays irregular shaped birthmarks, the traces of lava flowing on the surface. These flows, some of which can be traced to volcanic vents, often extend for thousands of miles, much farther than comparable features on Earth..."

Rivers of lava, some of them, thousands of miles long, have been observed on Venus' surface. The enormity of Venus being literally volcanic from pole to pole and on every region of the planet has led one *Magellan* investigator to say, "Everyone says *Olympus Mons* on Mars is the biggest volcano in the solar system... It isn't. Venus is. The entire planet is one big volcano." [Henry S.F. Cooper Jr., *The Evening Star Venus Observed*, (NY 1993), p. 180] A completely volcanic crust is precisely what a new-born planet that was completely molten a few thousand years ago would exhibit as Velikovsky's theory demands.

The actual picture of Venus is diametrically opposed to what the astronomers had anticipated. Yet its volcanic nature is in complete harmony with Velikovsky's concept. Scientists like Sagan can deny the theory of Velikovsky only by ignoring this volcanic evidence.

THE SUPER ROTATING ATMOSPHERE

One of the enigmas of Venus is the rapid flow pattern of its upper atmosphere. On Earth, the atmosphere flows in the same direction as the Earth rotates taking weeks to circle our planet at the equator. Venus' rotation period is 243 days; it possesses an atmosphere with clouds 39 miles above the surface which flows at 100 m/sec circling that planet in only about 4 days. Other measurements indicate the circulation takes about six Earth days in the direction Venus turns. Young and Young say,

"In the Earth's atmosphere, such winds are encountered only in narrow jet streams. Jet streams could not, however, account for the rapid atmospheric movements observed on Venus, since the Venusian winds seem to involve large regions of the planet...

"Theoretical attempts to explain the generation of the winds have produced several possible mechanisms, such as convection caused by the uneven heating of the day and night sides of the planet. None of them, however, has been shown to be capable of explaining velocities greater than a few meters per second."[26]

Velikovsky's hypothesis fits the bizarre atmospheric behavior quite well. Since Venus was a comet-like body, its tail gases, that is its atmosphere, would still have great momentum after Venus entered its present orbit. This momentum still exists in the Venusian atmosphere and causes it to move with great velocity around the planet in the upper regions where the density of the gases are thin and turbulence of these gases do not act to halt this motion.

R.E. Newell in reviewing the data of space probes sent to Venus, has come to the conclusion that Venus is at least 3 billion years younger than the Earth.[27] This makes Venus 1.6 billion years old, far older than the age claimed by Velikovsky. Nevertheless, the picture that has been emerging indicates that some scientists are beginning to speculate and suggest that Venus is a young planet.

CATASTROPHISM TO THE RESCUE

To save the concept that Venus was formed at the same time as the Earth, an *ad hoc* hypothesis is ready should Venus' heat be found to be coming from below the surface as Velikovsky predicted and not from the greenhouse effect. This theory is that Venus once had a moon — Adonis—which circled Venus in a retrograde orbit and fell into Venus. This created Venus' heat, put it into retrograde rotation and the impact created such great heat that it boiled away Venus' water while producing immense volcanism to change the nature of the atmosphere. There is, however, a major problem with this idea. The moon would not, based on uniformitarian theory, be born in a retro-grade orbit. An additional catastrophe is required to change its revolution around Venus from prograde to retrograde. Thus, a catastrophe is needed to get Adonis to revolve in a retrograde orbit. A second catastrophe is needed to get Adonis to fall into Venus, to get Venus to rotate in a retrograde direction. It seems the doctrine of uniformitarianism demands some such *ad hoc* theories to save the day. If this argument is advanced, based on Talbott's fully quantitative analysis, Adonis fell into Venus about 3,500 years ago. Simply stated, two major catastrophes must occur in the Venus-Adonis system to save uniformitarian theory. But, of course, Sagan implies that Velikovsky's catastrophic hypothesis is unnecessary.

There is also the problem of accounting for the Moon in its present orbit, and a catastrophic theory of the Moon's birth has been offered by the scientific establishment. The theory states that the Earth was struck by a large body which knocked a portion of the crust into space, forming a circular ring of debris which coalesced over time to form the Moon. A similar hypothesis is also offered to explain the retrograde rotation of Uranus and the reason that Uranus is tilted on its side by over 90 degrees. This hypothesis does not explain why Uranus' moons travel around the planet at its equator in retrograde orbits. In each of these instances, an assumed body crashed into Venus, Earth and Uranus. The bodies must be quite massive to produce the observed results. A body the size of Earth had to strike Uranus. A body the size of Mars had to strike the Earth. Why bodies of planetary dimensions had the highly elliptical orbits or were in retrograde orbits is never explained except by a series of catastrophes, each dependent upon the other.

To explain that Mercury's iron core is 75 to 80 percent of the planet's diameter, Clark R. Chapman in "Mercury's Heart of Iron", in *Astronomy* for November, 1988, p. 32, states,

"The protagonist for the new (catastrophic) view (of Mercury) is George Wetherill of the Carnegie Institute in Washington, D.C. Wetherill has been making calculations about how the planets can accumulate from their smaller precursor, asteroid-like planetesimals. His calculations show that planets do not grow in an orderly way. In fact, in Wetherill's view, Mercury's formation was wildly disorganized and catastrophic.

"Here is one possible way that Mercury might have formed, according to Wetherill's computer runs. About two dozen bodies, each with a third the mass of the Moon and made of rock and metal, accumulated to form proto-Mercury about twice as massive as Mercury is today. The bodies came either from near Mercury's present orbit or from beyond the Earth's orbit. After about six million years, when proto-Mercury was orbiting at about Earth's distance from the Sun, it happened to hit another small, but growing planet. That violent collision was so energetic and released so much heat that the proto-Mercury melted throughout, and its heaviest elements—such as iron—sank toward the center and the lighter elements floated upward to form a skin-like rocky mantle at the surface.

"About 10 million years later, when the melted and differentiated proto-Mercury was located halfway between the modern day Earth and Mars, it collided with an even larger body—greater than half its own size—which impacted at a velocity of 8 miles per second. This enormous collision tore proto-Mercury asunder, stripping away and crust and leaving the core with only a thin, rocky covering. The ejected material was shot at high velocity into interplanetary space, never to return.

"In one shattering collision, Mercury thus became a metal rich planet. Over the ensuing 18 million years, in Wetherill's calculations, Mercury wandered around the inner solar system. It was occasionally deflected by the gravity fields of the other growing planets—including proto-Earth and proto-Venus—until it settled down into its present orbit close to the Sun.

"This gripping scenario is only one of several that Wetherill has devised to account for Mercury's formation. Impacts are chance events and the reality may not have happened exactly that way. The important point, according to Wetherill, is that giant collisions were common in the early history of the solar system...

"From looking at Mercury as it is today, planetary scientists believe that the planet did get blown apart at a critical point in its formation. Proto-Earth, on the other hand, was lucky enough to grow to a larger size before it suffered a chance collision with a Mars sized body that many scientists now believe created the Moon... Was a giant collision responsible for nearly stopping Venus' rotation? Or for tipping Uranus on its side?

Catastrophes in series are required to salvage uniformitarianism; if this isn't a contradiction in terms, then nothing is. None of this has ever been observed. However, to "save the appearances"—that is, make the phenomena fit the theory—scientists like Sagan somehow attempt to make their particular brand of planetary catastrophism consistent with planetary physics. Who's kidding whom? Sagan has both plate tectonics and numerous ancient impact craters on Venus— a total contradiction. Plate tectonics removes craters. This is pointed out by Isaac Asimov in *Frontiers,* (NY 1989), p. 213, wherein he writes,

"...radar photographs...show craters on Venus, which...seem to be half a billion to a billion years old. This speaks against plate tectonics, for on Earth, the plate tectonics keep renewing the surface."

But if Venus does not have plate tectonics, it presents a further con-tradiction to Sagan since to rid itself of heat without plate tectonics, Venus would have to have an enormous amount of volcanism. On page 228 of *Frontiers,* Asimov informs us that,

"If Venus were as volcanic as Earth is, most of the craters would be filled in and covered by lava... Venus has a thicker crust, not broken into plates. Its internal heat can escape only by way of volcanic action. If we suppose that Earth and Venus, being of nearly equal size, should have the same amount of internal heat, and should lose it at the same rate, then Venus ought to be much more volcanic than Earth, perhaps a hundred times as volcanic, because there are no plate joints from which it can lose heat." [Venus is over 1,000 times more volcanic than Earth.]

A hundred times the amount of volcanic action on Venus would, therefore, erase all of Venus' impact craters. With or without plate tectonics, Venus should not possess ancient impact craters. Sagan never discusses such outstanding contradictions in his criticism probably because there is simply no answer to this enigma other than the concept that Venus is a volcanic cauldron, which is precisely what one would expect in terms of Velikovsky's views.

NOTES & REFERENCES

[1] B.B., p. 120.

[2] Ellenberger, C.L.; "Heretics, Dogmatics and Science's Reception of New Ideas", Part 4. *Kronos* VI, 4, p. 84, Note 209.

[3] Asimov, Isaac; *Venus, Near Neighbor of the Sun*, (NY 1981), p. 121.

[4] SCV, pp. 83–84; B.B., p. 118.

[5] Asimov, I.; *Venus, Near Neighbor*, op. cit., pp. 120–121.

[6] Greenberg, Lewis M.; *Kronos* V, 2, p. 91.

[7] Gore, Rick, *National Geographic*, (Jan. 1985), p. 36.

[8] Sagan and Druyan; *Comet*, op. cit., p. 258.

[9] *Science Digest*, (May 1984), "Violence on Venus", p. 21.

[10] Marov, M.Y.; "Results of the Venus Missions", *Annual Review of Astronomy and. Astrophysics*, Vol. 16, (1978), p. 145.

[11] Gore, Rick; *National Geographic*, op. cit., pp. 42–43.

[12] B.B., p. 120.

[13] Burgess, Eric; *Venus, An Errant Twin*, op. cit., p. 141.

[14] Asimov, I.; *Venus Near Neighbor*, op. cit., p. 120.

[15] Burgess, E.; *Venus, An Errant Twin*, op. cit., p. 141.

[16] SCV, p. 75; B.B., p. 111.

[17] Young, Andrew and Young, Louise, *The Solar System*, op. cit., p. 53.

[18] Ibid.

[19] Ibid., p. 54.

[20] Beatty, J.K., O'Leary, B., Chaikin, A, ed. *The New Solar System*, op. cit., p. 50.

[21] Colin, Lawrence' "Venus the Veiled. Planet", in *The Planets*, op. cit., p. 280.

[22] SCV, p. 85; BB pp. 119-120

[23] Zeilik, M.; *Astronomy*, 4 ed. op. cit., p. 160.

[24] Ibid., pp. 160-161.

[25] *Age of Velikovsky*, op. cit., p. 127.

[26] Young, Andrew and Young, Louise; *The Solar System*, op. cit., p. 51.

[27] Newell, R.E.; *Speculations*, op. cit., pp. 51-57.

SAGAN'S TENTH PROBLEM

THE CIRCULARIZATION OF THE ORBIT OF VENUS

NON-GRAVITATIONAL FORCES IN THE SOLAR SYSTEM

Sagan admits that the odds against the possibility of comet Venus achieving a planetary orbit in a short period are not "overwhelming," but he maintains, "...that the field strengths implied are unreasonably high...[and] they are counter-indicated by studies of rock magne-tization."[1] Since it has already been shown that Sagan is wrong regarding the strength of rock magnetization on Earth, (the strength of remanent rock magnetization is over a thousand times stronger than the field the Earth produces), we must deal with electromagnetic forces. These forces Sagan terms "non-gravitational"; however, I shall deal with this non-gravitational force and call it by its proper term, "electromagnetic" force.

PENDULUM EXPERIMENTS

In essence, Sagan claims electromagnetic forces play practically no role in celestial mechanics. He had earlier stated, "Newtonian dynamics and the laws of conservation of energy and angular momentum are on extremely firm footing. Literally millions of separate experiments have been performed on their validity..." However, it only takes one experiment to put this concept in doubt. Velikovsky posed the fol-lowing respecting the motion of charged and uncharged pendulums in an article titled "The Pendulum Experiment" in which he cited Albert Einstein's *Relativity,* 11th edition, p. 59, regarding fundamentals of gravity. Einstein stated,

> "In contrast to electric and magnetic fields, the gravitational field exhibits a most remarkable property, which is of fundamental importance... Bodies which are moving under the sole influence of a gravitational field receive an acceleration, which does not in the least depend either on the material or the physical state of the body."

In other words, electrical forces should not influence the motion of a celestial body, a pendulum or a body falling to the Earth. Velikovsky, after citing Einstein, states,

> "This law is supposed to hold with great accuracy. The velocity of the fall is generally explored with the help of a pendulum; it appears to us that a charged object may fall with a different velocity than a neutral

346

object. This is generally denied. But the denial is based on the observation that there is no difference in the number of swings of a pendulum in a unit of time, in the case where a charged or neutral bob is used. This method may produce inaccurate results. *In an accurate method, the falling time and the time of ascent of the pendulum must be measured separately.* In the case of a charged body, the increase in the velocity of descent of the pendulum may be accompanied by a decrease in the velocity of ascent or vice versa and thus the number of swings in a unit of time would remain the same for charged and non-charged bobs."[2]

If, as Sagan assures us, that "Literally millions of separate experiments have been performed on [Newtonian law] validity," then the following experiments should also be another confirmation of the validity of the laws of gravitation. Unfortunately, they are not. The experiment of the late Erwin J. Saxl was published in *Nature* in 1964 ten years prior to Sagan's first presentation of his criticism.

Saxl, a post-doctoral student of Albert Einstein, carried out experiments with torque pendulums that were charged, grounded and uncharged. Here are the results of his experimental work:

1) Charged pendulums accelerate differently than uncharged or grounded ones.

2) Reversal of charge on the pendulum produced an additional variation of acceleration.

3) Increased charge caused further variation of acceleration.

4) The experiment when carried out during the different seasons of the year caused variations of acceleration in 1, 2 and 3 above.

5) Eclipses of the Sun and Moon produced more variations of acceleration.

Because the result differed so greatly from gravitational expectations, Saxl stated,

> "The physicist hesitates to form a working hypothesis. Having accounted reluctantly to the conclusion that there may exist variations in *g* (gravity) even if such cannot be noted with grounded quasi-stationary instruments. ...When working as a post-doctoral student with Einstein, we discussed the possibility that there were interrelations between electricity, mass and gravitation. These experimental results make me wonder whether they may be properly interpreted."[3]

Thus, it seems clear that electromagnetism plays a role in the motion of pendulums. There is no doubt that these were the most careful experiments of gravity using charged and uncharged pendulums, and it is certainly odd that Sagan, who cites, "Literally millions of experiments" performed on their validity prior to 1974 when he delivered his paper somehow overlooked this historically significant work.

Another aspect of gravity is that it is considered to be a force which cannot be shielded as can electromagnetism. In this respect, there are two experiments with pendulums in which one pendulum not electrically charged, but which apparently showed that gravity, like electromagnetism, can be shielded. Maurice F.C. Allais, in a paper titled "Should the Laws of Gravitation be Reconsidered" wrote about an experiment with a paraconical pendulum that showed:

> "...an abnormal lunar and solar influence...[which] became apparent in the form of a remarkable disturbance of the motions of the paraconical pendulum (which gave the very definite impression of a screen effect) during the total solar eclipse of June 30, 1954. The plane of oscillation of the paraconical pendulum shifted approximately 15 centesimal deg. during the eclipse. The forces involved were of the same order of magnitude as those which corresponded to the Foucault effect."[4]

At Harvard, Erwin J. Saxl and Mildred Allen wrote a paper on this shielding effect titled "1970 Solar Eclipse as 'Seen' by a Torsion Pendulum". They state,

> "During the solar eclipse of 7 March 1970, readings were taken and recorded electromagnetically of the times requires for the torsion pendulum to rotate through a given fixed part of its path, involving both clockwise and counterclockwise motions, on its first swing from rest. Significant variations in these times were observed during the course of the eclipse as well as in the hours just preceding and just following the eclipse itself. Between the onset of the eclipse and its midpoint there is a steady increase in the observed times. After the midpoint, the times decrease suddenly and level off promptly to values considerably greater than those observed before the eclipse. Furthermore, before the eclipse there is a periodic variation in these times. This strange periodicity was essentially repeated two weeks later [when the moon is on the other side of the Earth] at the same hours, though the actual values were somewhat greater than the earlier ones. These increases in actual values exceed by a factor of 10^5 [100,000] those that can be explained by the attraction of the moon due to its change in position relative to the sun and earth. All this leads to the conclusion that classical gravitational theory needs to be modified to interpret these experimental facts."[5]

Thus, the millions of experiments to test the validity of gravitation are all called into question by these pendulum experiments which show electricity affects the motion of a pendulum and that, like electromagnetism, gravity can be shielded.

THE COUNTER FORCE TO GRAVITY

In fact, the inadequacy of gravity by itself to explain several celestial phenomena is also known. Karl Darrow, in the March, 1942 issue of *Scientific Monthly* wrote a paper to prove that there is indeed a counter force to gravity. Dewey Larson, in his book *The Universe of Motion*, (Portland, OR 1984), p. 31, discussed Darrow succinctly,

"One of the few authors who has recognized that an 'antagonist' to gravitation must exist is Karl Darrow. 'This essential force has no name of its own,' Darrow points out in an article published in 1942. 'This is because it is usually described in words not conveying directly the notion of force.' By this means, Darrow says, the physicist 'manages to avoid the question.' In spite of the clear exposition on the subject by Darrow (a distinguished member of the Scientific Establishment), and continually growing number of cases in which the 'antagonist' is clearly required in order to explain the existing relations, the physicists have 'managed to avoid the question' for another forty years."

NEWTONIAN MECHANICS
Sagan states,

"We can approach the problem empirically. Straightforward Newtonian mechanics is able to predict with remarkable accuracy the trajectories of spacecraft—so that, for example, the *Viking* orbiters were placed within 100 kilometers of its designated orbit; *Venera 8* was placed precisely on the sunlit side of the equatorial terminator of Venus; and *Voyager 1* was placed in exactly the correct entry corridor in the vicinity of Jupiter to be directed close to Saturn. No mysterious electrical or magnetic influences were encountered. Newtonian mechanics is adequate to predict, with great precision, for example, the exact moment when the Galilean satellites of Jupiter will eclipse each other."[6]

What Sagan has not told us is that is that on long space flights course corrections are made as needed and that the *Mariner 2* spacecraft, to the chagrin of the space navigators, missed its target area near Venus by a figure of almost 12,000 miles. Why didn't Sagan tell us that? Furthermore, these spacecraft did not carry a significant charge nor high magnetic fields compared to their mass; thus, they could not interact with the Sun's electromagnetic field. How can a non–charged, non–magnetic body interact with one that is charged and magnetic as did E.J. Saxl's charged pendulum which moved differently than the uncharged pendulum? Clearly Sagan's argument begs the question. Like his biased chart of the temperature of Venus, which says nothing regarding Venus' cooling rate, his statement regarding spacecraft says nothing about motion of celestial bodies based on electromagnetism.

Therefore, let us examine a few examples of bodies in space that offer evidence that is contrary to the view that "Newtonian mechanics is adequate to predict with great precision the exact" motions of bodies in space. For example, David E. Smith in *Nature*, Vol. 304, (1983), p. 15, discusses the erratic descent of *Lageos,* an artificial Earth satellite, which descended much faster than can be explained by aerodynamic drag or other forces such as gravity. *Lageos* is an acronym for 'Laser Geodynamic Satellite' which was placed in orbit in May 1976. This two foot diameter aluminum sphere possessed hundreds of mirrors that permitted it to be tracked by laser beams from Earth. *Lageos'* core of brass weighed 386 pounds and therefore, its orbital descent was expected to behave with great precision based on gravity and aerodynamic drag. The space scientists were, nevertheless, upset when *Lageos* began losing altitude at the rate of 1/25 of an inch each day. "This descent rate is about ten times greater than it should be at 3,700 miles above the Earth's surface. However, its descent became even more perplexing because the rate of fall changed. For about one year—late 1978 through late 1979, its descent rate was 60 percent greater than that of the 1982 value which was 1/25 inch per day. But then in 1983, *Lageos* appears to have stopped descending and seemingly against all laws and logic to possibly have begun to ascend..." This behavior does not show the precision of gravitational law. However, let us turn to a similar phenomenon with which Sagan is intimately acquainted. Martin Caiden, in his book *Destination Mars,* discusses this:

> "Ever since their discovery, Phobos and Deimos [the moons of Mars] have been regarded as 'dynamical nightmares.' They simply didn't behave as should any ordinary or conventional moons. They moved in what was regarded as 'a most peculiar fashion.'
>
> "Phobos, [like *Lageos*] is only 3,700 miles above Mars, astonishingly close to its parent body...
>
> "No other planet has such tiny moons...[states Russian physicist I.S. Shklovsky], and he has attempted to find a natural cause of the unique characteristics of Deimos and Phobos... Shklovsky, stressing point after point, noted that Phobos has not remained in the orbit long ago, calculated for its passage about Mars. American and German astronomers, notes Shklovsky, predicted with great accuracy the theoretical positions of Phobos in orbit. *But Phobos astounded astronomers by ignoring the rules of celestial mechanics.* It moved in its orbit by 2.5 degrees, which in astronomical terms is a tremendous figure. As Shklovsky stated, 'In just a few decades, Phobos moved as much as 2.5 degrees away from the point in orbit, where, according to calculations, it should have been. *An incomprehensible fact—simply a scandal in celestial mechanics'*... [emphasis added]
>
> "Astronomers note that the orbital changes of Phobos are apparently not matched by the outer moon Deimos. More than one astronomer has wondered aloud if the decay in Phobos' orbit might not be caused by dust in space. But if Phobos, why not Deimos? Scratch one theory. And there is no tidal movement on Mars to account for the change in orbit.
>
> "Shklovsky (and other scientists) presented his theory about Phobos before the flight of *Mariner 4*. At that time some scientists speculated that

perhaps a magnetic field was affecting the behavior of Phobos. Shklovsky, even before the flight of *Mariner 4,* did not place much faith in a major magnetic field of Mars, and rejected (correctly, as it turns out) this theory...

"The rate at which Phobos drops toward Mars allows scientists to calculate the mass of the little moon, and the astounding answer is that Phobos must have an average density of only 1/10,000 that of water. As far as a solid body is concerned, that's impossible. That would be a density lower than air."[7]

However, *Viking 1* orbiter, according to *New Scientist,* "measured for the first time as...[it] swung within 97 km of it [Phobos] in the February series of flybys. [Phobos' mass] about 2.1 g cm^{-3}."[8] This density is many orders of magnitudes greater than the density calculated for Phobos based on Newtonian laws of gravity. Thus, the motion of Phobos is still, as Shklovsky stated, "a scandal in celestial mechanics." Sagan was a close collaborator of Shklovsky and is fully aware of this "scandal" which completely denies the precision of gravitational law.

Furthermore, J. Kelley Beatty holds out the possibility in the *McGraw-Hill Encyclopedia of Science and Technology,* Vol. 8, (1982), p. 223, that Deimos and Phobos may be captured asteroids. However, William K. Hartmann in *The Solar System,* (San Francisco 1975), p. 114, tells us that,

"The remarkable photographs made by *Mariner 9* provides further evidence of interplanetary collisions. Both chunky moons are liberally cratered... Their circular orbits around Mars are not of the kind one would expect if they were true asteroids, gravitationally captured by the planet from the main asteroid belt nearby. Several of my colleagues and I have, therefore, proposed that they may be the shattered, and perhaps, partially reconstituted remains of an earlier single large satellite of Mars."

Thus, Mars' satellites are the remains of a collision. But if this is so, why are the orbits so circular? It is extremely improbable that the destruction of a body in orbit around Mars would permit its pieces to fall into only highly circular orbits. The capture model for Deimos and Phobos demands, based on gravitational law, that they have highly elliptical orbits. That is why Caiden called them "dynamical night-mares." Basically, there is no solid explanation of the orbits of these small moons, though there are theoretical *ad hoc* ones.

JUPITER'S SATELLITES

Sagan has stated, "Newtonian mechanics is adequate to predict, with great precision, for example, the exact moment when the Galilean satellites of Jupiter will eclipse each other," which is not quite true.

This is because Newtonian mechanics cannot predict, with great precision, the exact movement of Jupiter. In *Science News Letter* appears
this remarkable piece of information.

> "Jupiter sometimes appears to be ahead of where it should be, sometimes
> behind. The difference changes regularly with time through a complete
> cycle once every 12.4 years. The magnitude of the effect is small; Jupiter
> never gets out of place in its orbit by more than 600 miles... The cycle
> in which Jupiter departs from its predicted position has repeated every
> 12.4 years for the past 160 years. This is slightly longer than 11.9 years
> required by Jupiter to revolve once around the sun. Since these two time
> periods are nearly equal, the error may be considered due to steady, but
> gradual change in the shape and orientation of Jupiter's orbit. However,
> there is at present no known reason for such a gradual change to
> occur."[9]

Since this phenomenon is so well observed, it appears that the precise
forces of gravity between Jupiter and the Sun get stronger for a while
and Jupiter moves ahead of its calculated orbit, but then gravity gets
weaker for a while and Jupiter falls behind its calculated orbit by 600
miles and this, of course, is the precision that empirically supports
Sagan's view of gravity. Again theoretical *ad hoc* concepts are proposed
to explain this.

According to W.P.D. Wightman's *The Growth of Scientific Ideas,*
(New Haven 1963), p. 114, "The law of gravitation...boils down to
this: Given certain masses of certain known distances from each other,
and whose relative velocities are known, we can calculate their subsequent motions." Now, what is apparent is that we know the masses of
Jupiter and the Sun around which it orbits; we also know the distances
and relative velocities of these bodies with respect to one another.
Therefore, we should accurately be able to calculate rather precisely
the orbit of Jupiter. These equations are quite brilliant, one could
almost call them beautiful and according to Sagan, they are correct.
Wightman, above, p. 114, states, "...the law of gravitation was
absolutely true when it was enunciated, and is very nearly true today
when the precision of measurement was greatly improved. Nor is it
conceivable that it will be otherwise." The only problem is that
Jupiter somehow doesn't understand these elegant equations of the
laws of gravitation and absolutely refuses to obey them. If, as Sagan has
stated, "Newtonian mechanics is adequate to predict, with great
precision...the exact moment when the Galilean satellites of Jupiter
will eclipse each other," why can't Newtonian mechanics adequately
predict, with great precision, the exact orbital behavior of Jupiter? To
affect the orbit of Jupiter, another body of great mass must regularly
affect Jupiter. There is no body inside the orbit of Jupiter (closer to

the Sun) that can affect Jupiter in this way. Neither can there be a body outside the orbit of Jupiter that induces this effect because it would have to also affect the orbits of Saturn, Uranus, Neptune and Pluto. Some force has to be affecting Jupiter, but what is it? Only two forces reach beyond the realm of the atom; one is gravity, the other is electromagnetism. Since gravitational forces cannot be operating to produce this phenomenon, perhaps electromagnetism is operating.

CHANGES IN THE SOLAR SYSTEM

Based on what the astronomer concludes about bodies of the solar system, briefly let us list the numbers of bodies that have had their orbits changed, or been captured by other bodies in the solar system and then had their orbits circularized.

According to R.A. Lyttleton, Mercury, Venus, the Earth, the Moon, Mars, Jupiter's four Galilean satellites (Io, Europa., Ganymede, Callisto), Saturn's large satellite Titan and Neptune's large satellite Triton, as well as Pluto, are all terrestrial bodies that were produced as comets and that were captured by the Sun or planets, and that their orbits were then circularized. This constitutes twelve bodies made up of five terrestrial planets and the seven largest terrestrial moons of the solar system.

J. Kelley Beatty in the *McGraw-Hill Encyclopedia of Science and Technology,* Vol. 8, 5th ed. (1982), p. 223, claims the two tiny moons of Mars—Deimos and Phobos—are captured asteroids that were then circularized. One of the major theories of the Earth's moon is that it is a captured body that had its origin inside the orbit of Mercury or elsewhere. J.G. Hills in "The Origin and Dynamical Evolution of the Solar System" Ph.D. Thesis (Michigan University, Ann Arbor 1969) held that Uranus, Neptune and Pluto, had their orbits changed by encounters with other planets. This is based on Hills' analysis of the solar nebula hypothesis which indicates that at first, Saturn and Jupiter were the outermost planets.

According to most astronomers, the small outermost satellites of Jupiter are probably captured asteroids as are the Trojan asteroids that follow and precede Jupiter.

According to T. Gold, Mercury has been in its present orbit no more than 400,000 years; that is, it has changed its place of orbit. While, according to Reginald E. Newell, Venus is no more than 1.6 billion years old. Thus, it was captured by the Sun and its orbit circulated. Tobias C. Owen in the *McGraw-Hill Encyclopedia of Science and Technology,* 5th ed. (1982), Vol. 14, p. 561, tells us Pluto is an escaped moon of Neptune. Since Pluto has a moon, Charon, it too must be captured; it is hardly conceivable that if Pluto was a moon of

Neptune, it also had a satellite orbiting it. In Vol. 12, p. 61, Owen tells us Phoebe, a moon of Saturn, because of its irregular orbit must be a captured satellite. He writes about Saturn's smaller moons, "it has frequently been suggested that these small, icy satellites are similar in composition to the nuclei of comets" and thus, they would be captured bodies. In Vol. 14, p. 276, Owen informs us that the five major satellites of Uranus "are thought to be captured objects" because of their retrograde orbits. In Vol. 19, p. 79, Owen states Neried, the small moon of Neptune, "may well be a captured object." Thomas van Flandern also offered the view that Pluto is an escaped moon of Neptune. All the short period comets are believed to have had their orbits changed from long period to short ones. Many of the planets, moons and comets also had their orbits circularized. In all, nearly every major and minor body of the solar system is believed by some astronomer, or other, to have had its orbit changed, or to have been captured by another body, and then have its orbit circularized. But Velikovsky's ideas, which require the same changes, are implied by Sagan and the scientific establishment to be absurd. Sagan should read this material carefully.

Not only is this circularization suggested by astronomers for bodies inside the solar system, but it is well established that stars also change their orbits from elliptical to circular. Ivars Peterson, in *Science News,* Oct. 29, 1988, p. 281, relates:

> "Within a million or so years after their birth, young [binary] stars appear almost indistinguishable from their more mature brethren. *The only systematic difference noted so far is the shape of certain orbits.* When young stars have periods of less than four days, their orbits are circular. Binaries with longer periods have highly elliptical orbits. In old binary systems, the dividing line between circular and eccentric orbits is roughly 10 days."
> [emphasis added]

When a new system of binaries is born in which both stars orbit each other in from 5 to 9 days, the orbits of the two stars will strongly tend to be elliptical; however, by the time the stars have aged onto the main sequence (where they will spend most of their lives as mature stars), the orbits of both stars will become highly circular. Once a binary system is formed, it will stay together through the eons. Yet, *every* newborn binary system, like the one just described, will inexplicably change its orbit from an elliptical to a circular one. G.O. Abell, in *Realm of the Universe,* 3rd ed., p. 376, informs us that stars larger than the Sun age from newborn to main sequence stars in a few thousand years. If this orbital change can only be made by gravitational forces, then a third star is needed to affect the orbits of the binary system. This is, however, not the case. Peterson, who reported this enigma in *Science News* adds that such large rapid changes in the orbits of binary stars is "...telling us something about the evolution of orbits... Something has to be making that happen." The fact of the matter is that gravitational theory absolutely forbids such a *large rapid change*. This clearly illustrates that orbits can be, and are, circularized in only a few thousand years in accordance with Abell's evidence.

COMETS AND THE ROCKET EFFECT

Sagan states, "Comets, it is true, have somewhat less predictable orbits, but this is almost certainly because there is a boiling off of frozen ices as these object approach the Sun, and a small rocket effect." [10] David W. Hughes, in *Nature,* discusses this small "rocket" or "jet" effect and shows that Encke's comet with a period of 3.3 years, has arrived earlier than calculated by about 2.5 hours. [11] The observations prove Encke's cometary period is decreasing. David A. Seargents explains,

> "The consensus nowadays is that these 'non-gravitational effects' arise from the evaporation of cometary ices, the resulting gas emission giving thrust to the nucleus. The ejection of gas will occur on the sunward side and in the case of a non-rotating nucleus, result in a push in the opposite direction to the Sun.
>
> "When the nucleus is rotating (as, presumably, most will be) there will be a delay in the maximum emission in the sense that it will not be at the subsolar point, but rather in the afternoon of the nucleus' day. The thrust will now be at an angle to the radius vector. For those comets whose nuclei rotate in a retrograde direction, [opposite to the direction of the Earth's rotation] the thrust will be such as to decrease their orbits... Direct [or prograde] rotation [the same as that of the Earth's rotation] give the opposite result..." [12]

Thus, one would expect that the jets from the nucleus of a comet would fire on the afternoon side of the comet that faces the Sun. The observations of Halley's comet seemed at times to do this, however, at other times this did not appear to be the case.

Accordingly, J. Kelley Beatty writes in *Sky and Telescope,* for May 1986, p. 442,

> "Another major surprise, says *Giotto* project scientist Rudeger Reinhard, was just how strongly discrete jets dominate Halley's outpourings of dust and gas. Models of the comet had assumed that it would release material uniformly from its entire sunlit hemisphere."

This, however, is not the case. The nucleus apparently holds heat. Beatty informs us on the same page that,

> "Not only is the nucleus big and black, but it is also hot. A French built infrared spectrometer on Vega 1 registered a temperature of about 330 degrees Kelvin [135 degrees Fahrenheit]. 'What we see is very hot' says

investigator Jean-Michel LaMarre, underscoring that the surface of the Halley's nucleus is not covered with ice. If that were true, LaMarre notes, the comet would be at least 80 degrees to 100 degrees K colder." [13]

Hence, the jet emissions appear to come from discrete openings in the crust of the comet. The crust conducts heat over its surface and into the heart of the nucleus where frozen volatiles melt and evaporate causing pressures to build up. These pressures force the gases to shoot out the openings in the comet to which these gases have access. Thus, the jets are not uniform in action as is required by the "rocket effect" agent.

However, if this rocket effect actually operated and changed the size of the orbits of comets just as Sagan has suggested, then the rotation of the comets would also be affected by these jets. Tom van Flandern explains:

"...in the dirty-snowball model, the observed 'jets' in Comet Halley are actively transferring material away from the nucleus. Calculations show that these [jets] should produce torques that affect the rotation of the nucleus in a short time. But long-term observations clearly show that the rotation of the nucleus has not been affected significantly by torques." [14]

Thus, the rotation of comets clearly invalidates the jet action concept that is supposed to affect the orbital motion of comets. If a jet is powerful enough to change the orbital velocity of an entire comet around the sun it is also powerful enough to affect the rotation of that body as well. This is clearly not observed.

In fact, G.B. Marsden in "Non-gravitational Effects in Comets; The Current Status", *The Motion, Evolution of Orbits and Origin of Comets*, International Astronomical Symposium No. 4 (Dordrecht Holland 1972), p. 136, states,

"Attempts to relate non-gravitational effects in the motions of a comet to direct observation of the influence of non-gravitational forces on the comet's physical appearance have failed miserably according to the calculations by Cunningham...and by Herget... P/Schwassmann-Wachmann 1—a comet that quite frequently throws off shells of matter and suddenly increases in brightness a hundred fold—shows in its motion no non-gravitational effects whatsoever."

If, as Sagan has claimed, the changes in cometary orbits that are non-gravitational in nature are due to "boiling off of frozen ices," then the comet P/Schwassmann-Wachmann 1 would change its orbit during the times when it throws off matter in jets and becomes brighter to the observer. Even though this brightening effect caused by these jets is well observed "quite frequently," there is no change "whatsoever." This knowledge was reported in 1972, two years prior to the AAAS

Symposium on Velikovsky. It is interesting that Sagan, an astronomer, should seem to be ignorant of this evidence.

However, it is quite clear that the rocket effect mechanism that Sagan invokes that "almost certainly" explains the non-gravitational forces that affect the motions of comets, according to Marsden, has "failed miserably."

ELECTROMAGNETISM

What other force exists besides gravity in the universe which reaches beyond the confines of the atom? It is electromagnetic force and as hard as Sagan tries to ignore its influence, it is the only other known force left to explain the many incongruencies that exist in celestial mechanics. Velikovsky claims that electromagnetic force plays a role in the motions of celestial bodies. Although these forces have not been tested on artificial satellites in space as yet they can be. Velikovsky has stated,

> "Lately I lecture frequently for physical and engineering societies and faculties, and I challenge those in the audience who believe that a magnetic body can move through a magnetic field without being affected by it to lift their hands. Can Jupiter with its immense magneto-sphere move in the magnetic field centered on the Sun, if only a few gammas, without being affected by it? Can the satellites plow through the magnetosphere of the giant planet without being affected by it? On no occasion I saw a hand raised." [15]

What is clear is that no scientist can argue (as we will show below) with the fact that magnetic bodies moving in an electromagnetic field must be influenced in their motion by that field.

Physicist James M. McCanney depicted the behavior of astronomers and astrophysicists when evidence of electromagnetic action is discussed, stating,

> "The magnetic fields of Jupiter and Saturn...discovered by *Pioneer 11* have created an unmentionable dilemma in the astrophysics community. In short, magnetic fields do not self-generate and sustain themselves for billions of years. Maxwell's equations [on electromagnetism] have been swept under the theoretical rug by traditional theorists who still maintain that gravity is the only force acting in the cosmos.
>
> "For magnetic fields to form, current must flow. For current to flow, a potential difference must be maintained. Herein lies the downfall of traditional astrophysical theory which does not allow for any electro-magnetic interaction.
>
> "When I pointed this out to a theorist in the physics department at Cornell University, he exclaimed that since both positively and negatively charged particles were moving in the recorded electric current between Io [the moon of Jupiter] and Jupiter, no potential difference was

required. This shows that too many scientists will say anything, no matter how absurd, to uphold traditional theory. Irrational behavior of this sort is found to be commonplace in astrophysics circles."[16]

Astrophysicists, if they admit that electromagnetism affects the motion of celestial bodies, know the concept that only gravity plays such a role will have to be amended, perhaps greatly. Einstein understood the dilemma that Velikovsky had raised for establishment science. When Velikovsky showed Einstein his memoirs to *Worlds in Collision,* Einstein read this statement by Velikovsky. "When I wrote: 'The real cause of indignation against my theory of global catastrophes is the implication that celestial bodies may be charged' he [Einstein] wrote in the margin: 'ja' ('Yes')."

The astronomer James Warwick specifically wrote a paper to prove that magnetic forces play so insignificant a role in celestial mechanics that Velikovsky's view must be in error. Because Warwick's attack is fundamental, let us examine it. C. Leroy Ellenberger presented this material, which is supposedly a devastating rebuttal to Velikovsky's views in *Kronos,* Vol. X, No. 3 (1985), pp. 10-11.[17]

The following is taken from my paper in *Aeon,* Vol. 1, No. 2 (1988), pp. 79-80:

"Immanuel Velikovsky claimed that two magnetic stars orbiting at close quarters would not only be influenced by gravitational interaction, but also by magnetic interaction. [*Earth in Upheaval* (NY 1955), "Forum Address", p. 297] Radio astronomer James Warwick calculated the influence of two magnetic stars, each of 10,000 gauss separated by three solar radii, and aligned with their dipole moments directed towards each other to maximize the magnetic interaction. Based on his careful calculation he could claim that magnetic interaction between the magnetic stars was about a billion times smaller than gravity—far too small—to influence the orbits of these bodies.

"We wish to challenge this conclusion based on mathematics and physics and offer observed evidence that denies the validity of Dr. Warwick's *calculation and conclusion.* Magnetic stars fall into two classes. The first type of magnetic star is called an Ap star.[18] Jean-Louis Tassoul, in his *Theory of Rotating Stars,* (Princeton, NJ 1978), p. 430, informs us that, 'Almost all high intensity magnetic fields known in non-degenerate stars are found in the peculiar A-type (*Ap*) stars.' Therefore, these stars are on the main sequence of stars with the largest magnetic fields. The other class of magnetic stars are the Am stars which if they have magnetic fields they are not nearly as powerful as the Ap stars with the strongest fields. What is quite unique is that the magnetic axes are at [a] great angle with the rotational axes and thus, the magnetic poles are observed to be close to the equator. (Tassoul, *Theory of Rotating Stars,* pp. 431-432) When stars are in close binaries, tidal forces drive the two stars into a position in which their rotational axes move toward parallelism. Thus, the point where very close companions would be in orbit around a magnetic star is at the rotational equator where in Ap magnetic stars, the gravitational forces will be in competition with the dipole moment's magnetic forces, in order to form a close binary system. There is also another phenomenon observed with magnetic stars. 'The distribution of the obliquity angles (of the magnetic axes) appears to be random...but becomes increasingly bimodal as evolution proceeds.' [P. North, *Astronomy and Astrophysics*, Vol. 148, (1985), p. 165] Therefore, as the star ages, the magnetic axis will move to align itself with the rotational axis. The magnetic axes of stars with the strongest magnetic fields will, based on this data, be found nearer to the rotational equator because that is where the dipole moments will tend to be situated. Thus, any star that forms a binary that is close to these magnetic stars, especially those with the strongest fields (namely the Ap type) must do so against at least one of the magnetic dipole moments.

"It is well observed that both types of magnetic stars—*Ap* and *Am*— form binary systems, but there is a significant difference that appears to deny Dr. Warwick's view that electromagnetism is an inconsequential force in these star systems. Spectroscopic binary systems are double stars that are so close together that they can only be distinguished by spectroscopic analysis. Nearly all main sequence stars can, and do, form spectroscopic binary systems. Based on Dr. Warwick's calculation, there is absolutely no reason for the highly magnetic Ap stars not to be members of such spectroscopic systems... The fact of the matter is that Ap stars are rarely found to have spectroscopic binary companions. And we are specifically told that, 'Ap stars are generally slow rotators, but they differ (from Am stars) in being *single* rather than *spectroscopic binaries*.'" [19] [emphasis added]

Since all other main sequence stars have such binaries, one astronomer remarked, "There seems to be a severe shortage of such (spectroscop - ic) binaries among the Ap stars." [*Magnetic and Related Stars*, (Baltimore 1967), p. 536] While M. Floquet in "Les Etoiles Ap Binaries" in *Astronomy and Astrophysics Abstracts*, Vol. 30, Part I (1983), p. 439, has come to the conclusion that "The magnetic fields seem to play an important role in the relation between binarity and the Ap phe - nomenon." That is, gravitational theory does not explain why these highly magnetic stars seem to be single rather than form binary systems. But what of the extremely weak magnetic Am stars? Interestingly, we are informed that they are "almost all short period spectroscopic binaries." [20] Clearly the only force that distinguishes Ap stars from Am stars is the strength of their fields and, perhaps, the inclinations of their magnetic axes to the rotational axes. There is no other force known to account for the differences in the nature of the stars in binary systems. The only principle which coherently explains the fact that Ap stars with high magnetic fields are overwhelmingly single and not spectroscopic binaries, is an electromagnetic repulsion that is too strong to be overcome by gravity at close distance. The only force which coherently explains the fact that Am stars with perhaps weak or tiny magnetic fields are, by and large, almost all binaries in spectroscopic systems, is an electromagnetic repulsion that is too weak and is thus overcome by gravity at close distance.

In fact, there is a basic experiment which can test this question. If a spacecraft containing a very small mass—such as a hollow steel ball— which is magnetized to say 35,000 gauss and highly charged, were sent on a long trajectory, say from 10,000 miles above the Earth, around the Sun and back, we would very quickly discover whether or not electromagnetic fields affect the motion of this body. Because the astrophysical community is so convinced that electromagnetism plays little or no role in the behavior of celestial bodies such a basic simple experiment has never been conceived nor undertaken. This experiment will most probably be undertaken as a lark or by some accident in space before astrophysicists will seriously attempt to honestly disprove their own theory. Those who have the most to lose will most probably oppose this experiment by all means simply because of the politics of science.

HALLEY'S COMETARY ORBIT

Sagan states, B.B. p. 121, "Halley's comet, which has probably been observed for two thousand years, remains on a highly eccentric orbit and has not been observed to show the slightest tendency toward circularization; yet it is almost as old as Velikovsky's 'comet.'" This statement is dead wrong!

According to Guy Ottewell and Fred Schaaf in *Mankind's Comet : Halley's Comet* in the past, the future, and especially the present, (Greenville SC 1986), p. 15:

> "The most recent and probably most accurate study of Halley's motion is that of [Donald K.] Yeomans and [Tao] Kiang in 1981…[Ottewell and Schaaf go on to explain] Finally in the year 1404 B.C. [the approximate time of one of Velikovsky's catastrophes] was reached. It turned out that there was then another approach to the earth, even closer than that of [the closest known approach] of A.D. 837. But this was before the age of known observations: there was none for this or earlier appearances, therefore no way for correcting to the true course of the comet before this [1404 B.C.] 'narrow place.' So the integration was stopped. This is as far back as the history of Halley can go, at least with any exactitude."

In essence these comet researchers deny Sagan's argument that we can go back in time beyond Velikovsky's catastrophe and know the orbit of Comet Halley.

The measure of the extent to which an elliptical orbit departs from perfect circularity is termed the eccentricity of the orbit, and it is designated by the symbol e. For a perfectly circular orbit, $e = 0$. The greater the orbit of a body departs from a circular orbit, the greater the e number becomes. For example, Venus with the most circular planetary orbit has an eccentricity where $e = 0.006787$, while Pluto, with the most elliptical planetary orbit has an eccentricity where $e = 0.249$. In the *Comets* by Donald Yeomans (NY 1991), pp. 268-269 he presents the parameters of Halley's cometary orbit. He presents a chart of the orbital elements since the 5th century. The chart shows that in 1301 A.D. e was 0.9689307; in 1378 $e = 0.9683723$; in 1456 $e = 0.9679974$; in 1531 $e = 0.9677499$; in 1607 $e = 0.9673329$; in 1682 $e = 0.9677987$; in 1759 $e = 0.9675306$; in 1835 $e = 0.9673962$; in 1910 $e = 0.9673038$; in 1986 $e = 0.9672769$. The pattern is quite clear; over the past, almost 700 years the orbit of Halley's comet has been unmistakably becoming more and more circular. Sagan stated, "Halley's comet, which has probably been observed for two thousand years remains on a highly eccentric orbit and has not been observed to show the slightest tendency toward circularization..." This statement is completely contradicted by the facts. Its orbit has unmistakably been changing slightly to become more and more circular over the years.

On the other hand, during "the past few decades have seen the discovery of comets with almost circular orbits which cannot be explained by capture."[21] That there are short period comets with highly circular orbits is very highly improbable. Stan Gibilisco, in his book *Comets, Meteors and Asteroids, How They Affect Earth*, informs us that,

> "The longevity of a particular comet depends on several factors. The more massive comets would live longer than the less massive ones. More dense comets would survive longer than the less dense ones. Comets with extremely small perihelia (sun grazers) would deteriorate more quickly than those that stay far from the Sun. Short period comets, attaining perihelion relatively often, would die sooner than those that have extremely long periods."[22]

COMETS WITH HIGHLY CIRCULAR ORBITS

When a comet is captured, according to Sagan, it is placed into the inner part of the solar system on a highly elliptical orbit. Because the comet is closer to the Sun and approaches its perihelion point—its nearest distance to the Sun—far more often than when it was a long period comet, it must be destroyed quite rapidly in astronomical terms. Sagan is well aware of this; in his book *Comet,* he states,

> "Very small comets in the inner solar system might be impossible to detect; they also would have an extremely brief lifetime as their ices vaporized and the comet dissipates. As comets split and break up, there are surely fragments released that are the size of a house or smaller, but they do not long survive."[23]

According to gravitational theory, it takes many millions of years for a body to change from a highly elliptical to a highly circular orbit. Short period comets in the inner part of the solar system would be completely dissipated long before they achieved their highly circular orbits. Nevertheless, there are comets with highly circular orbits inside the inner solar system. This can only occur if their orbits changed quite rapidly. Thus, there are comets that have achieved circular orbits, while others, such as Halley's comet, have not. What we do not know regarding this question is the electrical charge and the magnetic strength of the different comets, as well as perhaps other phenomena. For example, does Halley's comet have a solid carbonaceous, chondritic or metal core which is highly magnetic, while Encke has no such solid core at all? Therefore, comets may furnish proof that electromagnetic fields are responsible for their orbital behavior. The "rocket effect" model has been shown to be wrong.

MORE CHANGES IN THE SOLAR SYSTEM

Is Velikovsky's claim of significant changes in the orbits of large bodies of the solar system outside the realm of science? Apparently not, since respected scientists have proposed just such changes for the capture of the Moon and its conversion form an elliptical orbit to a circular one. C.J. Ransom pointed out that,

> "In 1970, S.F. Singer published in *Science,* [Vol.] 170, page 438, an article titled "Where Was the Moon Formed?" He mentioned some of the properties discovered about lunar rocks and a previously published opinion about how these properties might have occurred. Singer then made some calculations relating to the acceleration process for material in Earth orbit and for material accreting elsewhere and later being captured as one body by the Earth. He stated that, 'the conclusion can be drawn that the Moon accumulated not in Earth orbit, but as a separate planet, and that it was later captured by the Earth.'
>
> "Later, A.G.W. Cameron expanded on this concept in a publication in *Nature,* Vol. 240, (1972), page 299. Cameron reasoned that the natural place for the Moon to form with the described characteristics would be inside the orbit of Mercury, and that the relative difference of the orbital radii of the Moon and Mercury would be less than other adjacent planets. Thus, gravitational perturbations of the orbits of the two bodies

would probably accumulate until a close approach [between the Moon and Mercury] took place, at which a very large modification in the elements of the Moon's orbit would be possible. If the modified orbit of the Moon were sufficiently great to allow it to approach the Earth, then gravitational capture of the Moon by the Earth would become possible, if not probable."[24]

Here we can see that the change in the orbit of the Moon offered by both Singer and Cameron is quite similar to the changes in the orbits of Venus and Mars offered by Velikovsky. In these examples the entire change of orbits was achieved without recourse to electromagnetic forces. Sagan's attack should, therefore, also be directed against Singer and Cameron. By the way, this orbital change by Singer and Cameron is based on Newtonian dynamics, which Sagan argues is a well established scientific law.

Furthermore, Ransom goes on, also as reported in *The New York Times,* 16 April 1965, an astronomer, Gold, stated,

"When it was found that Mercury is not fixed with one and the same face toward the sun, Gold concluded that the planet cannot have been in its orbit for more than 400,000 years—possibly less—which is 1/10,000th the assured age of the solar system, and thus, dates its arrival during the age of man."[25]

The conclusion by Gold that the planet Mercury arrived in its present orbit 400,000 years ago or less, is quite similar to Velikovsky's view that Venus arrived in its present orbit less than 3,500 years ago. Gold's conclusion is also based on Newtonian dynamics. Therefore, why is Velikovsky singled out for attack on a point for which other astronomers proposing similar types of changes in the recent history of the solar system, are not attacked and disparaged?

Early on, attempts were made to discredit Velikovsky along these lines. Velikovsky remarks that Dr. Waldemar Kaempffert was going to write a negative review of *Worlds in Collision* in *The New York Times.* Velikovsky states,

"On the occasion of an additional consultation I had with Professor [Lloyd] Motz, he told me that Kaempffert, who wanted to check several points concerning my theory with an astronomer, had called at Columbia University and chanced to come to Motz. The latter could tell him only that he had gone carefully through the pages of the Epilogue [of *Worlds in Collision*] that dealt with celestial mechanics and that he could not tell him of anything methodologically wrong with my [Velikovsky's] hypothesis, as expressed in the Epilogue. Kaempffert left without finding a useful point to attack and therefore, omitted the astronomical side of the story in this review."[26]

ANCIENT EVIDENCE OF VENUS' ORBIT

Nevertheless, there is a way to determine whether or not the orbit of Venus has changed. Currently, Venus possesses the most circular orbit of all the planets. The difference between its perihelion distance (closest point) to the Sun and its aphelion distance (farthest point) from the Sun of its orbit, differs by only a few thousand miles. Hence, if evidence exists that shows Venus pursued a different orbit, then at present one can only conclude that Venus was on a less circular (more elliptical) orbit. There is clear evidence that Venus' orbit was quite different in the past. Currently, as Venus travels along its orbit, it disappears behind the Sun (superior conjunction) for about 60 days. It also disappears from naked eye vision when it passes in front of the Sun (inferior conjunction) for about eight days. Livio C. Stecchini discusses ancient Babylonian records of the disappearance of Venus stating,

"the Venus tablets of [Babylonian king] Ammizaduga [is] the most striking document of early Babylonian astronomy. These tablets, of which we possess several copies of different origin, report the dates of the heliacal rising (appearances) and setting (disappearances) of the planet Venus during a period of 21 years."[27]

The tablets do not correspond to our modern observations of Venus' disappearances. Stecchini goes on to say,

"Since the first effort at explanation of Archibald Henry Sayce in 1874, these figures have challenged the wit of a score of experts of astronomy and cuneiform philology. [Franz Xaver] Kugler [1862-1929, a recognized major authority on Babylonian and biblical astronomy, chronology and mythology], opposed the contention of those who claim that these documents must be dismissed as nonsense."[28]

Professor Lynn E. Rose and Raymond C. Vaughon, in a paper titled "Analysis of the Babylonian Observations of Venus" state,

"...the patterns of motion given on the tablets do not fit the present observed motions of Venus. Neither the lengths nor the spacings of the invisibilities are compatible with the present orbits of Earth and Venus."[29]

Therefore, to save the present theory of astrophysics regarding the stability of the solar system

"Uniformitarians, who believe that the orbits of the planets have remained substantially unchanged for thousands, millions or even billions of years, will take it upon themselves to 'correct' what they perceive to be 'scribal errors' of the tablets."[30]

The question arises about the accuracy and competence of the Babylonian astronomers. Stephen Toulmin and June Goodfield in their book, *The Fabric of the Heavens,* state

"In two things, particularly, the Babylonian astronomers were the masters of us all: they kept continuous, dated records of celestial events from at least 747 B.C., and their best mathematical techniques were not excelled until quite recently."[31]

After discussing the many brilliant accomplishments of the Babylonian astronomers, they say,

"Looking back at Babylonian astronomy from the twentieth century, one is struck by two things: the care with which the records were kept, and the mathematical brilliance of the predictive techniques."[32]

Thus, the records of Babylonians of Venus' disappearances do not match Venus in its present orbit and only by changing or discarding a large part of this data, which is known in science as "culling evidence" can these Babylonian tablets be force fitted into the present system. However, when an ancient astronomical record happens to coincide with the present astronomical system, it is hailed as evidence that proves the validity of the present theory respecting astronomy. Hence, the only evidence that is acceptable is that which validates uniformitarian astronomy, and it is good, reliable evidence from an astronomically advanced culture. When the evidence contradicts the uniformitarian astronomy, it is bad, unreliable evidence from the very same astronomically advanced culture.

But what if two cultures have records of Venus' disappearance that do not coincide with our present astronomical system? On the other side of the Atlantic, the Mayans of Mexico also observed and recorded Venus' appearances and disappearances. The Mayan record of Venus is found in the Dresden Codex. E.C. Krupp in his book, *In Search of Ancient Astronomies* says,

"On pages 46 to 50 of the Maya Dresden Codex, we see a complete record of the apparitions of Venus as morning and evening star. Several Venus years are recorded... These...represent the elapsed time between successive heliacal risings or initial appearance of the planet as morning star. The Maya divided this 584 day period into four subintervals, or stations, representing the appearance of Venus as morning star and evening star and its disappearance intervals in front of and behind the Sun."[33]

Evan Hadingham in *Early Man in the Cosmos,* states this about Venus' appearances and disappearances in the Dresden Codex, "the Venus pages bear little resemblance to a modern astronomical table."[34] Since Hadingham, like the astronomers that dealt with the Babylonian tables, cannot conceive nor accept this evidence that Venus' orbit was different in the past, an analysis is created to dispose of this information. This is so in spite of Hadingham asserting the following regarding the accuracy of Maya astronomy,

> "The precision of the observations documented in the few surviving hieroglyphic books is astonishing. For instance, one book contains a scheme for the correction of Venus observations which ensures an accuracy of approximately two hours in five hundred years... How were they able to score such phenomenal success in their observations?"[35]

In *The Frame of the Universe* by F. Durham and R.D. Purrington, (NY 1983) p. 41 we learn that instead its present two periods of disappearance of eight and fifty days, according to the Dresden Codex of the Maya, Venus had disappearances of eight and ninety days. That is, Venus had a ninety day disappearance as described in the Babylonian Venus tablets of Ammizaduga. Instead of having about two equal appearances of 260 days, as at present, the Maya had two unequal appearances of 236 and 250 days. This flies in the face of Venus' present orbit. Modern astronomers, it seems, cannot tolerate such data. Anthony Aveni in the *Empires of Time,* (NY 1989), p. 229 states,

> "paradoxically, while Maya time keepers were concerned about accurate prediction, they seemed to be getting away with grossly distorting the Venus dates for religious purposes... This is but one of a host of Maya Venus mysteries that have yet to be solved."

According to Diego de Landa's *Yucatan, Before and After the Conquest,*

> "The astronomical observations of the Mayas were so precise that in computing the solar year, they arrived at figures not only more accurate than the Julian year, but also more accurate than the Gregorian year, introduced in Europe in 1582, ninety years after the discovery of America, which is our calendar year today."[36]

And again, all the Mayan records that conform with the present astronomical system are proof of its validity, but the Venus records are merely an attempt to reconcile two separate calendars—one religious with one astronomical. But what if three cultures have observations of Venus' appearances and disappearances that contradict the present system? John S. Major's *Heaven and Earth in Early Han Thought,* (NY), p. 76 has a section on "The Motions of Venus" which shows that instead of having two 263-day appearances, it had two 240-day appearances. Instead of 8 and 50-day disappearances, it had 35 and 120-day disappearances. There were corrected by Wuxingzhan to two 224-day appearances and 120 and 17-day disappearances. None of these measured observations fit the present orbit of Venus.

Apparently the records that validate our present system are accurate, while those that don't are not. As science, such a view is intriguing. Venus on its present orbit cannot rise very high above the horizon at night. However, in the *Chicago Assyrian Dictionary*, Vol. 16, *Sallummu* is interpreted to mean "fireball meteor" (p. 75). We also find that "[if Venus] rises very high and constantly has a red glow (explanation) constantly (SAG US = kunny) red fireball moves across variant: *at its zenith* (?) it is altogether red-hued..."[37] [emphasis added] The zenith would be, of course, the point in the sky directly above Assyria or Mesopotamia. Needless to say, Venus at night does not ascend to the Mesopotamian zenith; thus, why did the Mesopotamians write that it did? Furthermore, in *Nature* is an article on Halley's comet as recorded on Babylonian tablets which tells us a variant interpretation of *Sallummu*. "A meaning 'comet' is suggested by evidence from the diaries which show the *Sallummu* to remain in the sky for an extended time (several days or even weeks) so that a meteor explanation is excluded."[38] Thus, the Assyrians say they saw Venus as a *Sallummu*—a comet. This evidence stems from astronomy, not mythology.

In dealing with this evidence respecting Venus, astronomers either ignore the data or invent systems to explain it away so that it will conform to their uniformitarian view. By employing a sledge hammer, they smash the tablets of Ammizaduga to pieces and then reassemble the fragments to prove that Venus' orbit has never changed. Although Velikovsky does not explain the precise cause for the circularization of Venus' orbit except to invoke electromagnetic forces, the evidence of the ancient astronomers shows that Venus' orbit was different, and more elliptical than its present, almost perfectly circular orbit and thus, there must exist a force that circularized it.

DOGMA AS EVIDENCE

A good example of how astronomers deal with evidence from ancient astronomy that contradicts their theories is given by Patrick Moore in *The New Guide to the Stars*, (NY 1974), pp. 117-118. Moore states,

"...I cannot resist saying something about a problem which has always fascinated me. The spectrum of the primary [of the double star system Sirius] is of type A and the colour is pure white... Yet Ptolemy, in the second century A.D., distinctly stated that he saw Sirius as red; so did some of the ancient Egyptians and Assyrians. Ptolemy was a careful and reliable observer... Certainly Sirius is not red today, and it has been colourless ever since the tenth century A.D. when it was described by the great Arab astronomer Al-Sufi.

"This is a true puzzle. There seems no reason why Ptolemy and the other ancient observers could make so obvious a mistake. Sirius itself cannot have been red two thousand years ago; there is no conceivable way in which a red star could change into a main sequence star of type A. The fascinating suggestions has, therefore, been made that it is the Pup [the white dwarf companion of Sirius] which is responsible, and that it used to be a Red Giant.

"Outwardly, this may sound plausible, and it is true that if our modern ideas are correct, the Pup [Sirius' companion] must have gone through the Red Giant stage before collapsing into a White Dwarf. Unfortunately, there are major difficulties. The time scale is wrong, and a Red Giant shining together with the present day Sirius would make an object as bright as Venus, which is equally at odds with the old descriptions. My own view, for what it is worth, is that we must admit an error by Ptolemy."

Again, we see that any evidence from ancient astronomy that cannot be explained by the accepted concepts of modern astronomy is solved by saying the ancient astronomers were in error. On the other hand, when evidence of ancient astronomy supports the accepted concepts, these are hailed as evidence for these views. After calling Ptolemy a careful observer, Moore says he could not distinguish between red and white. It appears that there are no limits to the evidence that one may discard to save the established views.

Kenneth Hsu words regarding geologists' adherence to uniformitarianism could well apply to the mind-set of the astronomers. "I finally became aware that some of my fellow scientists are possessed by a mentality not much different from that of a fundamentalist in religion. A theory [or recorded measurement] cannot be valid [or accurate], I seemed to be hearing, if it is contrary to the gospel according to Charles Lyell [or Isaac Newton]."[39]

GRAVITY AND CRATERS

Before leaving this topic, let us return to Sagan's views respecting the ages of craters on Mercury, Venus, the Moon and Mars. Sagan has stated that Mercury, Mars and the Moon bear eloquent testimony to the fact that there have been abundant collisions during the history of the solar system (p. 85 of B.B., p. 47 of SCV.) He tells us the cratering of Venus, like that of the Moon, Mercury and Mars, may have occurred more rapidly in the very earliest history of the solar system (p. 119 of B.B.; p. 84 of SCV). In spite of all the evidence against this view, the establishment scientists maintain that the craters of Mercury, Mars, the Moon and Venus formed about four billion years ago. Let us use gravity to evaluate the age of the craters on Mercury, Mars, the Moon and Venus. All structures on the surfaces of these bodies are subject to the force of internal gravity, internal heat, and heat from the Sun, and over time rocks and craters will deform. The study of the deformation and flow of matter is called *rheology*. [40] William R. Corliss informs us based on an analysis of the data for the Moon,

> "The equations and data of rheology seem to indicate that the lunar craters should have disappeared long ago under the influence of gravity; but they have not!... Either the lunar craters are very, very young or, the viscosities of lunar rocks are about 10^5 [100,000 times] higher than those of similar terrestrial rocks... The lunar craters are youthful, belying radiometric dating..."[41]

The surface gravity of the Moon is about one-sixth that of the Earth. However, its surface gravity is smaller than that of Mercury, Mars or Venus; therefore, if rheology shows that craters should have long ago disappeared on the Moon, then the four billion year-old craters on these other bodies should have disappeared as well. Mercury and Venus possess very high surface temperatures compared to the Moon. The higher the temperature, the more rapidly the craters will deform. Thus, Mercury's and Venus' ancient craters should have deformed and disappeared even more rapidly than those on the Moon. J. Kelley Beatty, in *Sky and Telescope* for Feb. 1982, p. 135, informs us that, "One scientist estimates that over geologic time the heat could eradicate 300 km diameter craters simply by making rocks deform plastically back to local sea level." Mars' "surface temperature" at noon at the equator is not as high as that of the comparable area of the Moon. However, its surface gravity is about 2.3 times greater than that of the Moon. This should cause Mars' craters to have deformed and disappeared long, long ago. Mars' craters would have been removed by erosion. According to Corliss, the equations of rheology show that the Moon's craters should flatten in less than one million years.

As we pointed out earlier, according to S.K. Runcorn in "The Moon's Ancient Magnetism" in *Scientific American*, Vol. 257 for Dec. 1987, p. 63, the force of gravity was so strong on the Moon that it ought to remove its primeval bulge. He writes that, "if solid state creep occurred at one-trillionth the rate at which we now know it occurs in laboratory materials at modest temperatures, such a primeval bulge (as that found on the Moon) would have disappeared long ago." If the lunar bulge would disappear by a process that is one-trillionth its normal rate, what would occur to craters that are not nearly as large being deformed by gravity at a trillion times that rate? The answer is, they would disappear. Thus, the craters of the Moon, Mars, Mercury and Venus could not be even one million years old—unless one repeals the laws of gravity.

Sagan's views regarding this aspect of gravity are nowhere seen in his criticism of Velikovsky. Since he seems to maintain that theories based on gravity are the strongest form of evidence, it is interesting that he somehow overlooked this evidence which "eloquently" indicates that the view he offers of ancient craters is simply impossible, unless one rewrites the laws of gravity to suit a particular prejudice. But that would not be sloppy thinking, according to astronomers like Sagan.

According to G. Harry Stine's "Sol III–The Twin Planet," in *Closeup: New Worlds,* (NY 1977), p. 16,

> "On Luna the surface has lain bare to space for...billions of years, acted upon only by the erosion of the temperature cycle of lunar night and day, the action of solar wind, and impacts of meteorites.
>
> "But each of these three types of lunar erosion was grossly misjudged by lunar experts before the Space Age. For example, conceptual paintings of the lunar surface before the *Apollo* landings showed ranges of steep, sharply-spired mountains. We now know from on-the-spot photographs that the highest and most rugged lunar mountains are gently-sloped, softly-rounded hills. At one time, they must have been rugged. But some three billion years of erosion...plus billions of years of countless impacts have reduced those once-majestic mountains to ant hills."

Thus, since this could erode high mountains down several thousand feet, this erosion would have also degraded all the ancient impact craters. In conjunction rheological deformation and erosion over the eons would have not left the lunar surface appearing as it currently does. Sagan never discusses this in his criticisms of Velikovsky.

NOTES & REFERENCES

[1] SCV, p. 85; B.B., p. 120.

[2] Velikovsky, I.; "The Pendulum Experiment", in *Kronos* IV, 4, p. 59.

[3] Saxl, E.J.; *Nature*, Vol. 203, (1964), pp. 136–138.

[4] Allais, Maurice, F.C.; "Should the Laws of Gravitation be Reconsidered?", *Aerospace Engineering*, Vol. 18, (Sept. 1959), p. 46 and (Oct. 1959), p. 51.

[5] Saxl, Erwin J., Allen, Mildred; "1970 Solar Eclipse as 'Seen' by a Torsion Pendulum", *Physical Review*, D Vol. 3, (1971), p. 3.

[6] SCV, p. 85; B.B., pp. 120–121.

[7] Caiden, Martin; *Destination Mars*, (NY 1972), p. 236.

[8] *New Scientist*, Vol. 74, (1977), p. 394.

[9] *Science Newsletter*, "Query Newton's Theory", Vol. 75, (1959), p. 291.

[10] SCV, p. 86; B.B., p. 121.

[11] Hughes, David W.; "Spinning Comets and Planets", *Nature*, Vol. 273, (1973), p. 100.

[12] Seargent, D. A.; *Comets*, "Vagabonds of Space", (NY 1982), p. 87.

[13] *Sky and Telescope*, (March 1987), p. 245.

[14] Van Flandern, Tom, *Dark Matter Missing Planets & New Comets*, (Berkeley CA 1991), p. 222.

[15] *Pensée*, Vol. 4, No. 2, p. 13.

[16] McCanney, James M.; "The Nature and Origin of Comets...", Part I, *Kronos* IX: 1;p. 33.

[17] Ellenberger, C. Leroy; "Still Facing Many Problems", Part II, *Kronos* X, 3, pp. 9-10.

[18] *The Fact on File Dictionary of Astronomy*, 2 ed., ed. V. Illingworth, (NY 1985), p. 214.

[19] Ibid., p. 22.

[20] Ibid., p. 11, p. 360.

[21] Vsekhsviatskil, S.K.; "Comets Small Bodies and Problems of the Solar System", *Astronomical Society of the Pacific Publication*, Vol. 74, (1962), p. 106.

[22] Gibilisco, Stan; *Comets, Meteors and Asteroids How They Affect Earth*, (Summit, PA, 1985), p. 70.

[23] Sagan and Druyan; *Comet*, op. cit., p. 131.

[24] Ransom, C.J.; *Kronos*, I, 3, pp. 47-48.

[25] Ransom, C.J.; "How Stable is the Solar System?", *Pensée*, Vol II No.2 (May 1972), p. 16.

[26] Velikovsky , I.; *Stargazers*, op. cit., 121.

[27] Stecchini, Livio C.; *The Velikovsky Affair*, pp. 146-147.

[28] Ibid., p. 148.

[29] Rose, Lynn E, Vaughn, Raymond C.; "Analysis of the Babylonian Observations. of Venus", *Kronos* II, 2, pp. 3-4.

[30] Ibid., pp. 3-4.

[31] Toulmin, Stephen, Goodfield, June; *The Fabric of the Heavens*, (NY 1961), p. 25.

[32] Ibid., p. 41.

[33] Krupp, E.C.; *In Search of Ancient Astronomies*, (NY 1977), p. 169.

[34] Haddingham, Evan; *Early Man in the Cosmos*, op. cit., p. 225.

[35] Ibid., p. 219.

[36] W in C, p. 196.

[37] *Chicago Assyrian Dictionary*, Vol. 16, (1962 Chicago), p. 75.

[38] Stephenson, F.R., et. al.; "Record of Halley's Comet on Babylonian Tablets", *Nature*, Vol. 314, (April 1985), p. 588.

[39] Hsu, Kenneth; *The Great Dying*, op. cit., p. 43.

[40] *Dictionary of Geological Terms*, 3 ed. eds. R.L. Bates, J.A. Jackson, (NY 1984), p. 431.

[41] Corliss, W.R.; *Moon and Planets*, (Glenarm MD 1985), p. 118.

SAGAN'S OTHER PROBLEMS

DEIMOS AND PHOBOS

Sagan states:

> "on page 280, the Martian moons Phobos and Deimos are imagined to have 'snatched some of Mars' atmosphere' and to thereby appear bright. But it is immediately clear that the escape velocity on these objects— perhaps 20 miles per hour—is so small as to make them incapable of retaining even temporarily any atmosphere; close-up *Viking* photographs show no atmosphere and no frost patches; and they are among the darkest objects in the solar system."[1]

What is the truth about Sagan's statement? Sagan tells us that Velikovsky claimed on page 280 that Phobos and Deimos retain a remnant atmosphere of Mars and, therefore, they were bright. Let us, therefore, contrast Sagan's claim with what Velikovsky actually wrote in (*Worlds in Collision*, pp. 279-280).

> "Actually Mars has two satellites, mere rocks... They were discovered by Asaph Hall in 1877. With the optical instruments of Swift [in the early 1700's or about 150 years before their discovery] they could not be seen, and neither Newton nor Halley...nor William Herschel in the eighteenth, or Leverrier in the nineteenth century suspected their existence..."

> "When Mars was very close to the Earth [2,700 years ago], its two trabants were visible. They rushed in front of and around Mars; in the disturbances that took place, they probably snatched some of Mars' atmosphere, dispersed it as it was and appeared as gleaming names."

All that Velikovsky claimed was that Deimos and Phobos gleamed as they passed with Mars close to the Earth 2,700 years ago. Velikovsky made no claim that they will "appear bright" to use Sagan's words. In fact, Velikovsky made it clear that these satellites could not still be bright by showing that Halley nor William Herschel nor Leverrier in the preceding centuries could see them with the telescopic instruments of their time. John Nobel Wilford in *Mars Beckons*, (NY 1990), p. 72 tells us about Asaph Hall, the discoverer of the Martian moons.

> "Asaph Hall, an astronomer at the Naval Observatory in Washington, decided this was the year [1877] to renew the search [for possible Martian satellites]. He had available the observatory's new 26-inch telescope, one of the best in the world then and especially equipped to establish positions of the fainter satellites in the solar system."

Sagan merely distorted and by inference, misrepresented the statements of Velikovsky. Thus, it is interesting to note that Sagan, along with certain other astronomers, held that Iosif Shklovsky's conclusion regarding Phobos was correct. What was this conclusion? Shklovsky held that Deimos and Phobos were extraterrestrial bodies created by a race of superior beings and that they were hollow. In other words, Sagan supported the view that Deimos and Phobos were alien space stations in orbit around Mars. In his co-authored book *Intelligent Life in the Universe,* (NY 1966), p. 373 Sagan states,

> "The idea that the moons of Mars are artificial satellites may seem fantastic, at first glance. In my opinion, however, it merits serious consideration. A technical civilization substantially in advance of our own would certainly be capable of constructing and launching massive satellites. Since Mars does not have a large, natural satellite such as our moon, the construction of large artificial satellites would be of relatively greater importance to an advanced Martian civilization in its expansion into space. The launching of massive satellites from Mars would be a somewhat easier task than from Earth, because of lower Martian gravity."

In our discussion of "Sagan's Fifth Problem" above, Sagan ridiculed Velikovsky's concept of the possibility of life in space. Yet Sagan had created an entire civilization on Mars complete with the capacity of orbiting satellites! Patrick Moore in *The Planets,* (NY 1962), p. 96 states,

> "Not so long ago considerable interest was aroused by the suggestion...that Phobos and Deimos might be artificial satellites, and were probably hollow. Amazingly enough, some newspapers treated this weird idea seriously, and a paper on the subject was solemnly read before the British Interplanetary Society. At any rate, it is an attractive theory, even if it is about as likely as the age-old hypothesis that our Moon is made of green cheese."

Sagan states,

> "Beginning on page 281, there is a comparison of the Biblical Book of Joel and a set of *Vedic* hymns describing 'maruts'. Velikovsky believes that the 'maruts' were a host of meteorites that preceded and followed Mars during its close approach to Earth, which he believes is described in Joel. Velikovsky says (page 286): 'Joel did not copy from the *Vedas* nor the *Vedas* from Joel.' Yet, on page 288, Velikovsky finds it 'gratifying' to discover that the word 'Mars' and 'marut' are cognates. But how, if the stories in Joel and the *Vedas* are independent, could the two words possibly be cognates?"[2]

Sagan implies that the story of Mars and its maruts is a diffused common tale related by both Joel and the Vedas. However, again Sagan somehow failed to complete Velikovsky's citation so that the question he raises is answered. In *Worlds in Collision*, (p. 286), if we complete the citation introduced by Sagan, this becomes clear:

"Joel did not copy from the *Vedas*, nor the *Vedas* from Joel. In more than this one instance it is possible to show that peoples, separated even by broad oceans, have described some spectacle in similar terms. These were pageants, projected against the celestial screen, that, a few hours after they were seen in India, appeared over Nineveh, Jerusalem, and Athens, shortly thereafter *over Rome and Scandinavia*, and a few hours later *over the lands of the Mayas and Incas*. [emphasis added]

"The spectators saw in the celestial prodigies either demons, as the Erinyes of the Greeks or the Furies of the Latins, or gods whom they invoked in prayers, as in the *Vedas* of the Hindus, or the executors of the Lord's wrath, as in Joel and Isaiah."

Therefore, using Sagan's brand of semantic logic, it would seem that if stories are similar, but the objects described are not cognate words, then there was no diffusion of a story at all.

Since 'maruts' are not cognates of the Greek Erinyes or of the Latin Furies, the description by the Greeks and Latins of the events in the sky must be interpreted as common observation. But if the Greeks and Latins truly saw something in the sky, using Sagan's logic then the Scandinavians and Mayas and Incas, who described these same events in which the bodies that accompanied Mars are called by non-cognate names must also be interpreted as common observation. Therefore, all the peoples of the Earth saw Mars preceded and followed by a host of meteorites or comets independently except the Hindus and Ancient Hebrews. Stories certainly can contain cognates referring to the same event and clearly need not be diffusionist in origin. Sagan has really taken the observations and stood them on their heads.

If Velikovsky's analysis of the Mars maruts observation is correct, there should be evidence on at least one and possibly both tiny moons that circle Mars. Since the maruts and debris were generally preceding and following Mars and moving in the same general direction, it is highly probable that they would have encountered one of the Martian satellites and left grooves that generally run parallel to each other and leave craterlets all along these grooves. In *Astronomy*, for Jan. 1977, is just such evidence:

"*Viking* has discovered another mystery in the most unexpected place— on one of the two small Martian moons. *Mariner 9*'s mapping of both Phobos (12 x 14 x 17 miles, or 20 x 23 x 28 kilometers) and Deimos (6 x 7 x 10 miles or 10 x 12 x 16 kilometers) showed many craters and left most investigators with the impression that they were merely rocky chunks that bore the scars of meteorite impacts. There was a puzzling feature on Phobos that a few analysts noticed, but without better data, could say little about.

"At the limit of resolution were a few small crater pits that seemed to align in one or two chains. This was unusual, because crater chains on the moon are traditionally explained as volcanic pits—small eruption sites strung along fracture lines. Yet, Phobos apparently is too small to generate heat and conventional volcanic activity.

"*Viking's* high resolution photos have revealed that the crater chains are real and part of an extensive system of parallel grooves, a few hundred yards wide. There may be a tendency for the grooves to lie parallel to the direction of the satellite's orbital motion, although there appear to be several swarms with somewhat different orientations. Scientists are at a loss to explain them. Theories being discussed include: grooves left by much smaller satellite debris also orbiting Mars (though the grooves seem to follow contours of Phobos' surface too closely for this to be tenable); fractures radiating from an impact crater...[such as the largest crater Stickney]...or fractures created in the body of the Martian satellite when it was part of a hypothetical larger body that spawned both Martian moon, perhaps during a catastrophic impact."[3]

What is most interesting is that although other unproved analyses of the grooves with craterlets inside them may be correct, these analyses are in full agreement with Velikovsky's views of Mars, the maruts or the Martian moons.

Sagan is well aware of this evidence, but wished to argue about the evidence based on a confused semantic analysis and not based on this astronomical evidence.

The many smaller bodies traveling with Mars were probably rocks from Mars that were ejected from it during its near collision with Venus. Velikovsky describes planetary discharges from Venus to Mars that were of such strength that they ejected a great deal of Martian debris. Therefore, when Mars approached the Earth, some of this trailing debris should have fallen on the Earth. In *Science,* Vol. 237, for Aug. 14, 1987, pp. 721-722, is an article on Martian meteorites that are being found on Earth.

The headline states that "...eight SNC meteorites found on Earth are probably from Mars, most researchers now agree, but how they ever got off their home planet remains a question." The article states,

"It seems too good to be true. After spending $25 billion to obtain rocks from the nearby moon, rocks from Mars are falling out of the sky to be picked up by anyone that happens by. 'We think [the existence of meteorites from Mars] is a very good working hypothesis,' says geochemist Michael Drake of the University of Arizona, who was not that quick to warm to the idea. 'It's probable, but not proven; it's not likely to be incorrect. But short of going to Mars, no one will be absolutely convinced.'

"An increasing number of reputable researchers are convinced enough of the reality of the Martian meteorites to begin inferring the nature of Mars from them, even though there is a lingering problem—no one is quite sure how the meteorites ever got off Mars. In this issue of *Science*, (p. 738), Ann Vickery and Jay Melosh of the University of Arizona propose one way that it may have happened. The catch is that more needs to be known about Mars to show that their mechanism for interplanetary rock transport."

There are several pieces of evidence that lead scientists to conclude that the eight SNC meteorites are from Mars. The major problem is the ejection mechanism. Velikovsky's description of Mars and its maruts passing near the Earth is supported by this evidence. Once again the existing scientific evidence does not contradict his research.

Only a very few meteorites that fall to Earth are ever found. This means that if, at present eight Martian meteorites are found, untold thousands had to have fallen over the Earth recently. Seventy percent of the Earth's surface is ocean, and the Arctic and Antarctic regions have few inhabitants. Why would thousands of Martian meteorites have fallen? Velikovsky's theory explains this evidence.

EVIDENCE ON MARS

If Mars was involved in these events, wouldn't it also have shown an excess of grazing-incidence craters? That is, Mars should exhibit many more craters that are more elliptical than other highly cratered bodies like Mercury and the Moon. In fact, Mars has approximately ten times as many grazing-incidence craters than is predicted for bombardment of heliocentric projectiles. In *Scientific American*, Vol. 249, for October 1983, an article titled "Star Grazers", p. 88 discussed the fact that about 5% of the Martian craters were found to be elliptical with butterfly-wing patterns of ejecta thrown perpendicular to the craters long axes. This type of elliptical crater is produced by objects striking the planet at a grazing-incidence angle. The 176 such craters represent 10 times the number expected from projectiles all in heliocentric orbit. Like the grooves on Phobos, the scientists who conducted this study—P.H. Schultz and A.B. Lutz-Garihan—claim that projectiles that made these elongated craters had to be in orbit around Mars. This is exactly what the maruts would do to Mars. Phobos would be grooved while Mars would have an unusual number of elliptical craters. Again, Velikovsky's theory finds support.

MORE OF DEIMOS AND PHOBOS

What then of Mars' other satellite Deimos? Does it show evidence to support Velikovsky's views that Mars was followed by maruts (debris of all sizes)? William K. Hartmann, in *Moon and Planets,* 2 ed., (Belmont, CA 1983), p. 380, points out that, "The terrain on Deimos seems blanketed with boulders protruding here and there..." J. Guest, et. al., in *Planetary Geology,* (NY 1979), p. 192, informs us that many of these boulders are "house sized boulders." Where did these boulders come from? One would suspect they are the debris produced from impact cratering. But this is extremely improbable because to produce craters large enough to eject house sized rocks requires impact with forces and velocities that impart escape velocity from Deimos. The escape velocity of Deimos is about 10 meters (31 feet) per second. A person standing on Deimos could literally throw a stone from it into space and it would never return because Deimos' gravitation pull is so tiny. Thus, to produce craters, practically all the debris would be ejected at the escape velocity. Michael H. Carr in *The Surface of Mars,* (New Haven 1981), p. 198 writes, "The presence of ejecta (on Deimos and Phobos) is surprising, since the low gravity...must allow almost all the ejecta to escape." He does contend that some material would be retained, but the probability of this occurring is quite small. Thus, the large debris blanket on Deimos and the much smaller debris blanket on Phobos requires an explanation because Phobos with the larger mass has less debris than Deimos with the smaller mass. Gravitational and impact cratering theory require just the opposite. Based on Velikovsky's description of Mars and its maruts, Phobos and Deimos' orbits were at different distances from Mars. Thus, they traveled at different rates of speed. It is thus a reasonable assumption that Deimos' velocity permitted it to collect and retain much more minute debris than Phobos. About the space scientist's suggestions to explain the differences in the debris blankets of the two satellites, Carr says, "none of these suggestions is well formulated, and the cause of the difference remains uncertain." However, Velikovsky's hypothesis does indeed fit and explain the evidence.

ANCIENT ASTRONOMY

Sagan states,

"On page 307 we find Isaiah making an accurate prediction of the time of the return to Mars for another collision with Earth 'based on experi-ence during previous perturbations.' If so, Isaiah must have been able to solve the full three-body problem with electrical and magnetic forces thrown in, and it is a pity that this knowledge was not passed down to us in the *Old Testament*."[4]

We have previously cited that, "Babylonian astronomers were masters of us all: they kept continuous, dated records of celestial events from at least 747 B.C."[5] That is, by having an event recur periodically, it is relatively easy to make predictions and one need not have recourse to gravitational theory. In their chapter titled "Celestial Forecasting", which deals with the period that Sagan is discussing, Stephen Toulmin and June Goodfield inform us that by keeping careful records the ancient Babylonians,

> "were able to spot more and more of the cyclical variations which combine to determine where precisely in the sky the Moon, the Sun, or a particular planet will appear on a given day or night."[6]

The precision of Babylonian astronomers, for example, Kidinnu who lived 400 years before the Christian Era, has been summed up as follows:

> "Hansen, most famous of lunar astronomers, in 1857 gave the value for the annual motion of the sun and moon 0.3″ in excess; Kidinnu's [of Babylonia] error was three times greater. Oppolzer in 1887 constructed the canon we regularly employ to date ancient eclipses. It is now recognized that his value for the motion of the sun from the node was 0.7″ too small per annum; Kidinnu was actually nearer the truth with an error of 0.5″ too great. That such accuracy could be attained without telescopes, clocks, or the innumerable mechanical appliances which crowd our observatories, and without our higher mathematics, seems incredible until we recall that Kidinnu had at his disposal a longer series of carefully-observed eclipses and other astronomical phenomena than are available to his present-day successors."[7]

Velikovsky, citing various experts on ancient astronomy tells us in *Worlds in Collision*, that,

> "In the eighth century in Babylonia, the planet Mars we called 'the unpredictable planet.'...the movements of Mars were extremely important in Babylonian astrology—its rise and setting, its disappearance and return...its position in relation to the equator, the change in its illuminating power, its relationship to Venus, Jupiter and Mercury." [pp. 242-248]
> "But we ask for a direct statement that the planet Mars-Nergal [its other name] was the immediate cause of the cataclysm in the eighth and seventh centuries, when the world, in the language of Isaiah was 'moved exceeding' and 'became removed from its place.'" [p. 243]

Mark Washburn in *Mars at Last*, p. 25 states, "A surviving fragment of Babylonian text informs us that 'When Nergal (Mars) is dim it is lucky, when bright unlucky.'" When Mars was closer to Earth it naturally appeared brighter.

Thus, Isaiah states that he had observed Mars-Nergal and saw it move differently than earlier and approach the Earth periodically every 15 years. The perturbations occurred in 687, 702, 717, 747 B.C. All Isaiah, who experienced a few of their earlier events, had to do was observe Mars growing larger and brighter in the night sky every 15 years to foretell what might befall the Earth soon thereafter. When the summer sky becomes very overcast during the day, one can often predict that it will soon rain, and one does not need a barometer or weather chart or a complete understanding of meteorology to do this. Isaiah and the Babylonian astronomers of that time did not understand the three-body problem, with electrical and magnetic forces thrown in, to know that as Mars grew much brighter as it approached the Earth every 15 years that the same events that occurred 15, 30 or even 45 years earlier could happen again. They did not have telescopes, clocks or the innumerable appliances which crowd our observatories. But they could observe, count and even keep records. And although their achievements are spectacular compared with what has only been achieved in the past century, Sagan maintains such a basic observation is not only sloppy, but can only be achieved by full knowledge of the three-body problem. What he does not tell the reader is that Mars and the Earth are now nearest to each other between 15 and 17 years.

If Mars had a few very close approaches to the Earth every 15 years as described by the ancients, then that is in full accord with its present orbit, based on gravitational theory. Taken together with all the other evidence, it seems clear that Velikovsky's views respecting Mars are in full accord with Mars' surface topography, its atmosphere, it's erosion data, its moons—Deimos and Phobos—and its gravitational interaction with the Earth. None of Sagan's establishment views are so fully in accord with the evidence found by modern astronomy.

MARS' ATMOSPHERE

This brings us to Sagan's statements about the composition of Mars' atmosphere. He states, "Velikovsky argues some other constituent of the Martian atmosphere must be derived from the Earth. The argument, unfortunately is a *non sequitur*."[8] Sidney M. Willhelm, however, states that,

> "The statement by Sagan is, however, a *non sequitur* because Sagan himself reversed Velikovsky's words that it was *Mars* that gave to *Earth* certain gases. Sagan made no slight mistake; his distortion must be considered either deliberate, his effort incompetent, or his ethics of honesty were tossed to the winds in light of the fact that upon the two page

length discussion (under the title "The Atmosphere of Mars") to which Sagan's comments are addressed. Velikovsky [in *Worlds in Collision* (1950), pp. 366-367] clearly writes—on three separate occasions—that the shift in gases took place from Mars to Earth."[9]

It is not, as Sagan states, some constituent of "the Martian atmosphere must be derived from the Earth." These citations are given by Lewis M. Greenberg who informs us,

"...Sagan erroneously concludes that Velikovsky argues for a terrestrial origin of Martian [in his AAAS paper, page 62] argon when in fact, Velikovsky claimed just the opposite. In *Worlds in Collision* ("The Atmosphere of Mars"), Velikovsky wrote: "The main ingredients of the atmosphere of Mars must be present in the atmosphere of the earth. Mars, 'the god of war' *must have left part of his property on his visits*. As oxygen and water vapor are not the main ingredients of the atmosphere of Mars, some other elements of the terrestrial atmosphere must be the main components of its atmosphere. It could be nitrogen, but the presence of nitrogen on Mars—or its absence—has not been established. [emphasis added]

"Besides oxygen and nitrogen, the main components of the terrestrial atmosphere, argon and neon are present in detectable quantities in the air. These rare gases excite spectral lines only when in a hot state; consequently, they cannot be detected through lines of emission from a comparatively cool body such as Mars. The absorption lines of argon and neon have not yet been investigated. When a study of these lines will make possible a spectral search for these rare gases on planets, Mars should be submitted to the test. If analysis should reveal them in rich amounts, this should answer the question: *What contribution did Mars make to the Earth* when the two planets came into contact?"[10] [emphasis added]

Sagan has argued that evidence based on gravitational theory is most valid. In this respect his view that Mars got its atmosphere from the Earth during a near collision also violates Newtonian theory. In brief, larger (more massive) bodies have a greater gravitational pull than smaller (less massive) bodies. Mars is only about 10% the mass of the Earth. During a close passage of Mars to the Earth, the pull of gravity from the Earth will be much greater than the pull of gravity from Mars. Therefore, Mars' atmosphere would be pulled by Earth's gravity a far greater distance than Mars could gravitationally pull the atmosphere of the Earth. The Earth would, therefore, tend to remove the atmospheric gases of Mars, but Mars would not remove atmospheric gas from the Earth.

Sagan distorts not only Velikovsky's statement, but also a basic gravitational concept. This supports Sidney M. Willhelm's statement regarding Sagan's competence and/or ethics.

In Sagan's AAAS paper, he claimed Velikovsky wrong with respect to oxygen in the atmosphere of Venus. He claimed that there was no atmospheric oxygen on Venus. When space probes found oxygen, as was pointed out earlier, Sagan edited out his error and raised the following argument. "*Mariner 10* found evidence of oxygen in the atmosphere of Venus, not massive quantities of molecular oxygen in the lower atmosphere.

"The dearth of O_2 on Venus also renders untenable Velikovsky's belief in petroleum fires in the lower Venus atmosphere."[11] We have previously discussed in Sagan's Seventh Problem the breakdown of carbon dioxide by ultraviolet radiation in the upper and middle Venus atmosphere. There, oxygen and carbon monoxide are constantly being produced and it was expected that CO_2 would break down too, in the upper atmosphere into CO and O_2 "within weeks, and from the entire middle atmosphere in a few thousand years." Based on Velikovsky's model that Venus is heated from below its surface, it follows that the cooler upper and middle atmosphere reaches the surface where it interacts with whatever chemical gases remain after 3,500 or so years of burning and produces "fires". This analysis is based on the second law of thermodynamics which requires the hot gases above the surface of Venus rise and the cooler gases of the upper and middle atmosphere descend. This also explains why the carbon dioxide in the upper and middle atmosphere has not been completely photolyzed into carbon monoxide and oxygen. These heavier products are constantly descending. Therefore, whatever oxygen decends will be consumed as burning occurs. Sagan is unable to bring himself to see this basic process because, to do so, he must give up his greenhouse theory which explains nothing about Venus' atmosphere.

In his original AAAS paper (p. 62), Sagan argued that Velikovsky was wrong that there were any quantities of neon on Mars. After reports from space probes suggested that there was neon in Mars' atmosphere, Sagan wrote, "More than trace quantities of neon are now excluded; about one percent argon was found by *Viking*."[12]

Sagan then tries to explain the quantities of argon on Mars predicted by Velikovsky with the following,

"But even if large quantities of argon have been found on Mars, it would have provided no evidence for a Velikovskian atmospheric exchange—because the most abundant form of argon, 40 Ar, is produced by the radioactive decay of potassium 40 which is expected in the crust of Mars."[13]

Let us, therefore, subject this conclusion to a uniformitarian analysis. If this is correct, Mars would possess about 10 percent of what the Earth possesses of potassium-40. This is because Mars has a mass which is 10 percent of that of the Earth. Mars' atmosphere is about 1/100th that of the Earth. Actually, according to J.K. Beatty, B. O'Leary and A. Chaikin in *The New Solar System,* (NY 1981), p. 85, Mars contains one-fifth the proportion of potassium as compared with potassium found in Earth's rock. Argon should be, as we shall discover, a considerable percent of Mars' atmosphere. On Earth, argon–40 is about 1/100 of one percent of the atmosphere. One tenth of that is 1/1,000 of one percent. But Mars' atmosphere is 100 times smaller than that of Earth, therefore, if we multiply 100 by 1/1,000, Mars would have 10 percent of its atmosphere made up of argon–40 if it has one-tenth as much potassium as does the Earth. But on Mars, argon–40 is 16 thousandths of one percent.[14] Mars would have to have 625 times as much potassium–40 as it originally contained to have 10 percent of its atmosphere as argon–40. Now, Mars does not have as much potassium–40 as 10 percent of that of Earth, but even at 5 percent of that of the Earth, Mars would still have about 322 times as much argon–40 as it currently possesses; at 2 1/2 percent that of the Earth, Mars would still have about 161 times as much argon–40, and at 1 1/4 percent that of Earth, Mars would still have 80 times as much argon–40. For Mars to possess its present amount of argon–40, it would have to have about 1/100 of one percent of potassium–40 as the Earth had. This may or may not be the case, but Sagan's analysis is based on conclusions that can have interesting results. This conclusion is based on Sagan's uniformitarian view that the great changes on Mars were ancient and that Mars' atmosphere is also extremely ancient.

Sagan also states that Harrison Brown was the first to predict that neon and argon are constituents of the Martian atmosphere and not Velikovsky.[15] It seems that since Velikovsky was correct regarding these gases in Mars' atmosphere and Sagan was wrong, Sagan attempts to deny Velikovsky priority for his claim. Lewis M. Greenberg, however, informs us that,

"The first person to contend publicly that Mars' atmosphere must contain argon, and also neon, was Immanuel Velikovsky (in a lecture titled "Neon and Argon in the Atmosphere of Mars", copyrighted in 1945; test requests addressed in 1946 to Harvard College and Mount Wilson observatories; *Worlds in Collision* in 1950, "The Atmosphere of Mars")...

"At least one prominent individual, Arthur C. Clarke, was positively aware of Velikovsky's claim. In a letter (dated January 4, 1973) which was sent to the journal Pensée, Clarke expressed his belief that Velikovsky had been mistaken in his expectation of Martian argon and neon. However, three years later, in a personal communiqué (dated 30th January 1976) to Dr. Velikovsky, Clarke remarked, '...when I saw the

Russian announcement of argon in Mars' atmosphere, I immediately said to myself, 'Dr. Velikovsky strikes again!' However, I think we should wait until the *Viking* landings next July, which hopefully may settle a great many matters about Mars.'

"In sharp contrast to Clarke's position, is the attitude of Carl Sagan who has attempted to deny Velikovsky's priority be crediting Harrison Brown, an atomic scientist, with the argon prediction. Brown did not make his conclusions concerning argon known, however, until 1949 (*The Atmospheres of the Earth and Planets*, ed. Kuiper, 1949, p. 268; also see *The Velikovsky Affair*, p. 238), a good four years after Velikovsky.

"Yet, in his revised AAAS paper (p. 62), Sagan improperly claims that, 'the first [*sic*] published argument for argon and neon was made by Harrison Brown in the 1940's.'"

In all three of his criticisms of Velikovsky (the AAAS paper pp. 62-3, *Scientists Confront Velikovsky*, p. 88, and *Broca's Brain*, p. 123) Sagan claims,

"A much more serious problem for Velikovsky is the apparent absence of N_2...from the Martian atmosphere. The gas is relatively unreactive, does not freeze out at Martian temperatures and cannot readily escape from the Martian exosphere. If such an exchange of gases (between Earth and Mars) occurred, where is the N_2 on Mars? These tests of the assumed gas exchange between Mars and Earth, which Velikovsky advocates are poorly thought out in his writings; in fact, the test contradicts his thesis."[16]

Lewis M. Greenberg answered this attack laconically.[17]

"With the discovery of 2 to 3 percent nitrogen in the Martian atmosphere (as reported in *Science News*, Vol. 109, June 5 and 12, 1976, p. 363), yet another of Sagan's assaults on *Worlds in Collision*, is rendered impotent."[18]

Professor Greenberg sums up Sagan's views of the Martian atmosphere as follows:

"In point of fact, where Velikovsky's thesis is concerned, tests involving Mars now contradict only Sagan. Wrong about argon, wrong about nitrogen, unsure about neon. It is Sagan's cosmological theories which [upon analysis of the evidence] are poorly thought out—if thought out at all—and found severely wanting in the balance."[19]

SANTORINI—ATLANTIS

Sagan states,

I should also point out that a much more plausible explanation exists for most of the events in *Exodus* that Velikovsky accepts, an explanation that is much more in accord with physics. The Exodus is dated in *I Kings* as occurring 480 years before the initiation of the construction of the Temple of Solomon. With other supporting calculations, the date of the

Biblical Exodus is then computed to be about 1447 B.C. (Covey, 1975). Other Biblical scholars disagree, but this date is consistent with Velikovsky's chronology, and is astonishingly close to the dates obtained by a variety of scientific methods for the final and colossal volcanic explosion of the island of Thera (or Santorini) which may have destroyed the Minoan civilization in Crete and had profound consequences for Egypt, less than three hundred miles to the south."[20]

Sagan further tells us that one of the consequences for Egypt was,

"The amount of volcanic dust produced is more than adequate to account for three days of darkness in daytime, and accompanying events can explain earthquakes, famine, vermin and a range of familiar Velikovskian catastrophes."[21]

Allan Chen's article, "The Thera Theory" in *Discover,* for Feb. 1989, p. 79, discussed the ash from Thera stating that,

"...Charles Vitaliano of the University of Indiana and Dorothy Vitaliano of the U.S. Geological Survey...after years of collecting and analyzing [ash] samples...found ash all right...but the ash fall was not very heavy. By looking at samples of the Mediterranean sea floor, oceanographers from the Lamont-Doherty Geological Observatory determined that most of the ash from the eruption actually fell to the east of Thera, not to the south [toward Egypt]. Thick ash falls were later found on Rhodes...and on Kos...but neither island seems to have been seriously affected by the ash fall."

The evidence on the subject thoroughly contradicts Sagan's conclusions about Thera. In *Science News* an article titled "Ancient Crete: Double Dose Destruction" states,

"The Bronze Age eruption of Santorini, roughly between 1200 and 1700 B.C. has been credited with an impressive list of feats. Scholars have tied it tentatively to the plagues during the Exodus from Egypt, to the disappearance of that mysterious realm, Atlantis, and to the demise of the sophisticated Minoan culture on Crete... In the June 7 (1984) *Nature,* W.S. Downey and D.H. Tarling described their efforts to establish relative dates for deposits of eruptive products such as ash and denser volcanic debris from Santorini, and for artifacts from late Minoan sites on Crete 120 kilometers to the south... Downey and Tarling support earlier findings that the ash fall did not contribute greatly to the downfall of Minoan culture. At most, five centimeters of ash fell on the island, and this shower was mostly on eastern Crete where civilization continued."[22]

Five centimeters is about two inches; if two inches of ash fell 120 km to the south of Santorini, even less ash fell on Egypt which was some 300 miles south. To reach the Nile River to the west of Cairo which is another 500 miles away, the dust would become extremely thin in the atmosphere. Probably less than half an inch of ash fell over a period of several weeks on Egypt and this would, by no credible stretch of the imagination, have occluded the Sun for three days.

In *Worlds in Collision*, Velikovsky wrote a chapter entitled "The Darkness" in which he claims that the darkness encompassed the entire globe. Thera could not darken Egypt for three days, thus, it could not darken the entire globe; but a comet could. Kenneth Hsu has informed us on this matter, "A large comet need not even hit the earth to produce that much dust; a near miss would leave enough debris in earth's atmosphere to produce a complete blackout."[23] And there is evidence for this dust debris in the Greenland ice cap which, in the Ice Age ice, is a hundred times greater than in present-day ice.

Velikovsky held that Venus had two near collisions with the Earth around 1447 and 1395 B.C. or about fifty years apart. Even if the date for the events should subsequently be moved either up or down, there should be evidence of these two events. But we have pointed out in our earlier discussion that the evidence presented by Downey and Tarling from the University of Newcastle-upon-Tyne showed from magnetic signatures that the destruction of Minoan sites on Crete occurred in two separate events. Furthermore, two West German scientists have shown from ceramic analysis of the pottery found among the Cretan ruins, that the events were separated by a period of fifty years. The magnetic evidence shows different magnetic intensities and directions for these fields during the Thera events. Thus, Thera's eruption supports Velikovsky's hypothesis and conflicts with Sagan's views on it. Sagan adds the following respecting the Thera event.

> "It also may have produced an immense Mediterranean *tsunami*, or tidal wave, which Angelos Galanopoulos 1964)—who is responsible for much of the recent geological and archaeological interest in Thera—believes can account for the parting of the Red Sea as well."[24]

Again, Allan Chen's article in *Discover* for Feb. 1989, p. 80, informs us that the "...claim that giant waves set off by the eruption of Thera...[at first] was at least plausible; eruptions of island volcanoes have been known to trigger such waves or tsunami. The problem was there was no clear evidence that a Theran tsunami actually occurred; no one had found the distinctive type of sedimentary deposit that the wave would have left on the coast. What's more, as tsunami experts achieved a better understanding of the physics of giant waves, their estimates of the potential size of the Theran wave came down dramatically from a terrifyingly destructive 600 feet to an eminently surfable 30 feet." A 30 foot wave can neither account for great destruction nor the parting of the Sea of Passage.

Bernard Newgrosh deals with the violence of the Thera eruption:

"Over Thera…[astronomers] come into difficulties with stratigraphy, for they see the eruption of Thera as both a part of and a consequence of a world-wide disturbance that took part at the time of the Exodus. If they had read their own references carefully they would have noticed that one of them states quite clearly that the Thera eruption took place in the Late Minoan IA period. Not only that but pottery of Late Minoan IA straddles the ash fall, and the disaster did not cause any alteration in the style of occupation… Major world-wide catastrophes cause hiatuses, breaks in occupation, *et cetera* and are notable because things like pottery styles are changed. Nothing of this sort happened as a result of the Thera eruption, confirming that it was little more than a big volcanic eruption with local effects, and formed no part of such a world-wide [or major geographic] event such as the Exodus was. Late Minoan IA continues with hardly a visible hiccup as a result of Thera."[25]

Though Sagan does not believe Thera is responsible for a world-wide catastrophic event, he does imply it was responsible for the destruction and plagues that occurred during the Exodus from Egypt. If the tidal wave did swamp Egypt and part the Red Sea, why was it barely able to change anything in the area nearest to where it occurred? In *Science News* for Jan. 13, 1990 p. 22 we are told that Minoan culture survived the Theran eruption. Jeffrey S. Soles of the University of North Carolina at Greensboro who conducted the excavation on Crete states, "We have conclusive evidence for the survival of Minoan civilization after the Santorini blast."

Thera does not explain the number of ancient ruins in the Near East that, according to Claude F.A. Schaeffer (cited in the chapter, *Sagan's Fourth Problem*) were damaged by catastrophic events of such magnitude that entire cultures were destroyed.

"In the ruins of excavated sites throughout all lands of the ancient East, signs are seen of great destruction that only nature could have inflicted. Claude Schaeffer, in his great recent work, discerned six separate upheavals. All of these catastrophes of earthquake and fire were of such encompassing extent that Asia Minor, Mesopotamia, the Caucasus, the Iranian plateau, Syria, Palestine, Cyprus and Egypt were simultaneously overwhelmed. And some of these catastrophes were in addition of such violence that they closed great ages in the history of ancient civilizations.

"The enumerated countries were the subject of Schaeffer's detailed inquiry; and recognizing the magnitude of the catastrophes that have no parallels in modern annals or in the concepts of seismology he became convinced that these countries, the ancient sites of which he studies represent only a fraction of the area that was gripped by the shocks."[26]

Thus, Sagan's views do not tally with the evidence. Sagan adds,

"In a certain sense, the Galanopoulos explanation of the events in *Exodus* is even more provocative than the Velikovsky explanation, because

Galanopoulos has presented moderately convincing evidence that Thera corresponds in almost all essential details to the legendary civilization of Atlantis. If he is right, it is the destruction of Atlantis rather than the apparition of a comet that permitted the Israelites to leave Egypt."[27]

Sagan appears to be quite uninformed regarding basic facts about Atlantis. Israel M. Isaacson in an article titled "Some Preliminary Remarks About Thera and Atlantis" states,

"With specific reference to Atlantis, Plato's account *might* retain a kernel of the memory of the Theran explosion, but even this is open to serious doubt. According to Plato, Atlantis was a tremendous land mass in the Atlantic Ocean, supporting a semi-fabulous civilization as well as elephants; it sank in floods and earthquakes *long* before the flood of Deucalion, in fact, some 9,000 years before Solon lived (i.e., ca. 9600 B.C.); and was submerged and invisible by the 6th century B.C. In order to equate this account with the eruption of Thera, one would need to discount the semi-fabulous part; shrink a huge island to a tiny one; physically transport it from the Atlantic Ocean to the East Mediterranean; change its mode of destruction from quake and flood to massive volcanic eruption; divide the 9,000 years by 10 without any textual justification and despite its "ring of truth" (not as real history, but as the very type of thing that Egyptians of the day told the Greeks), then, still subtract 500 years from this figure to fit the...chronology; and explain why an island which is and always has been perfectly visible, and which in the 6th century B.C., supported a vigorous population of seafarers was said to be totally submerged."[28]

Velikovsky claimed that the date for the sinking of Atlantis, if it occurred, would have been about 3,500 years ago—about 1,400-1,500 B.C. However, the story told by Plato has Solon claiming this event occurred 9,000 before Solon's time, not 900 years. James W. Mavor Jr., in *Voyage To Atlantis*, (Rochester, VT 1990), p. 21 explains:

"To explain this discrepancy [between 9,000 and 900 years] Dr. [Angelos] Galanopoulos...came up with a startling and original concept...he told me, he found that all numerical references in the thousands seem to be implausible, while those involving sums under 1,000 were always entirely reasonable [regarding his attempt to fit Atlantis into the framework of Minoan history]."

On page 26 Mavor shows that the symbol for 'one hundred' in Cretan Linear script of 1,500 B.C. is a circle (O) but 'one thousand' is a circle with lines emanating up and down, and right and left from the circumference (⊕). Thus the discrepancy of time may be explained by saying that when Solon said Atlantis was destroyed 9,000 years ago, it could have been a scribal error, making the meaning, Atlantis was destroyed 900 years ago.

All in all, Sagan's arguments are without substance. His attempt to explain the immense destruction in the ancient East wrought by the volcanic eruption of Thera is lamentable. Sagan's glaring mistakes bring to mind what N.J. Mackintosh wrote in *Nature,* for April 27, 1978 about Sagan's, *The Dragons of Eden,* pp. 768-769. Mackintosh remarks that Sagan's book is, "profoundly unscientific" and goes on to say, "It is inaccurate, full of fanciful and unilluminating analogies, infuriatingly unsystematic and skims hither and yon over the surface of the subject, unerringly concentrating on the superficial and misleading." I conclude that Sagan's analysis of *Worlds in Collision* and the rest of Velikovskian writings fare no better.

NOTES & REFERENCES

[1]SCV, pp. 86-87; B.B., p. 121.
[2]SCV, p. 87; B.B., pp. 121-122.
[3]*Astronomy*, Vol. 5, (1977), p. 55.
[4]SCV, pp. 87-88; B.B., p. 122.
[5]Toulmin and Goodfield; *The Fabric of the Heavens*, op. cit., p. 25.
[6]Ibid., pp. 37-38.
[7]Ibid., pp. 39-40.
[8]SCV, p. 88; B.B., pp. 122-123.
[9]Willhelm, Sidney M.; *Kronos* III, 2, pp. 55-56.
[10]*Kronos*, Vol. II, 1, pp. 107-108.
[11]SCV, p. 88; B.B., p. 122.
[12]B.B., p. 123.
[13]B.B., p. 123.
[14]*The New Solar System*, op. cit., p. 57.
[15]B.B., p. 123.
[16]B.B., p. 123.
[17]*Kronos* II, 1 p. 107.
[18]Ibid., p. 109.
[19]Ibid.
[20]SCV, pp. 89-90; B.B., p. 124.
[21]Ibid.
[22]*Science News*, Vol. 125, No. 24, (June 16, 1984), p. 374.
[23]Hsu, K.; *The Great Dying*, op. cit., p. 109.
[24]SCV, p. 90; B.B., p. 124.
[25]Newgrosh, Bernard; *Kronos*, VIII, 4, p. 70.
[26]*Earth in Upheaval*, op. cit., p. 177.
[27]B.B., pp. 124-125.
[28]Isaacson, Israel M.; "Some Preliminary Remarks About Thera and Atlantis", *Kronos* I:2; p. 95.

SAGAN'S APPENDICES

"Mathematics is only a tool, though an immensely powerful one. No equation, however impressive and complex, can arrive at the truth if the initial assumptions are incorrect. It is really quite amazing by what margins competent, but conservative scientists and engineers can miss the mark, when they start with the preconceived idea that what they are investigating is impossible. When this happens, the most well-informed men become blinded by their prejudices and are unable to see what lies directly ahead of them. What is even more incredible, they refuse to learn from experience. They will continue to make the same mistakes over and over again."

Arthur C. Clarke, *Profiles of the Future,* (NY 1984), p. 21

A scientific appendix is added to a paper to prove the material already presented. The author must employ the proper physics and the proper mathematical equations. Furthermore, the physical and mathematical models presented must also conform to those physical and mathematical models which have been accepted by science as properly applicable to the problem under discussion. The appendix should prove mathematically that the conclusions raised by the author are valid based on indisputable scientific evidence.

This is what is required, but as we shall observe, this is not what Sagan has delivered.

APPENDIX I
Simple Collision Physics of the
Probability of a Recent Collision With Earth
By a Massive Member of the Solar System

Here, Sagan proposes to prove his conclusion in "Problem II" that the probability of comet-Venus having collisions with the Earth is, "7.3 x 10^{-28}, or about a trillion quadrillion to one." Sagan says,

"We here consider using the most elementary treatment which preserves the essential physics, the probability that a massive object of the sort considered by Velikovsky to be ejected from Jupiter might impact the Earth. Velikovsky proposes that a grazing or near-collision occurred between this comet and the Earth. We will subsume this idea under the designation 'collision' below among other objects of similar size. Collision will occur when the centers of the objects are 2R distance. We may then speak of an effective collision. Cross section of $\pi(2R)^2 = 4\pi R^2$; this is the targeted area which the center of the moving object must strike in order for a collision to occur."[1]

Here, Sagan informs us that the Earth and comet-Venus had to actually touch each other for there to be a collision. Sagan assures us that "Velikovsky proposes that a grazing or near collision occurred" This is a contradiction because Sagan claims that Velikovsky proposed that the collision was a "hit", but also at the same time a "miss". However, Sagan wishes to concentrate his mathematics on a grazing "hit" because he demands that the "effective collision cross section of $(\pi 2R)^2 = 4\pi R^2$; thus, the target area which the center of the moving target must strike in order for collision to occur." Based on Sagan's analysis, if the bodies do not strike, his mathematical model becomes invalid.

Does Sagan actually believe that Velikovsky proposed that comet-Venus struck the Earth 5 or 6 times? The answer is that he knew that this conclusion was not Velikovsky's. In *Scientists Confront Velikovsky,* p. 67 and *Broca's Brain,* p. 121, we are able to quote Sagan admitting that Velikovsky never proposed a grazing collision. Sagan let slip this remark, "...in fact, *if these planets* (Earth and comet-Venus) *were even tens of thousands of kilometers away,* AS HE [Velikovsky] SEEMS TO THINK..." Here, Sagan completely contradicts the entire thrust of his model by allowing that "Dr. Velikovsky thinks that the Earth and comet-Venus were tens of thousands of kilometers apart." Thus, they do not "hit" each other.

When we pursue Sagan along this line of inquiry, it becomes obvious that if the bodies do not strike each other, then they must miss one another. The effective collision cross section of the target area changes and becomes larger and thus, the probability also changes. Sagan is well aware of this, but offers two sentences that debunk the validity of *his* grazing collision at the end of the appendix. Sagan casually adds "that in both calculations, an approach to within N Earth radii has N^2 times the probability of a physical collision. Thus, for N=10, a miss of 63,000 km, the above values of T must be reduced by two orders of magnitude. This is about 1/6 the distance between the Earth and the Moon."[2]

We have shown that Shulamith Kogan used Sagan's physical model and mathematical calculation for N = 100 Earth radii. What she discovered was that the Earth and comet-Venus would collide more than once for each close approach on their orbits. At N = 1,000, using Sagan's equations, the Earth and Venus collide over 10 times as they pass by each other once. It is the same as a batter swinging his bat at a ball, and with one swing missing it, based on Sagan's calculation, several times. At N = 500, the batter misses the ball with one swing a few times. One begins to get the impression that Sagan's analysis can, and does have impossible results. Unlike "Casey at the Bat", Sagan's first appendix strikes out with one swing. Furthermore, Sagan's probability can only be calculated if we repeal the laws of gravity; since Venus has to move inside the Earth's Roche limit 5 or 6 times; according to Asimov, it must disintegrate at each close approach, but somehow maintain its integrity; that is, it must not disintegrate inside the Roche limit. The probability of Sagan's analysis being correct is zero.

What then is wrong with Sagan's mathematical model? Shulamith Kogan in a published letter in *Physics Today* for Sept. 1980, pp. 97-98, has correctly informed us that the model Sagan employs is "only a good approximation when short time periods or small distances are considered." Robert Bass has also informed us of the same when he wrote Sagan's model was derived from Opik and Urey "to obtain apparently reasonable statistics about meteoric collision with the Moon, Mars and Venus; but in such calculations, it is assumed (as an approximation)..." [3] The model employed by Sagan was never meant to be applied in the way he used it. Therefore, it can give 10 collisions for one near approach.

In simple terms, Sagan's model does not conform to those models accepted to analyze gravitating bodies on long orbits over long time periods. His model applied to small bodies over short distances and short time periods. As was pointed out, Sagan's expertise is not in the field of dynamical astronomy. Therefore, he was taken to task by Robert Jastrow and it was further shown that "Sagan, in effect, did not appear to know the differences between kinematics...and dynamics. Kinematics is an analysis based on the use of velocities while dynamics is the use of accelerations. These two models are quite different." [4] Apparently, it seems Sagan was out of his depth when he undertook this calculation.

However, Sagan introduces a new set of parameters. He states,

"The objects which have since the earliest history of the solar system produced impact craters on the Moon, the Earth and the inner planets are ones in highly eccentric orbits; the comets and, especially, the *Apollo* objects—which are either dead comets or asteroids... If, therefore, Velikovsky's proto-Venus comet were a member of some family of objects like the *Apollo* objects or the comets, the chance of finding one Velikovskian comet 6,000 km in radius would be far less than one-millionth of the chance of finding one some 10 km in radius. A more probable number is a billion times less likely, but let us give the benefit of the doubt to Velikovsky.

"Since there are about ten *Apollo* objects larger than about 10 km in radius, the chance of there being one Velikovskian comet is then much less than 10,000-to-1 odds against the proposition." [5]

Here, Sagan raises the argument that comets and Apollo asteroids which are on highly elliptical orbits are known to be small. Thus, what is the chance of there existing a comet or Apollo asteroid as large, say as Venus? Sagan apparently forgot the argument he raised to explain the retrograde rotation of Venus and Uranus. According to Sagan, Venus and Uranus rotate in the opposite direction because they were struck by large bodies which produced these changes in their rotation. [6] For Uranus to be knocked on its side, it had to be struck by an object on a highly elliptical orbit with the mass about the size of Earth. In *Mercury* for July/August 1975 is an article, "The Trouble with Uranus". In discussing the axial inclination, the author writes,

> "The most extraordinary thing about Uranus is the fact that the pole of the ecliptic plane is 98 degrees. Uranus and Venus are the only two planets with retrograde rotation. The rotation period of Uranus...is apparently normal, being very similar to that of Jupiter, Saturn and Neptune, all of which have their axes of rotation nearly perpendicular to the ecliptic, and it is very difficult to understand why the direction and not the rate of rotation is anomalous. The only explanation that has been suggested is that, at the very end of the accretionary phase of the formation of the planets, a chance collision with another protoplanet (with perhaps 10% of the mass of Uranus) knocked the axis of rotation awry." [7]

Uranus is some fourteen times more massive than the Earth. Therefore, the body that tilted Uranus, according to the scientific explanation offered, was about 1.4 times the mass of Earth. To be that massive, the body had to be at least as large as the Earth. The Earth is larger than Venus and therefore, the body that struck Uranus would be larger than even Venus. Hence, it is generally accepted, contrary to Sagan's odds, that large bodies existed and were on highly elliptical orbits and actually struck a planet. Oh! but one forgets Sagan's odds and it seems strange that such evidence exists in the literature at least a year after Sagan's paper was presented.

Finally we showed that William K. Hartmann in his book *Moon and Planets,* 2 ed. 1983, p. 238, claimed that planetesimals from the early history of the solar system could be as large as 10 percent of the planet in their zone. Some of these would have been ejected and become part of the Oort cloud. A planetesimal one-tenth the diameter of Jupiter would be about 8,000 miles in diameter, the approximate size of Venus. Sagan's odds and discussion are really foolish since his apparent motivation to disprove Velikovsky seems to have outrun his scientific good sense. C.J. Ransom informs us that,

> "Sagan also cluttered the calculations with unnecessary details in order to make the odds against collisions higher. If enough specific restraints are placed on an idea, anything can be 'proven' to be improbable. For

example, it is known that a 10 pound meteorite fell through the roof of a house and hit Mrs. Hewlett Hodges in her left side just below the ribs. Obviously a person being hit by a meteorite is a rare event. An estimate of about how rare the evidence is, might be obtained by calculating a 'probability' that the event would happen. This might be done by using estimates about the number of meteorites hitting populated portions of the earth per year, and the average number of people per unit area in populated areas. However, Sagan's technique would more closely resemble the following. He would calculate the probability that someone named Hewlett Hodges would be born and eventually get married. Then he would calculate the probability that a house would be hit by a meteorite and that Mrs. Hewlett Hodges would be home at the time. If it were a 10 room house, he would calculate the probability that she is in the room that was hit. He would then assume a certain size room and calculate the probability that she was in the part of the room that was hit. Included would be a fifty-fifty chance that she was facing the proper direction so that she would be hit in the left side. He would also add an estimate of the square inches on her body and the area just below the ribs to get a probability that this is where she was hit. Combining all of this, his final number would be used as 'proof' that this did not happen because the odds are improbably high."[8]

In spite of Sagan seemingly proving the odds improbably high by stating,

"Since there are about ten Apollo objects larger than about 10 km in radius, the chance of there being one Velikovskian comet is then much less than 100,000-to-1 odds against the proposition" [and then piling on the following probabilities or should one say improbabilities], "The steady state abundance of such an object would be (for r = 4 a.u. and 1.2 degrees) n

$$n = (10 \times 10^{-5})/4 \times 10^{40} = 2.5 \times 10^{-45} \text{ Velikovskian comets/cm}^3.$$

The mean free time for collision with Earth would then be

$$T = 1/(n \, \sigma v) = 1/[(2.5 \times 10^{-45} \text{ cm}^{-3}) \times (5 \times 10^{18} \text{ cm}^2) \times$$
$$(2 \times 10^6 \text{ cm sec}^{-1})] = 4 \times 10^{-21} \text{ secs} \approx 10^{14} \text{ years}$$

which is much greater than the age of the solar system (5×10^9 years). That is, if the Velikovskian comet were part of the population of other colliding debris in the inner solar system, it would be such a rare object that it would essentially never collide with the Earth."[9]

This, however, flies in the face of what we know of the collision of an Earth-sized body with Uranus. Uranus is located in the outer part of the solar system: therefore, the area in which the body that is believed by the scientists to have hit it would be enormously larger than the smaller, inner area of the solar system. We are further informed that the collision occurred "after the major time period when the planets had already formed" and thus, there was little debris available to cause collisions. Based on the enormous restraints Sagan has placed on the probability (or improbability) of this event happening for the Earth and comet-Venus, it becomes even more improbable that Uranus could have been struck. But the scientific community has maintained for a long time, against the views of Sagan, that it probably did occur.

The event that supposedly knocked Uranus on its side is assumed to have occurred billions of years ago. Nevertheless, it is believed that there may currently exist a planet called Planet X that is 5 times more massive than the Earth and is on a highly elliptical orbit beyond Pluto. It is also assumed that Planet X's orbit is at high inclination to the plane of the ecliptic. David M. Raup in *The Nemesis Affair,* (NY 1986), p. 143, states,

> "Yet another explanation for the extinction…[of the dinosaurs and other life forms] has surfaced. D.P. Whitmire and J.J. Mattese published a paper in *Nature* in January, 1985 suggesting that the comet showers could also be produced by an unseen planet, Planet X, lying beyond the orbit of Pluto… The idea of a missing planet in our solar system has been kicking around for a long time because of possible (but debatable) discrepancies between observed and predicted motions of the outer planets."

Apparently, because the uniformitarian astronomers need this large body to fulfill their views, it is perfectly plausible for it to exist, while Sagan denounces the possibility of a large body one-fifth the size of Planet X based on comparison of masses. Sagan does not seem to understand the nature of the evidence he offers as refutation.

In fact, this is exactly what the opponents of Galileo did when they argued with him about the validity of the Copernican theory which claimed that the Earth was in orbit around the sun. That is, they use the technique of adding unnecessary calculations as considerations for rejecting the heliocentric theory. Galileo in *Dialogue Concerning the Two Chief World Systems,* 2 ed., (translated by Stillman Drake), (Berkeley, CA, 1967), p. 294 tells us Galileo said that, "…the truth of the matter is that I was much astonished to see how this [anti-Copernican] author goes to such lengths, [to disprove the view that the Earth moves] and puts into so many computations which are not in the least necessary to the question he is examining." Yet Sagan has done essentially the same with respect to the view that proto–planet Venus is an Apollo asteroid. Proto-planet Venus was not an Apollo asteroid or a comet; it was a planet on a cometary orbit.

With regard to Sagan's argument that Apollo asteroids are small while Venus is immense by comparison, I this time refer the reader to Galileo's book *The Assayer,* originally published in Rome in 1623 in *Discourses and Opinions of Galileo,* (translated by Stillman Drake), (NY 1957), p. 253. Here Galileo argues,

"...that in order to make a comet a quasi-planet, and as such to deck it out in the attributes of other planets, it is sufficient [by my opponents]...to regard it as one and so name it. If their opinions and voices have the power of calling into existence the things they name, then I beg them to do me the favor of naming a lot of old hardware I have about my house 'gold.'"

To make proto-planet Venus an Apollo asteroid, Sagan simply regards it as one and so names it. By naming it thus magically it becomes one. But Galileo long ago saw through such legerdemain when he said calling "old hardware" by the name "gold" changes it not one wit; it is still old hardware. How little has changed since Galileo's time.

Sagan does not discuss this, although I think he is well aware of this evidence. But somehow he overlooked this long-standing contra-diction in the literature. Nevertheless, not only is his high probability calculation disproven by the evidence, but even if we too overlook this error on his part, his calculation is irrelevant. If anyone sought clarity and an appropriate physical model in Sagan's first appendix, he would have to first get rid of the debris with which it is littered.

APPENDIX II
Consequences of a Sudden
Deceleration of the Earth's Rotation

Sagan here attempts to prove that if the Earth stopped rotating in one day and then restarted rotating the same day; the energy to do both jobs would have all been converted to heat energy. The seas would have boiled above the boiling point, and it follows that all life in the seas would have been cooked and there could not still exist the abundant life that thrives in the deeps. One suspects that Sagan thinks he has cooked Velikovsky's goose with this argument.

What Sagan failed to explain is that the Earth does not have to come to a full stop for the inhabitants to see the Sun appear to stand still for a few hours in approximately the same area of the sky. Let us be very specific about this process, especially respecting gravitational theory which Sagan maintains has great validity. David Morrison in his book, *Voyage to Saturn*, (NASA) (Washington D.C. 1982), pp. 167–168 discusses the "Co-Orbital Satellites", stating,

"Two satellites share nearly the same orbit circling Saturn in just under 17 hours... Both were first seen in 1966, at the time the rings were edge-on as seen from Earth, and again in 1980 under similar geometry...Officially, they are called 1980S1 and 1980S3; ultimately they will be named as the tenth and eleventh satellites of Saturn.

"The period of revolution of 1980S1 was 16.664 hours and that of 1980S3, 16.672 hours, both measured at the time of the *Voyager 1* encounter. The slower object has an orbit just 50 kilometers larger than the faster one. About once every four years, the inner [faster] one catches up with its slower moving sibling, but since the space between the orbits is smaller than the objects, there is no room to pass. What happens is that, just short of a collision, the two satellites attract each other gravitationally and exchange orbits. They then slowly move apart in opposite directions, and the four year cycle starts all over again. This strange orbital dance is unique, as far as we know, in the solar system."

To explain this "unique" gravitational interaction, the astronomers claim that the two satellites exchange angular momentum, that is, they transfer energy. These satellites are relatively similar in size and thus, the angular momentum energy lost by one satellite is gained by the other. As the Earth and Venus are also relatively similar in size, a similar transfer of angular momentum can occur. As the two bodies gravitationally move toward each other the angular momentum transferred to Venus would cause the rotation of the Earth to appear somewhat slower and the Sun would appear to stand still in the sky. In point of fact, moving the Earth closer to Venus as it came from the Sun would also move the Earth somewhat closer to, then away from, the Sun like Mercury at perihelian. Mercury because of its 59 day rotation period on its highly elliptical orbit shows just this effect as it comes closest to the Sun. Kenneth F. Weaver in *National Geographic* for August 1970, p. 165 informs us, "If you were on Mercury at dawn just at perihelion passage, you would see the sun come up, hang for a brief time in the sky, drop back below the horizon, then rise again. Because of this strange phenomenon Professor Bruce W. Hapke, of the University of Pittsburgh, calls Mercury 'the Joshua planet.' He refers, of course, to the *Old Testament* prophet who commanded the sun to stand still over Gibeon." According to Patrick Moore's *The Unfolding Universe,* (NY 1982), p. 65, "Near perihelion…an observer on the surface [of Mercury] would see the Sun move slowly backward for eight Earth-days…" Although this effect on the Earth would not be as dramatic as that at Mercury it would be quite similar: the observer on the Earth would see the Sun seem to stand in the sky for several hours instead of a few weeks. But what is most significant is that the energy Earth passed over to Venus would not generate heat in the Earth. Hence, based on the laws of gravity, the Sun could appear to stand still and the oceans would not have boiled. The energy that Sagan demands would not all go into the production of heat; it would go into decreased revolutionary angular momentum. The Earth's orbit would then be ever so slightly closer to the Sun. Velikovsky claims that after this collision the length of the year was shorter and this is reflected in the calendars of ancient peoples.

Furthermore, Sagan importantly has overlooked the question of the shifting of the polar axis during the 1500 B C. catastrophe that Velikovsky suggests. In *Earth in Upheaval,* p. 39 Velikovsky pointed out that, "the northern part of Greenland was never glaciated. 'Probably, then as now, an exception was the northernmost part of Greenland; for it seems a rule that the most northerly lands are not and never were glaciated,' writes the polar explorer Vilhjalmur Stefanson [*Greenland* (1949), p. 4] 'The islands of the Arctic Archipelago,' writes another scientist, [R.F. Griggs, *Science,* XCV, (1942) 2473] 'were never glaciated. Neither was the interior of Alaska.'

'It is a remarkable fact that no ice mass covered the low lands of northern Siberia any more than those of Alaska,' wrote James D. Dana [*Manual of Geology,* (4th ed.) p. 977] the leading American geologist of the last century." Why weren't these lands glaciated as the Ice Age theory requires? This is further corroborated in Charles Hapgood's *Maps of The Ancient Sea Kings,* (NY 1979), revised ed., pp. 177-178. Hapgood remarks that, "the freezing of up of Siberia [occurred] simultaneously with the thawing of North America." If as Velikovsky suggested the axis of the Earth tilted, or as Hapgood suggested the crust of the Earth slipped, then as Siberia moved into the polar regions it would freeze while as North America moved into the temperate regions it would thaw. It is quite strange that Siberia which was warmer during the Ice Age should become colder after the Ice Age ended. This strongly implies a polar shift.

Velikovsky has also shown that ancient man built and inhabited a city, Ipiutak, in Alaska, that bronze was cast in a metal shop unearthed in Siberia, both places well north of the Arctic Circle. There were human artifacts and cave paintings discovered on the New Siberian and Spitzbergen islands, and abundant trees which are currently found only in regions much farther south were growing on these Arctic islands on which herds of mammoth and other life forms fed. A pole shift responsible for such a change of climate would most certainly cause the Sun to appear to stand still in the sky.

In conjunction, the Earth moving on its orbit closer toward then away from Venus as the great planet-comet approached and departed, the tilting of the Earth's poles would have made the Sun appear to be unmoving. The tilting of the poles may have occurred either by the magnetic interactions of the two bodies, or by the sliding of the continents over the Earth's mantle, or some combination of both processes. The fact is that the rotation of the Earth need not have been greatly altered to make the Sun appear stationary. Charles Hapgood in *Path of the Pole*, suggested that the Ice Age ended with a polar shift of a tectonic nature. Velikovsky's thesis does not preclude such behavior and his theory can accommodate and fit in well with Hapgood's view. Hapgood asked Einstein to consider his concept. What was Einstein's evaluation of the idea that the Ice Age ended with a pole shift?

In the *Foreward* to Hapgood's book, p. XIV, Albert Einstein states,

> "I frequently receive communications from people who wish to consult me concerning their unpublished ideas. It goes without saying that these ideas are very seldom possessed of scientific validity. The very first communication, however, that I received from Mr. Hapgood electrified me. His idea is original, of great simplicity, and—if it continues to prove itself—of great importance to everything that is related to the history of the Earth's surface.
>
> "A great many empirical data indicate that at each point on the earth's surface that has been carefully studied, many climate changes have taken place, apparently quite suddenly. This, according to Hapgood, is explicable if the virtually rigid outer crust of the Earth undergoes, from time to time, *extreme displacement* [emphasis added] over the viscous, possibly fluid inner layers. Such displacements may take place as the consequence of comparatively slight forces exerted on the crust derived from the earth's momentum of rotation, which in turn will tend to alter the axis of rotation of the earth's crust."

Since Hapgood's thesis ends the last Ice Age by this crustal motion, the same would also pertain to Velikovsky's theory. The force that Velikovsky suggests could trigger just such a sudden violent movement. The rotation of the Earth need never have ceased completely and then started up again to make the Sun stand still. Based on this view very little of the energy of the interaction between the Earth and Venus would have gone into the production of heat.

Sagan merely chose the worst case scenario to attack Velikovsky and ignored totally the concept to which Einstein was eager to lend his consideration. S.K. Runcorn, a well respected geophysicist, in the Sept. 1955 issue of *Scientific American* stated categorically that, based on geomagnetic evidence respecting the Tertiary period, "the Earth's axis of rotation had changed." Runcorn claims that the Earth in the past rolled around on its axis. If Sagan wishes to dispute Velikovsky's concept, shouldn't he have dealt with evidence supported by both Einstein and Runcorn that agrees with Velikovsky's thesis?

Fred Warshofsky in *Doomsday The Science of Catastrophe,* (NY 1977), pp. 164-165 gives proof that in historical times the axis of the Earth did change. He states:

> "In 1940 a British archeologist named W.M. Flinders Petrie published the first detailed survey of the pyramids of Giza. He found that the Great Pyramid of Cheops and the smaller one of Chephren that flanks it were aligned to within four minutes (of arc) west of true north. Moreover

Petrie was able to prove that when they were originally built the Great Pyramids were virtually bang on an alignment with true north.

"What shifted these incredibly massive structures from their original position to a new alignment four minutes west of north? Petrie believed it was the result of a shift in the earth's poles. The idea been largely ignored, even unknown save for archeologists, but now a pair of geophysicists at Edinburgh University Scotland and Aarhus University in Denmark have examined the Petrie theory in the light of continental drift.

"'Continental drift,' they noted, 'can cause the direction of true north to vary with respect to the moving block. The Americas have been separating from Africa and Europe owing to the spreading of the sea floor. This movement has a hinge southwest of Iceland and is about 5 cm. per year between South America and Africa. If this causes only the latter [Africa] to rotate and if the rotation is uniform, in 4,500 years the pyramids would be rotated 0.1 [degrees] in the observed sense.'

"While the direction of rotation is right, the amount is far less than the actual four-minute movement made by the pyramids away from true north. But what about the other plate upon which the Arabian peninsula rides? Would its rotation, added to that of the South American and African plates, equal four minutes of arc? Again the answer is no. 'Africa and the Arabian peninsula are moving apart as if hinged near the north end of the Red Sea.' G.P. Pauley of Edinburgh and N. Abrahamsen of Aarhus point out. 'This suggests a rotation in the wrong sense, but again the magnitude is too small.'

"Some mechanism besides continental drift was clearly at work. 'Earthquakes are a possible mechanism for a local reorientation,' suggest Pauley and Abrahamsen, 'but a single quake of unprecedented magnitude would be needed to move the pyramids by strain release.'

"From their tone the two scientists clearly dislike the earthquake idea and throw the whole question open to speculation to the geological community. 'Expert geological opinion, 'they plead' would be worthwhile on this point'..."

In essence sometime after these pyramids were built there was a large shift in the Earth's plates that reoriented the African plate on which the pyramids ride away from true north. The event implies a pole shift and had to have occurred in the age of civilized man!

However, Velikovsky's catastrophe would cause the water in parts of the seas to be thrown into the atmosphere by meteorite impacts and greatly saturate the atmosphere. The water vapor would rise, cool and condense to produce great hurricanes in which the cool water fell in torrents. And Velikovsky's catastrophe is concurrent with ending the Ice Age, producing enormous volcanism. The great ice sheets would have broken apart, especially along the coast, adding great numbers of icebergs to the oceans. The clouds of dust would have halted solar insulation. Indeed, the mammoths seem to have been deep frozen in less than four hours, somewhat like Sagan's "nuclear winter".

Nevertheless, the fact remains that the Earth did not have to come to a complete halt to create the conditions that made it seem as if the Sun stood still.

Lastly, what does Sagan say regarding this subject? Sagan, in his original paper delivered at the AAAS symposium, explicitly stated the following about the temperature of the oceans during this catastrophe. He maintained that the oceans would be hot "but not lethal." Therefore, when Sagan states, "It is doubtful that the inhabitants [of the seas] would have failed to notice so dramatic a climatic change. The deceleration might be tolerable if gradual enough, but not the heat." [10] Hence, using Sagan's brand of logic, the heat would have killed the fish because the oceans would be hot "but not lethal."

APPENDIX III
Present Temperature of Venus
If Heated by a Close Passage of the Sun

Sagan states,

> "The heating of Venus by a presumed close passage by the Sun, and the planet's subsequent cooling by radiation to space are central to Velikov-sky's thesis. But nowhere does he calculate either the amount of heating or the rate of cooling." [11]

Sagan claims that Venus would be at least 6,000 degrees K or as hot as the Sun's surface. But after 3,500 years of radiative cooling, Venus would have cooled down to 79 degrees K, about the temperature at which air freezes. Thus, he concludes Venus cannot be called hot. However, another astronomer, W.C. Straka, also made a calculation respecting the cooling rate of Venus. C.J. Ransom reports,

> "Astronomer W.C. Straka wrote in 1972 that if Venus were incandescent (say 2,000 degrees K) only 3,500 years ago, then it would now be much hotter than the measured temperature. Later Sagan wrote that if Venus were 6,000 degrees K about 3,500 years ago, its temperature today would be lower than the freezing point on Earth." [12]

Thus, there are two astronomers calculating the cooling of Venus from a high temperature 3,500 years ago to the present. One—Straka—claims Venus would be hotter than it is today; the other, Sagan claims Venus would be freezing cold.

In fact, another astronomer has come up with an analysis that contradicts Sagan. That astronomer is Sagan. Based on Sagan's conclusion that Venus would be colder than freezing by "radiative cooling," it is obvious that the crust of the planet would also be frozen and thus must be very thick. If, as Sagan assures us, the surface of Venus would be frozen, then the temperature below the surface might be warmer, but would be at least the same or slightly cooler than the crust of the Earth. Based on Sagan's calculations, Venus would have a fairly thick crust. But this is the very point he disputes stating, "If, as Velikovsky believes, Venus were completely molten only a few thousand years ago—from planetary collision or any other cause—no more than a thin outer crust, at most [about] 100 meters thick, could since have been produced by conductive cooling." [13] Therefore, based on Sagan's views, the crust of Venus would be cool enough to support small continents because in 3,500 years Venus radiated its heat into space while at the same time the crust would be so hot it could not support small continents because in 3,500 years Venus could not conduct enough internal heat to its surface. The surface of Venus, according to Sagan, must be both hot and cold. Sagan states, "Velikovsky's mechanism cannot keep Venus hot, even with very generous definitions of the word 'hot.'" [14] Sagan's mechanism, however, can keep Venus both hot and cold since he seems to employ a generous definition for "hot" which also means "freezing". This contradiction casts extreme doubt on Sagan's calculation.

According to the *McGraw Hill Encyclopedia of Science and Technology*, (NY 1987), Vol. 10, p. 269, the temperature of magma (molten rock) ranges between 2,200 and 1,500 degrees Fahrenheit. Sagan has molten rock at near 2,000 degrees 100 meters below the Venusian surface but radiating so little heat to the surface that air should freeze. However, in *Broca's Brain*, p. 119, Sagan stated that,

> "Velikovsky has failed to understand two classic and basic geophysical results. Thermal conduction is a much slower process than radiation or convection, and, in the case of the Earth, primordial heat makes a detectable contribution to the geothermal temperature gradient and to the heat flux from the Earth's interior. The same applies to Venus."

On the Earth, magma originates in the lower crust about 100 kilometers (60 miles) under continents and 65 kilometers (about 40 miles) under ocean basins. Sagan claims Velikovsky was confused regarding the difference between convection, conduction and radiation of heat from Venus' surface while he has a 2,000 degree heat source only 100 meters below Venus' surface barely able to heat that surface above the freezing point of air. This, according to Sagan, would not be confused. Such a concept is not just confused; it is a scientific absurdity.

Nevertheless, this is not the only contradiction Sagan's calculations show. C.J. Ransom pointed out that,

"Sagan incorrectly implies that according to Velikovsky's theory, Venus was heated only by a close approach to the Sun. Sagan assumed that Venus was at a certain temperature (6,000 degrees K) 3,500 years ago, that it received no additional heat from the Sun, and radiated perfectly since that time. The temperature he then calculated for Venus today (79 degrees K) was naturally much lower than the actual present temperature of the planet. This enabled Sagan to claim that Venus would not have been hotter than 6,000 degrees K, 3,500 years ago. But by the same logic, Venus would have (had to have) been hotter than 6,000 degrees K, 3,500 years ago, even if Velikovsky had never presented his theory."[15]

Let us assume that Velikovsky never wrote anything about Venus. What was Venus' temperature 3,500 years ago? According to Sagan's calculation, Venus must have had a temperature hot enough to allow for its current 750 degrees K surface temperature. Since Sagan claims that if Venus was as hot as the surface of the Sun, its surface would now be frozen, it follows that Venus had to be much hotter than the surface on the Sun. How does Sagan explain that Venus' temperature was hotter than the surface of the Sun while it has been in its present orbit for billions of years?

Ransom pointed out that Sagan's calculation omits the radiation emitted by the Sun. This brings us to the third contradiction in this appendix. To prove that Venus must have a frozen surface, Sagan stopped the Sun from radiating. On the other hand, to maintain a greenhouse effect to heat Venus' surface, Sagan permits the Sun to radiate. Now, if the Sun does not radiate for 3,500 years to allow Venus to cool below the freezing point; how does Venus, over the same 3,500 years, maintain a greenhouse effect without sunlight?

By turning off the Sun's radiation, Sagan has shed his own light on science.

APPENDIX IV
Magnetic Field Strengths Necessary to
Circularize an Eccentric Cometary Orbit

In this appendix, Sagan calculates what he contends would be the necessary electromagnetic field strength that would circularize the orbit of Venus from a cometary orbit with high ellipticity to that of a planet which is highly circular. He claims that ten mega-gauss is re-quired. If a body passes close to the Earth, Sagan claims that rocks that were molten at that time would have remanent magnetism to reflect their immersion in such a powerful magnetic field. He states, "Had Earth experienced, even fairly briefly, a 10 Mg field 3,500 years ago, rock magnetization evidence would show it clearly. It does not."[16]

Once again, Sagan is quite wrong. For a magnetic field to affect rocks and leave traces of the field in the rocks on Earth, the field must exist for a very long time, but it must also be quite close to the molten rock. Thomas McCreedy states, p. 47,

"Any competent worker in the field of paleomagnetism would concede that the magnetic field reversals described by Velikovsky—which are of very short duration, of the order of a few hundred years—would be inordinately difficult to detect." [17]

The question arises, is there evidence for magnetic field excursion 3,500 and 2,800 years ago? McCreedy continues,

"As the paleomagnetic evidence accumulates, Velikovsky's position continues to be strengthened though evidence for a similar event (geomagnetic field) reversal in the 15th century B.C. is missing at present from the record. However, in view of the evidence put forward previously, detection in the geologic record is virtually impossible especially with regard to the close spacing 52 years between the Venus/ Earth encounters.

"Two reversals within such a short period of time would leave a negligible trace on the geological record while the paucity of archaeological samples from this period leaves the matter temporarily undecided. It is, nevertheless, more than of passing interest to note that there are magnetic delineation inflections ca. 1500 B.C. at both Loch Lomond and Lake Windermere of the type advocated by Warlow (P. Warlow, "Geomagnetic Reversals" in the *Journal of Physics,* A-11, No. 10, Oct. 1978, pp. 2107-2130) Taken within a catastrophic context, these indicate a double magnetic reversal occurring in quick succession."

For these brief magnetic reversals to be left in lake sediments during a 52 year period, 3,500 years ago, the strength of the magnetic field must have been very large.

Sagan has argued that the eruption at Santorini (Thera) in 1500 B.C. explains the evidence of catastrophism Velikovsky has promulgated. We have pointed out that according to researchers (Downey and Tarling) in *Science News* and in *New Scientist,* there occurred two phases of destruction separated by fifty years based on archaeomagnetism and ceramic dating. But back to circularization of Venus' orbit.

In the *Study of Comets,* Part 1, (NASA SP-393), (Washington D.C. 1976), pp. 450-452, Edgar Everhart discusses how long it takes for a long-period comet to have its orbit reduced after an encounter with Jupiter to become a short-period comet. He states that these comets must "also survive the solar thermal dissipation of *hundreds of thousands of returns at small perihelia.*" [emphasis added] However, comets cannot come relatively close to the Sun (small perihelia) as is required by the Jupiter capture model and survive these thermal events hundreds of thousands of times. Comets lose material during each close passage to the Sun and can survive perhaps a few hundred or so close approaches to the Sun, but not hundreds of thousands. The problem is that there are some short-period comets with highly circular orbits. To have this happen, the change from a highly elliptical orbit to one that is highly circular must occur quite rapidly or the comet will dissipate all its material. But the fact of the matter is that there are short-period comets with highly circular orbits and the only way they can exist is if they changed their orbital structures in less than a few thousand years. All this is well-known to astronomers.

Now, since comets change their orbits from highly elliptical to highly circular in a few thousand years, some mechanism has to be operating to accomplish this. The non-gravitational forces that Sagan suggests affect the orbits of comets requires tens of thousands of years and thousands of passages near the Sun to produce such changes, according to Everhart. Something else, some other force has to be operating to convert comets with elliptical orbits to circular ones. But, in fact, Sagan does not ever touch upon this problem.

Sagan's argues that at least ten million gauss fields are necessary to circularize the orbit of comet-Venus. Is there evidence that shows cometary orbits can be changed within in a few thousand years from highly elliptical orbits to circular orbits? Sagan has informed us that short-period comets with highly elliptical orbits will dissipate their materials in a few thousand years. This concept is well accepted in the scientific community. But there are currently comets in the solar system with highly circular orbits. Thus, their orbits had to be converted from elliptical ones in less than a few thousand years without ten million gauss fields. No such field strengths are known to exist in the solar system yet the orbits have changed. Why?

C.J. Ransom states that the highly circular orbits of comets,

"makes it clear that this (Sagan) appendix has little, if anything, to do with the total thesis of *Worlds in Collision*. Although Velikovsky clearly and correctly suggested in 1950 that electromagnetic fields played a greater role in the solar system than most scientists thought, his model for orbital changes was fully compatible with these forces responsible for the behavior of comet *Ortema III*. Before 1938, *Ortema III* had an orbit entirely between Jupiter and Saturn. During a near approach to Jupiter that year, it changed its orbit so that it was entirely between Mars and Jupiter."

"Furthermore, according to Bass, 'if one removed Venus from its present orbit and gave it the initial conditions of the comet *Ortema III*, the initial orbit of Venus works itself inward into an orbit lying entirely between the orbits of Mars and Jupiter...'

"Overall, Sagan's appendices exhibit the same critical sloppiness as the text which they pretend to support. They have little to do with Velikovsky's historical model of the recent events in the Solar System and are as error-filled as the rest of his paper. Anyone with a physics background who is familiar with Velikovsky's ideas should read these appendices since they demonstrate the depths to which one can go when one considers himself so great an authority that he does not need to think about what he writes."[18]

William Broad and Nicolas Wade in *Betrayers of the Truth,* (NY 1982), pp. 193–196 state:

"The essence of a scientific attitude is objectivity. The scientist is supposed to assess facts and test hypothesis while rigorously excluding his own expectations or desires as to the outcome. In the public's eye objectivity is the distinguishing feature of the scientist, for it keeps his vision pure from the distorting effects of dogma and allows him to see the real world as it is. Objectivity does not come easily and researchers undergo a lengthy training to acquire it.

"With some scientists, nevertheless, objectivity is only skin deep, not a sincerely felt attitude toward the world. Under its guise, a scientist can foist his own dogmatic beliefs on the world far more easily than could a plain demagogue...

"Science is meant to be a community of intellects, dedicated to a common goal. If one scientist falls prey to dogma and tries to promote doctrinaire beliefs in the name of science, won't his colleagues immediately perceive his error and take action to correct it? History shows that, to the contrary, a community of scientists is often ready to swallow whole the dogma served up to them as long as it is palatable and has the right measure of scientific seasoning...objectivity often fails to resist infiltration by dogma. [The authors then expose the dishonest work of Samuel G. Morton who used scientific data to prove the white race was superior to all others. They continue:] What is striking about the Morton case is how thoroughly he allowed his prejudices to permeate his scientific work. His dogma shaped not just his theory but the very data from which it was supposedly derived. He juggled the numbers to get the results he wanted. The juggling was done right out in public, on the open page of his scientific reports so that he was evidently doing it unconsciously. The final paradox in the extraordinary story is that no contemporary scientist spotted the errors glaring though they were...

"The finagling in Morton's data was never discovered by the scientists who relied on his results. His tables remained unchallenged in the scientific literature..."

Sagan's work has been preeminent in discrediting Velikovsky for twenty years; why then, has it not been exposed in full by his colleagues? The answer, according to Broad and Wade on pp. 210–211 of their book is that,

"Science is not self-policing. Scholars do not read the scientific literature carefully. Science is not a perfectly objective process. Dogma and prejudice, when suitably garbed, creep into science just as easily as into any other human enterprise, and maybe more easily since their entry is unexpected... Against such weapons the scientific community that harbored him [Morton] was defenseless. Against rhetoric and appearance, the scientific method and the scientific ethos proved helpless. Against dogma disguised as science, objectivity failed."

In *Bones of Contention,* (NY 1987), p. 68, Roger Lewin states, "It is, in fact, a common fantasy, promulgated mostly by the scientific pro-fession itself, that in the search for objective truth, data dictates conclusions... But we've seen...again and again, frequently this does not happen. Data are just as often molded to fit preferred conclusions." It seems clear that Sagan only presented evidence to support his version of the scientific truth. He has not only misinformed his readers about the physics of his appendix, but has understated the contrary evidence to ensure that the vision of science he believes will be seen as correct. Lewin adds, "the interesting question then becomes, 'What shapes the preference of an individual'...[such as Sagan] not 'what is the truth?'" This we will now explore in our "Conclusion".

NOTES & REFERENCES

[1] SCV, p. 94; B.B., p. 320.
[2] SCV, p. 98; B.B., p. 324.
[3] *Kronos* III, 2, p. 135.
[4] Ellenberger, C. Leroy; *Kronos* V, 4, p. 61.
[5] SCV, pp. 99-100; B.B., p. 322.
[6] Sagan and Druyan; *Comet,* op. cit., p. 204.
[7] "The Trouble with Uranus", *Mercury,* Vol. 4 (July/Aug 1975), p. 25.
[8] *Age of Velikovsky*, op. cit., p. 230.
[9] SCV, pp. 99-100; B.B., p. 322.
[10] SCV, p. 99-100; B.B., p. 325.
[11] SCV, p. 100; B.B., p. 326.
[12] Ransom, C.J.; "A Note on the Temperature of Venus", *Kronos* IV, 2, p. 26.
[13] B.B., p. 120.
[14] B.B., . 326.
[15] Ransom, C.J.; *Kronos* III, 2, p. 137.
[16] B.B., p. 327.
[17] McCreedy, Thomas; *Kronos* VI, 3, p. 44.
[18] Ransom, C.J.; *Kronos* III, 2, pp. 138-139.

CONCLUSION

"Almost all scientists think they know how science differs from philosophy, religion and plain old guess work. Or, at least, all scientists I have talked to think they know. But it is quite a different matter to find out just how science differs from other forms of inquiry. To some, science is defined by the use of experiments to test the predictions of hypotheses. To others, it is any kind of careful scholarship not burdened by a prior commitment to a particular answer or belief system. Most scientists argue that religion is not science because religion involves no experiments, tests, no hypotheses, and is committed before hand to a set of beliefs. Science and faith are antithetical. It follows that scientific research is objective because the scientist is not influenced by prior expectations and is willing to let the chips fall where they may. I think these statements contain a fair amount of bunk."

David Raup, *The Nemesis Affair*, (NY 1986), p. 194

"...establishment science itself has become a church, with its own dogmas, hierarchies and heresies. It has its popes and cardinals, and the power to excommunicate. This power has been used to pillory scientists who challenge prevailing scientific orthodoxy..."

Alvin Toffler, "Interview" *OMNI*, (Nov. 1978), p. 134

The above statements by Raup and Toffler reflect a fundamental problem with science's reception to new and revolutionary ideas. Raup wrote his book because he was also subjected to the slanderous reception system that currently pervades the halls of science and academia. Sagan has offered the naive view that science and scientists operate on high levels of integrity; that they approach radically new ideas without bias or malice. However, Raup, who experienced the behavior of the scientific system, wrote, "A great many controversies are waged as much by rumor as by formal means... No laws in science say hearsay has to be wrong... But this kind of communication can be extremely damaging to the truth just as it is in human affairs in general."[1] The reason Raup offers for scientific dishonesty is that "When a truly controversial idea or conclusion comes up, the whole neat [scientific debate] process tends to break down... A hot topic short circuits the system."[2] He further informs us that, "Science is basically an adversarial process."[3] Perhaps because "Working scientists are subject to many more prejudices and preconceptions than is generally thought."[4] And when "Given a choice, the scientific community invariably sticks with the conventional wisdom. Furthermore, the older ideas have usually been around long enough to have accumulated supporting evidence, whereas the new idea rarely has much going for it, at least at first. It is not a fair game."[5]

George Grinnel, an historian of science, explains, "It became a scientific heresy to believe in catastrophic theory... The reaction of the scientific community was one of repression, not because Velikovsky was wrong, but it basically feared he may be right."[6] Thus, if establishment scientists can portray by misrepresentation and disinformation to the rest of the scientists and public that Velikovsky and his ideas, as Harlow Shapley would have us believe, are insane, then no matter what Velikovsky or his supporters say on his behalf or in reply to criticism, they will not be listened to or believed. After all, they are cut from the same cloth. This is a political technique with which the scientific establishment has maligned and smeared Velikovsky, and rather tragically, this technique has worked. Sagan's criticism has been cited in a great many journals and other publications without the evidence of opposing scholars given in rebuttal.

Name calling has played a role in this affair also. Lawrence Lafleur called Velikovsky a "crank," Henry Bauer called Velikovsky a "crack pot," Patrick Moore called Velikovsky a "charlatan," Isaac Asimov called Velikovsky "C.P." [crack pot], Sagan called Velikovsky "ignorant," etc. Martin Buber, the humanist thinker, in his book *Thee and Thou* explains that when one wishes to destroy someone, that person must first be dehumanized, that is, in some way characterized as less human; that is the essence of name calling. Once a label is attached to a person, it is then possible to treat this person as that object, or by the name with which he has been labeled. The actions of the established scientific community of calling Velikovsky names and those who study his ideas a "cult", gives this community leeway to misrepresent Velikovsky and his work. Scientists who are truly secure in themselves and their knowledge and understanding of science have no need to resort to such tactics.

Sagan stated only in *Broca's Brain,* p. 127, "In the entire Velikovsky affair, the only aspect worse than the shoddy, ignorant and doctrinaire approach of Velikovsky and many of his supporters was the disgraceful attempt by some who call themselves scientists to suppress his writings. For this, the entire scientific enterprise has suffered... Scientists are supposed to know better, to realize that ideas will be judged on their merits if we permit free inquiry and vigorous debate." True, Sagan did not join with those who tried to suppress Velikovsky's work; instead he presented distortions of Velikovsky's statements and then argued with his straw men. Does Sagan really believe that this is not a shoddy, ignorant or doctrinaire approach? Is distortion and misrepresentation better than suppression? Doesn't the entire scientific enterprise suffer when either occurs? Should such tactics be acceptable as part of free inquiry and vigorous debate? I doubt it.

VELIKOVSKY AND EINSTEIN

In this respect, let us examine how establishment science dealt with the relationship that existed between Velikovsky and Albert Einstein during the last years of Einstein's life. In the 1950's, Velikovsky moved to Princeton, New Jersey and became reacquainted with Albert Einstein. In his book *Stargazers and Gravediggers,* Velikovsky devotes a portion of his memoir to his relationship with Einstein.

"On November 8, 1953 we were invited by Einstein to visit with him. The story of my relations and debates with Albert Einstein, from his first reading of the manuscript of *Worlds in Collision,* until his death, is related in a separate book, *Before the Day Breaks.* On that evening he greeted my wife and me—his long hair well groomed, his face lighted with a friendly smile—and started to move a chair with a very high back, which had already drawn my attention in the modestly furnished living room. As I helped him, he said, 'This is my Jupiter chair.'

"During our conversation, I took this lead and remarked: 'If one evening I should stop every passing student and professor on the campus and should ask which of the stars is Jupiter, it is possible that not even one would be able to point to the planet. How is it then, that Jupiter was the highest deity in Rome, and likewise, Zeus in Greece, Marduk in Babylonia, Amon in Egypt, and Mazda in Persia—all of them represented the planet Jupiter. Would you know why this planet was worshipped by the people of antiquity and its name was in the mouth of everyone? Its movement is not spectacular; once in twelve years it circles the sky. It is a brilliant planet, but it does not dominate the heavens. Yet, Apollo—the sun—the dispenser of light and warmth, was only a secondary deity.' After explaining that Marduk was the Babylonian name of the planet Jupiter and Mazda its Persian name, Einstein expressed his wonder. Then I told him in the *Iliad* it is said that Zeus can pull all the other gods and the Earth with his chain, being stronger than all of them together; and that an old commentary (by Eustathius, a Byzantine scholar) states that this means the planet Jupiter is stronger in its pull than all the other planets combined, the Earth included. Einstein admitted that it was really very strange that the ancients should have known this.

"When, after three quarters of an hour during which we were served tea, we rose to go. Einstein kept us, saying, 'We have only started.' In order not to appear a bore or a fanatic of one idea, I repeatedly changed the theme of conversation as is so easy with Einstein, whose associations are rich and whose interests are many. The conversation was vivid. We spoke again of the problem of time, which apparently occupied his mind then, and of coincidence and accident. He observed that it was an accident of unusual rarity that his chair should occupy its very position in space, but that it was no accident that we were sitting together, because *meshugoim* (Hebrew for 'crazy people') are attracted to one another.

"In the following weeks I put my lecture before the Princeton Graduate Forum in writing, and discussed it with Professor Motz of Columbia University. Then I sent a copy of it to Einstein. After a few days he invited Elisheva and me to come and discuss it.

"The problem he selected for discussion that evening, from a series of problems mentioned in my lecture, was the round shape of the sun. Because of rotation it should be somewhat flattened; and, in addition, the sun rotates at a greater velocity at its equator than at higher latitudes. We spent the evening talking about this and a few points in my lecture.

"In the meantime, I thought of calling Helen Dukas, Einstein's secretary, to say a few words of apology for our too long conversation, when the phone rang and Miss Dukas said: 'The professor would like to talk to you.' His voice sounded resonant and clear and, I thought, if one does not see Einstein but only hears him, he might imagine that he's speaking with a young man. Einstein said:

'After our conversation last night I could not fall asleep. For the greater part of the night, I turned over in my mind the problem of the spherical form of the sun. Then before morning I put on the light and calculated the form the sun must have under the influence of rotation, and I would like to report to you.' I mention this episode only to emphasize Einstein's attitude toward a scientific problem that intrigued him and, even more, his behavior toward a fellow man.

"Einstein made no secret of his interest in my ideas and his good personal feelings toward me; often he asked me not to go away when it was late, but to spend more time in discussion. He was surrounded by much love but was a lonely man. Not once and not twice he called me to follow his example and be content in isolation. 'Don't you feel fine being alone? I feel unconcerned being alone.' The fact was that most physicists of the younger generation, including those connected with the Institute for Advanced Study, opposed his later stand in physics that conflicted with the quantum theory, which requires the principle of chance or indeterminacy in natural events. On one occasion I answered his monition: 'Yes, there are two heretics in Princeton. Only one is glorified; the other, vilified.'

"His theory increased immensely the regard the general public has for science: If a scientist's theory can be understood only by a very few in the entire world, as it was in the beginning with Einstein's theory, what a supreme race are the scientists! But if one comes with a theory, which, if true, would let many reputed scholars appear in error before the public, what should he expect from them?

"One evening in May 1954, sitting with Einstein in his study, a few days after another ugly attack on me and my theory, I referred for the first time to the behavior of the scientists against me, and I showed him a file with some letters quoted earlier in this book. He read them with very great interest, and he was obviously impressed.

"He thought that the letters and other material must be put into a readable form, as a story, and that somebody with a talent for dramatic writing should be entrusted with the composition; and he was already concerned with the success of my defense. He wished to read more of the letters, but I was interested in taking up the problem that really occupied my mind: my theories.

"The same evening I left with Einstein Chapters VIII to XII of *Earth in Upheaval* in typescript, and we parted at close to midnight. Upon reading these chapters, he wrote me a long handwritten letter with criticism. In this letter he also inserted a few passages concerning the letters he had seen. He thought that Shapley's behavior could be *explained*, but in no way *excused* ('*erklären* aber keineswegs *entschuldigen*') and he added:

'One must, however, give him credit that in the political arena, he conducted himself courageously and independently, and just about carried his hide to the marketplace.'

'Therefore it is to some extent justified if we spread the mantle of Jewish neighborly love over him, however difficult that is.'

"Yet, Einstein did not change his opinion that the material pertinent to the suppression of my book must be made public.

"At the conclusion of my letter written several weeks later, as the next step in our debate, I returned to that issue:

'Too early you have thrown the mantle of Jewish compassion over Shapley; you have seen only the beginning of the file of the documents concerning the "Stargazers and Gravediggers" and their leader. His being a liberal is not an excuse, but an aggravating circumstance.'

"In the summer and fall of 1954, I wrote most of *Stargazers and Gravediggers*. Its first reader was professor Salvador de Madariaga of Oxford University, who visited me while he was a guest lecturer in Princeton. A few months later I gave the manuscript to Einstein; it was in March 1955, fully ten months after he had first read a few letters quoted in it. It was almost finished, the section 'Before the Chair of Jupiter' included. He supplied some of the pages of *Stargazers and Gravediggers* with handwritten marginal notes; some of the notes were very emphatic: 'mean' and 'miserable' to some letters, and 'bravo' to others, and which side commanded his sympathy is clearly discernible.

"Upon reading the first of the three ring files of *Stargazers* Einstein wrote me on March 17, 1955:

'I have already read with care the first volume of Memoirs to *Worlds in Collision* and have supplied it with a few marginal notes in pencil that can be easily erased. I admire your dramatic talent and also the art and the straightforwardness of Thackrey who has compelled the roaring astronomical lion to pull in to some extent his royal tail without fully respecting the truth. I would be happy if you too, could enjoy the whole episode from its humorous side.'

"Interesting is his note on the back of the page on which I tell about Larrabee's article in *Harper's,* which broke the story of *Worlds in Collision* to the public in 1950. He wrote:

'I would have written to you: The historical arguments for violent events in the crust of the earth are quite convincing. The attempt to explain them is, however, adventurous and should have been offered only as tentative. Otherwise, the well-oriented reader loses confidence also in what is solidly established by you.'

"This came very close to Atwater's judgment, in his capacity as a reader for Macmillan, and it sealed his fate.

"But this was a great stride away, on Einstein's part, from the view he once took that the events I have described could not have happened. Einstein said, not once and not twice, also in the presence of his secretary: 'The scientists make a grave mistake in not studying your book (*Worlds in Collision*) because of the exceedingly important material in contains.'

"In our debate, which spread over eighteen months, I drove ever closer to a point not necessary for the validation of *Worlds in Collision,* but of prime importance per se: the revision of celestial mechanics in the face of the accumulated data pointing to the charged state of celestial bodies. When I wrote: 'The real cause of indignation against my theory of global catastrophes is the implication that celestial bodies may be charged,' he wrote in the margin: 'Ja' ('Yes').

"My understanding of the nature of the sun and planets made me assume that these bodies are charged, or that, at least, their atmospheres are strongly ionized. I wished for several years that a check could be made on Jupiter. I took the opportunity of my lecture before the Graduate College Forum of Princeton University on October 14, 1953, and after presenting many reasons for believing that the members of the solar system—the sun, the planets, the satellites, the comets, the meteorites—are not electrically or magnetically neutral, made the following statement:

'In Jupiter and its moons we have a system not unlike the solar family. The planet is cold, yet its gases are in motion. It appears probable to me that it sends out radio noises as do the sun and the stars. I suggest that this be investigated.'

"The lecture was a discussion of my theory of 1950 'in light of the new discoveries in the fields of astronomy, geology and archaeology'; I presented a considerable collection of recent findings that support the theory given in *Worlds in Collision.* It was natural to offer, at the end of that register, some new tests. And this I did by asserting that Jupiter sends out radio noises. The radio noises coming from the sun were explained as the effect of its great heat, but Jupiter is a cold planet and therefore, nobody expected radio noises coming from it or from any other planet. In conventional astronomy, Jupiter is an inert body, neutral in charge; in my understanding it is the center of a powerful electromagnetic system.

"The summer of 1954, in a letter that I wrote to Einstein, I included these sentences: '...I question the neutral state of celestial bodies. There are various tests that could be made. For instance, does Jupiter send radio noises or not? This can easily be found, if you should wish.'

"It was a plea to help me convince others that this test should be performed. I did not doubt the result of such a test. Einstein did not respond in that instance. I have the original of my letter with many of Einstein's marginal notes.

"Eighteen months after my lecture and nine months after my letter to Einstein (written June 16, 1954), strong radio noises coming from Jupiter were discovered. They were detected entirely by chance, yet the discovery appeared of such importance that it was immediately reported to the scientific world in a dramatic manner.

"The spring 1955 semiannual meeting of the Astronomical Society met in Princeton. A very long list of papers was scheduled. The new discovery was presented to the meeting because of its importance, though it was not scheduled, having been made only a few weeks earlier. The next day the newspapers displayed the sensational discovery. *The New York Times,* (April 6, 1955) in a column-long story from Princeton reported the news: 'SOUND' ON JUPITER IS PICKED UP IN U.S.

> "Radio waves from the giant planet Jupiter have been detected by astronomers at the Carnegie Institution in Washington... No radio sounds from the planets in our solar system have been reported previously... The existence of the mysterious Jovian waves was disclosed by Dr. Bernard F. Burke and Dr. Kenneth L. Franklin... The two scientists said that they did not have an explanation for the observed radio emissions.'

"The press reported that the discovery was made entirely by chance, when the Carnegie Institution astronomers scanned the sky for radio noises from faraway galaxies. The noises were so strong that the discoverers thought they were caused by some experiments in a neighboring radio station. It was only after they found out that the noises were repeated every third day for six minutes, when the receiving antenna was directed toward the spot crossed at these minutes by Jupiter, that the astronomers came to the correct conclusion, unexpected and surprising as it was.

"In November, 1955, Harlow Shapley, reviewing the field of astronomy for the year that was coming to a close, selected a few 'highlights' as the most important events of the year. At the top of the discoveries he placed:

> "Detection of 'thunderbolts of Jove' of some similar strong electric effect in the atmosphere of the planet Jupiter...the first to be found from another planet in the solar system."

"Shapley did not know the true significance of the metaphor. Of the thunderbolts of Jupiter, the classic literature and the religious beliefs of the races of the Earth, speak without end. My own treatment of the subject will be resumed when I present the story of earlier cataclysms.

"When I brought Einstein the news he was obviously much taken by what he learned. He was also embarrassed, for not only had he disregarded my request for this test, but also at our previous meeting, he had stressed the great importance for the acceptance of a theory that it be able to generate correct predictions.

"He stood up and asked: 'Which experiment would you like to have performed now?' I asked him to help me obtain radiocarbon tests to check on my reconstruction of ancient history. He was very emphatic in his desire to help me in this. This was our last meeting; he died a few days later. In fulfillment of his wish, a letter went from his home after his death to the Metropolitan Museum of Art with the request that some of the relics of Egypt be submitted for radiocarbon analysis."[7]

[The letter sent by Einstein's secretary in full reads:]

May 25, 1955

"Dear Dr. Hayes:

"I am writing to you at the request of Dr. Immanuel Velikovsky and because, as the secretary of the late Professor Einstein, I feel that I should inform you of the following matter. "During the course of the last eighteen months, Professor Einstein had several discussions with Dr. Velikovsky—with whom he had friendly personal relations—about the latter's work. The last such discussion took place on April 8th. In the course of this conversation Professor Einstein said that he would write to you and suggest that you should give Dr. Velikovsky an opportunity to have his theory subjected to a radiocarbon test. As I was present at this discussion, I can assure you that Professor Einstein did intend to write that suggestion to you and, but for the lateness of the hour, the letter to you would have been written then and there.

"Yours sincerely,
Helen Dukas, Secretary to Albert Einstein[8]

"As I reported earlier, Bernard Cohen, the historian of science of Harvard, took a vacillating position in relation to me and my work at the 1952 symposium of the American Philosophical Society. In the abstract of his paper he took an objective stance regarding the ultimate value of my work, but in his oral delivery and especially in the subsequent published paper, he relied on Payne-Gaposchkin and put me down in one short sentence.

"Two and one-half months after Einstein's death the July 1955 issue of *Scientific American* carried an article by Bernard Cohen describing his visit and interview with Einstein on April 3, two weeks before the latter's death. It was Cohen's first and only meeting with Einstein. The recentness of Einstein's death made the interview appear like a testament, utterances of a now dead person spoken to a witness. Illustrated with pictures of Einstein's home and of the street he used to walk on his way to the Institute of Advanced Studies, the piece attracted much attention.

"Einstein and Cohen talked about 'the history of scientific thought and great men in the physics of the past.' As Cohen reported it, Einstein started by saying: 'There are so many unsolved problems of physics. There is much that we do not know; our theories are far from adequate.'

"They spoke of Newton, whom Einstein had 'always admired,' and of the fact that Newton would not give Hooke any credit for his priority in the discovery of the law of the inverse square in gravitation, to the extent that Newton had expressed his preference not to publish the third and most important part of the *Principia* at all rather than give credit to Hooke in the Introduction to the volume. In his feud with Leibniz over the precedence of the invention of the calculus, Newton had secretly directed the activity of the committee that had to decide between the two savants in order to have it declare Leibniz a plagiarist.

"According to Cohen's report, Einstein was dismayed by Newton's conduct and 'did not appear too much impressed when I asserted that it was the nature of the age to have violent controversies, that the standard of scientific behavior had changed greatly since Newton's day.'

"Then the conversation turned to Benjamin Franklin, who had prided himself for not engaging in polemics in defense of his ideas, believing that they must make their own way by proving their vitality. Cohen professed his admiration for this behavior. Einstein, however, disagreed. 'It was well to avoid personal fights,' he said, 'but it was also important for a man to stand up for his own ideas. He should not simply let them go by default, as if he did not really believe in them.'

"Then, almost inescapably, Einstein talked about me and my work. Though my name was not mentioned, it was obvious about what book and author he spoke. His opinion of the standards of scientific behavior and the obligation of a man to stand up for his ideas in science was a good introduction to my case. The fact that Einstein spoke of me and my work after talking about Benjamin Franklin and discussing Isaac Newton, did not surprise me. He was then very much taken by my work. He was reading the second and third files of *Stargazers and Gravediggers* and was reading *Worlds in Collision* once again, this time in German translation. However, in Cohen's presentation Einstein's comments went thus:

> "The subject of controversies over scientific work led Einstein to take up the subject of unorthodox ideas. He mentioned a fairly recent and controversial book, of which he had found the nonscientific part—dealing with comparative mythology and folklore—interesting. 'You know,' he said to me, 'it is not a bad book. No, it really isn't a bad book. The only trouble with it is, it is crazy.' This was followed by a loud burst of laughter. He then

went on to explain what he meant by this distinction. According to Cohen, Einstein said:

'The author had thought he was basing some of his ideas upon modern science, but found the scientists did not agree with him at all. In order to defend his ideas of what he conceived modern science to be, he had to turn around and attack the scientists.'

"I knew that Einstein could never have expressed himself this way about my work. In his report of the interview, Cohen made Einstein appear as my opponent, while Cohen allowed himself to seem open-minded and sympathetic—the reverse of the actual attitudes of the two men. Cohen continued:

'I replied that the historian often encountered this problem: Can a scientist's contemporaries tell whether he is a crank or a genius when the only evident fact is his unorthodoxy? A radical like Kepler, for example, challenged accepted ideas; it must have been difficult for his contemporaries to tell whether he was a genius or a crank. 'There is no objective test' replied Einstein.'

'Einstein was sorry that scientists in the U.S. had protested to publishers about the publication of such a book. He thought that bringing pressure to bear on a publisher to suppress a book was an evil thing to do. Such a book really could not do any harm, and was therefore not really bad. Left to itself, it would have its moment, public interest would die away and that would be the end of it. The author of such a book might be 'crazy' but not 'bad,' just as the book was not 'bad.' Einstein expressed himself on this point with great passion.'

"The rest of the conversation turned around Newton.

"That he was speaking 'with great passion' on the subject was true: Before that interview was also at my last meeting with Einstein five days after he spoke with Cohen, I heard him speak on the subject with great passion. But there was a wrong twist in Cohen's story: It appeared as if Einstein spoke with great passion against my book.

"The word 'crazy' may have various connotations—one meaning 'most unusual,' the way Einstein used the word *meshugoim* in referring to himself and myself in one of our conversations. Thus, he likened himself to me. (*Meshuga* is a Hebrew word; it means 'crazy' in both senses—like the English word—and more often its milder meaning. *Meshugoim* is the plural form.)

"It appeared in Cohen's report as if Einstein thought that suppression of a book was evil because a bad book left alone would not survive anyway. *This is true, but this was meant by Einstein, if he said it, in defense of my book, which he was reading again and again.* [emphasis added] Opposing the ways my book was suppressed, he could have said that *if worthless* and left alone, it would have died by itself—but the 'if worthless' fell out. Einstein, five days later, in his last conversation with me, said, and with passion, that the book contained much of importance; five days earlier he could not have said that the book would have died a quiet death if not suppressed.

"I was deeply hurt. In the five and a half years of vilification, distortion and abuse I had usually remained unperturbed; all the attacks that had taken place until then had not really stung. This time I was angered; Einstein, who obviously in the last weeks of his life was occupied with my book—it was he who raised the subject with Cohen—was made to appear my antagonist. Several years earlier, under the influence of the agitation among the scientists, Einstein may have felt hostile toward me, as so many other scientists did. But at the time of the interview with Cohen, his relation to me was at its highest and closest point. The manuscript of *Stargazers and Gravediggers* was on his desk when he spoke to Cohen—he finished reading its almost 400 pages—and his marginal notes better than anything else spoke of his feeling at that time. I could not bring together the attitude and the words Cohen ascribed to Einstein with the feelings Einstein had displayed during the hours he and I had sat side by side discussing my work; with his encircling my letters and pages of my manuscripts with numerous notes all around the margins; with his writing me by hand—a distinction he reserved for only a few select; with his saying to me before parting on March 11 that he thought it was a great mistake on the part of the scientists that they did not study my book for the useful information and fruitful problems it contains; with his writing me on March 17 the letter from which I quoted above, and my meeting with him on April 8, after his talk with Cohen; with his saying words of praise and offering to explain all in my book in the frame of accepted principles in science; and with his offering to help me with his authority so that a test of my theories could be performed.

"During his lifetime, the scientific establishment could not make Einstein express himself publicly against my work or myself, though it must have tried. Now, as soon as he died, his name was used to combat me and my work.

"I wrote Einstein's secretary, Miss Dukas, who knew of our meetings and correspondence, a letter for the record.

"Was it worthwhile to write a rebuttal to Cohen's article? The reader would have to decide where the truth was, and how could he know?

"I went for three days to the ocean shore to regain my peace of mind, watching the surf and the great expanse of water. I decided how to act. The only one who could revise what had been printed was Bernard Cohen himself.

"I wrote him this letter:

"July 18, 1955

'Dear Professor Cohen:

'In your published interview with the late Einstein, you refer to the great passion with which he spoke of my book. The reader may conclude that with great passion he opposed my work.

In the last eighteen months of his life, Einstein spent not a few long evenings with me discussing my work...also several...extensive, manuscripts, supplied them with marginal notes, in short, showed great interest in my ideas and gave me very much of his time. *On a manuscript containing the history of my first book, he wrote what he exactly thinks of Worlds in Collision—he wrote it in the very week you have seen him; it is in great disagreement with what I read in your interviews.* [emphasis added] In a letter of March 17, 1955, he made very clear what he thought of my adversaries and their methods of combating my book; and on margins of the pages containing copies of letters confidentially written by some scientists to my publishers with expressions similar to those you ascribe to him, he marked, 'miserable.'

I assume that with great passion he spoke against my opponents and their campaign. This does not mean that he agreed with my theories on all points; after many gradual agreements there remained between us a large area of disagreement, but our debate, orally and in writing, was carried on in the spirit of mutual respect and friendliness. Our last long conversation took place on April 8th, five days after your interview, and nine days before his death. He was rereading my *Worlds in Collision* and he said some encouraging sentences—demonstrating the evolution of his opinion in the space of 18 months.

I assume that the expressions that you mention were not used by Einstein in the meaning you have unintentionally given to them. I think that upon searching your memory, you will find that the predominant feature of his in speaking of my book was positive and not negative, sympathetic and not hostile. Would you like to write down a more complete version of that part of your conversation? I believe you would like to have a chance to rectify yourself.

"Einstein appears from the portion of your interview dealing with me as unkind and cynical—and these features were very far from him. And certainly he was not two-faced. It appears to me that the scene you describe is in a final count more damaging to Einstein's memory than to me.

Is not an historian of science, even more than any other scientist, kept under scrutiny by future members of his guild? There can be no greater mishap to an historian of science as when he unwittingly becomes the cause of a distortion of history at its source.

If I understand right, you have not yet made up your mind conclusively as to my position in science as it will find its evaluation by a future generation (see also the advance abstract of your lecture before the *Amer. Philos. Soc.,* April 1952). So why not to learn about a dissident from close? When in Princeton, you are welcome to visit me and read the letters Einstein exchanged with me, his notes on my manuscript, or any other material that may interest you. You are really welcome.'

"I did not hear directly from Cohen. Dr. Otto Nathan, executor of Einstein's estate, protested the fact that the interview had not been submitted to him before being printed, as it would have been submitted to Einstein for approval if he had been alive; Nathan's letter, the first part of which I quote here, was printed two months later in the September issue of *Scientific American*. He began:

"In 'An Interview with Einstein', published in the July issue of your magazine, I. Bernard Cohen quotes remarks which Albert Einstein allegedly made about a recently published book and its author. Professor Cohen represents Einstein has having said that both the book and its author were 'crazy,' but not 'bad.'

"As executor of Einstein's estate, and as one who has the responsibility to protect his scientific and literary interest, I feel compelled to say that I deeply regret Professor Cohen's statements. The article was not submitted to me before publication. If it had been, I should have made every effort to prevent it from being published in its present form. Professor Cohen would certainly not have published it without Einstein's approval had he been alive. Similarly, after Einstein's death, it was Professor Cohen's duty to seek permission for publication..."

"Bernard Cohen answered in the same issue and offered what was actually his response to my letter to him, though he did not mention it:

"The immediate cause of Dr. Nathan's concern is my reporting of the remarks made in my presence by Professor Einstein about a book. The remarks were evidently intended to illustrate two main points: (1) that any acts toward suppressing a book which contains heretical or unorthodox ideas (even in science) is evil; (2) that there is no objective test of whether notions that contravene accepted scientific ideas and theories are the work of a crank or a genius, nor whether such ideas will forever seem crazy or perhaps become the orthodoxy of the future. As an illustration there was a reference to Kepler and to a book which Professor Einstein had read and had found in part interesting. Professor Einstein did not mention terms about the above mentioned issue and was using the book only as an example of work that was sufficiently unorthodox to appear 'crazy' to a scientist. Thus, on the basis of the few words said, and reported by me in full, there is no basis for concluding that Professor Einstein might not have had a friendly feeling for the author in question or that he might not have had some interest in his work. As is plain from my article, Professor Einstein sympathized with the author when he was attacked and disliked the methods used by some of his attackers."

"Although Bernard Cohen, under pressure, wrote the above letter, I could only hear Einstein's words, 'Don't let the abuse discourage you; are you not happy in your isolation'?"[9]

There are many ways to destroy the reputation of a man as Raup showed. By misrepresenting Einstein's views of Velikovsky's work in

Worlds in Collision, we see that misrepresentation by an historian in the journals of organized science is not really any different than the way his views were handled at the AAAS symposium at which Carl Sagan first attacked Velikovsky.

SAGAN'S COMMITMENT

Toward the end of the symposium there was to be a question period in which Sagan, as one of the participants, could be questioned. However, this was not to be. Sagan left before he would have had to face his audience and answer their questions. The pressing engagement was an appearance on the Johnny Carson show. So far, he has never publicly faced and answered any of the questions raised by supporters of Velikovsky. Does Sagan lack the courage to face those scholars and answer their questions? This is a serious question and raises suspicion that when serious debate develops about revolutionary ideas, the scientists wage warfare by disreputable means. But what does Sagan discuss during his visits to Johnny Carson on the *Tonight Show.* According to Joseph Goodavage, *Storm on the Sun,* (NY 1979), p. 134,

> "During an interview on the *Tonight Show*...Professor Sagan casually told host Johnny Carson about the false 'Mars-moons-are-cities' myth [which he had presented in *Intelligent Life in the Universe*] and explained how he worked with NASA to change *Mariner's* trajectory so that its cameras would be trained on the mysterious Mars moons from close up to settle their real nature once and for all. 'There were, of course, no cities,' Sagan said coolly. 'In fact the pictures of Phobos look something like a pockmarked potato.' That seemed to be that. No big triumph for science, just another crack-pot theory shot down by self-correcting scientific research—*and* a fantastically good job of space technology.
>
> "At no point, did Sagan give the slightest indication on that televised show that he had been the co-author of that most interesting [Mars-moons-are-space cities] story."

HALTON ARP

As a case in point, let us briefly examine the case of an establishment scientist who presented a revolutionary concept about the red shifts of stars. It is generally accepted that the light from stars will be shifted toward the red end of the electromagnetic spectrum as a star moves away from an observer on the Earth. The greater the red shift, the farther away the star is considered to be. On the other hand, if the star's light is shifted toward the blue end of the spectrum, it is believed that the star is approaching the observer on the Earth. One important astronomer challenged this view. His name is Halton Arp. If Arp's views should turn out to be correct, much of present day astronomical theory would fall, and with that fall would go the reputations of several notable scientists. When Arp's views became impossible to face any longer, he was refused permission to use large telescopes and thus could not continue to make his case by producing new evidence to vindicate his hypothesis. Geoffrey Burbidge, a world renowned astrophysicist described the techniques used to destroy Arp.

> "The community of astronomers is totally polarized by this argument [about the meanings and interpretations of red shifts]. Most do not want to hear about it. The strong disbelievers hold that those who propose or believe in this hypothesis are variously naive, mistaken, ignorant of how to do statistics, overly zealous or worse. They claim that...[the results Arp claims as evidence] are not reproducible, that we have no theory to explain these phenomena, and that we should recant, and that in fact, the red shift controversy is over: that is, the *status quo* has been maintained. This last statement is often made in meetings to which the proponents of unorthodoxy are either not invited, or not allowed to speak."[10]

The reader will remember that at the AAAS symposium speakers from the scientific community were not allowed to speak on Velikovsky's behalf; and attacks in the journals of science on Velikovsky over the years could not be answered in these organs of science because the editors suppressed and still suppress evidence if it is clearly related to his theory. Thus, Sagan's remarks are nothing but rhetoric. Burbidge goes on to relate the rank depths to which the astronomical community sank to excommunicate Halton Arp. He writes,

> "The other part of the learning process has been unpleasant, probably because I have a strong instinct for fair play. It may be argued that this is no substitute for good judgment. But neither are the tactics that have been used by those who want to maintain the *status quo*. These include interminable refereeing [holding up the publication of papers for long periods of time and requiring endless corrections to frustrate publication], blackballing of speakers at meetings [withholding positions at universities for those who support Arp], distortion and misquotation of the written word, rewriting of history..."[11]

This begins to sound all too familiar. Against such venality what chance has any new idea?

THIS IS SCIENCE TODAY

The view that a comet may have struck the Earth and caused the extinctions of the dinosaurs has also led to vilification all around. *The New York Times* reports;

> "The impact of a large comet may or may not have killed the dinosaurs 65 million years ago. But as the debate over dinosaur extinction rages on, personal rancor is increasingly clouding scientific issues.
>
> "Scientists on both sides of the argument agree that it has taken an unusually harsh and personal turn, coloring published professional corre-

spondence, scientific meetings and books on the topic. The debate has crystallized into a conflict between opposing camps whose partisans rarely seem to change their minds or soften their position whatever the objective evidence may be."

'Lacking conclusive evidence one way or the other' said one scientist, 'opponents in this debate have been reduced to *name calling*.' [emphasis added]

"Charges and recriminations have also flowed through the informal grapevine that can make or break scientific careers. The bitterness of the debate is so much more intense than usual that several historians of science have begun detailed examinations of the sociology of the debate.

"All this ill will is a reflection of the deep disagreement among scientists over a crucial aspect of the Earth's history. But the personal venom in the debate, some scientists fear, is inhibiting rational scientific discourse."[12]

In this debate Sagan has a stake. His theory of nuclear winter is part of it, although the winter he postulates that caused the extinction of the dinosaurs was caused by a body striking the Earth, not by thermonu-clear bombs. Sagan and Stephan J. Gould, who also advocates ancient collisions such as one that killed the dinosaurs, have been in the forefront in attacking Velikovsky because they cannot allow *recent* catastrophism. Now their chickens come home to roost. Sagan is, perhaps, an old hand at dealing with this kind of debate. But in such an atmosphere, there is no objectivity and Sagan's platitudes about objective and vigorous debate are nonsense.

ATOMS AND SOLAR SYSTEMS

Sagan then raises another argument in his "conclusion":

"but on the next-to-the last page of the book, a breathtaking departure from the fundamental thesis is casually introduced. Here we read of a hoary and erroneous analogy between the structure of the solar system and of atoms. Suddenly we are presented with the hypothesis that the supposed errant motions of the planets, rather than being the results of collisions, are instead due to changes in the quantum energy levels of planets attendant to the absorption of the proton—or perhaps several. Solar systems are held together by gravitational forces, atoms by electrical forces. While both forces depend on the inverse square of distance, they have totally different characters and magnitudes: as one of many differences, there are positive and negative electron charges, but only one sign of gravitational mass. We understand both solar systems and atoms well enough to see that Velikovsky's proposed 'quantum jumps' of planets is based on a misunderstanding of both theories and evidence."[13]

Velikovsky wrote that if electromagnetism plays a role in celestial motion, then the solar system would, in this respect, be like the atom. Sagan states, "We understand both solar systems and atoms well enough..." By "we", one can only presume Sagan means scientists. Does he mean *all* scientists? Let us see if relationships exist in the solar system of a quantum nature that are similar to what is found in the atom. A.E. Caswell in *Science* has written about one aspect of the solar system which apparently behaves just like Niels Bohr's quantum physical analysis of the atom.

"The writer has discovered another relation between the planetary distances, and so far as he is aware this relation has not been reported hitherto. It suggests the possibility that the orbits of planets may be 'quantized' somewhat after the manner of the electronic orbits of the Bohr atom. For this reason, it may prove to have theoretical importance.

"*The distances of the planets from the sun are proportional to the squares of simple integral numbers.* The four innermost planets are represented by four successive integers, viz., 3, 4, 5 and 6. The space between the...asteroids and Jupiter to a difference of 3, that between Jupiter and Saturn to a difference of 4, and that between Saturn and Uranus to a difference of 6, that between Uranus and Neptune to a second difference of 6."[14]

William Malisoff in *Science,* Vol. 70, (1929), pp. 328–329, reported that there was a quantum relation between the mean period of planets and their mean velocity that appeared to exhibit a wave of velocity similar to the electron in the atom.

And J.B. Penniston in *Science,* Vol. 71, (1930), pp. 512–513 showed that all the planets and all their satellites obey these same "quantized" mathematical relations that reflect orbits in the Bohr atom.

John P. Bagby in an article in *Speculations in Science and Technology,* Vol. 2, (1979), p. 173, titled "A Comparison of the Titus-Bode Rule With the Bohr Atomic Orbitals", has written an equation that not only derived the Titus-Bode Rule of planetary bodies and their satellite distances but which can also "accommodate the atomic Bohr Law as well...[His results] show how the Constant of Gravitation, Planck's Constant, and the Bohr integer squared relationship can be combined with the mass and density of any primary body to define electrical and gravitational orbit(al)s. This should go a long way towards the ultimate goal of a Unified Field Theory."

Furthermore, J. Barnothy in an article titled, "Quantization of the Solar System" in *Nature,* Vol. 157, (1946), p. 808, has written a very provocative equation regarding the motions of the planets. The surprising feature is the appearance of 137 as the fine structure constant which is also the same for the Bohr atom. But most provocative is the fact that the Pauli exclusion rule also applies to the planets; no two planets can have the same numerical values unless they rotate in opposite directions. In the solar system only Venus and Uranus rotate in the retrograde (opposite) direction to that of the other planets; nevertheless, it is found that Saturn and Uranus have the same quantum numbers as do the Earth and Venus. If the scientific community had not been as dogmatic, as Sagan seems to be on this issue, then back in 1946 astronomers would have known that Venus rotates in the retrograde direction, whereas this was discovered almost 20 years later.

Petr Beckmann in *Einstein Plus Two,* (Boulder, CO 1987) using classical physics that apply to atoms was also able to use these same physical formulas to derive the distances of all the planets from the Sun. Furthermore, his last chapter, pp. 184-186 discusses the relationship between gravity and electromagnetism, but he comes to exactly the opposite conclusion Sagan has presented. Beckmann states,

"The fact that this hypothesis leads to agreement with observation for both electrons and planets, that is, regardless of whether the field propagating from its source is electric or gravitational, suggests that electric and gravitational forces might not be siblings, but the same phenomena. Thus suggestion gets some support from the macroscopic magnetic field, which is proportional to the velocity of a charge with respect to neutral matter.

"Since all macroscopic matter [matter above the size of atoms] (the only type where gravitation has been observed) is known to consist of positive nuclei and negatively charged electrons, the idea that gravitation is ultimately due to electric forces appears plausible; it has been proposed by several scientists of which Walther Ritz may not have been the first.

"Let us see what stands in the way of such a proposition:

"1) there are two types of electric charges, positive and negative, but only one type of mass;

"2) there is no charge without mass, but there is mass without charge;

"3) mass has inertia, charge does not;

"4) mass varies with velocity, whereas charge is invariant;

"5) electric force can be screened, but nothing is opaque to gravity

"6) we have always been told the two are different.

"Of these, only no. 6 is a serious obstacle; the others may be false, removable or of our own making."

No. 3, for example, [mass has inertia, charge does not] is false: inertia is the reaction to a force, the resistance to being accelerated. The phenomenon is observed with charges, or at least with their fields, just as surely as with masses... When a large number of charges—a current—is accelerated, we call the phenomenon "self-inductance"; when a single (point) charge is accelerated, the same resistance to acceleration sets in, but we have not been paying much attention to it in our textbooks. The classics of the late 19th century worked with the concepts of 'electromagnetic mass' and 'electromagnetic momentum'; the concepts have gone out of fashion, but they have not been refuted. In both cases, the resistance to acceleration is proportional to the acceleration, and to the mass or charge of the object that is resisting. 'Proportional to' is made by nature; 'equal to' when it depends on units, is made by man. Any difference between the inertia of mass and the inertia of charge is, therefore, of our own making.

"No. 4 [mass varies with velocity, whereas charge is invariant] statement from the point of view of the present theory is that inertial mass is velocity-dependent, whereas gravitational mass, like its analog electric charge, is invariant—or can we make it so, if we choose to regard velocity-dependent field and aberration effects as separate factors rather than incorporate them in the value of gravitational mass or electric charge.

"No. 2 [there is no charge without mass, but there is mass without charge] becomes at best trivial in view of these two realignments of concepts (gravitational mass the analog of electric charge and inertial reaction the equivalent of self-inductance). The nearest valid statement to no. 2 is then that neither gravitational mass nor electric charge is without inertial reaction, with which no one (even an Einsteinian, if he knows his electromagnetics) should have any quarrel. However, as an objection to the identity of gravitational and electric force, no. 2, as it stands, becomes an exhibit of flawed logic. Suppose one could overcome the experimental difficulties—say by charging and discharging the leaden balls in the Cavendish torsion scales. No one doubts that the balls would move; but how does that decide whether a force of an entirely different character was activated or whether the gravitational force of the leaden block was made to change its value?

"That leaves no. 1 [there are two types of electric charges, positive and negative, but there is only one type of mass; Sagan's argument] which says that an electric force may either attract or repel, but a gravitational force can only attract. Now we all know that positive and negative charges neutralize each other—but do they? Consider the field of an electric dipole...

"The force between two electric dipoles varies as the inverse *cube* of the distance between them, as is easily shown from the field of one dipole derived in virtually all textbooks. The same inverse cube would apply to two 'masses' of dipoles. ...Gravity, however, acts by the inverse *square* law just like single electric charges which appears to make the theory untenable.

"The reason why I have nevertheless mentioned it is to show what a comparatively near miss it is. It *fails only by the wrong dependence on distance*. However, this failed theory is not needed to remove the other objections, which are removed by no more than a conceptual realignment: charge and gravitational mass are twins not just because they appear in the same place in the Newton-Coulomb Law, but because both exhibit the same type of inertial reaction when accelerated, the measure of resistance to acceleration being given by electromagnetic and Newtonian inertial mass, respectively.

"Such a conceptual realignment from contemporary thinking, beyond leading to agreement with experiment without perverting space and time, will also convert electromagnetism and gravity from siblings to twins...

"However, since there is only one flaw—the wrong dependence on distance—to overcome, the failure of this particular theory is no refutation of the idea that the supposed twins (gravity and electromagnetism) may yet turn out to be one identical child."

Sir Arthur Eddington in *Fundamental Theory*, (Cambridge Univ. Press 1953), pp. 103-104 wrote formulas to derive the constant g of gravity. George R. Talbott in his book *Sir Arthur and Gravity*, (1993) pp. 26-32, expressed Eddington's formula as $g = \tilde{Q}$, times (the charge on an electron divided by the mass of the hydrogen proton squared). This formulation shows that to derive g, there is an electromagnetic component, *the charge of the electron,* which when applied, derives gravity.

Because some astronomers refuse to admit the electromagnetic quantum nature of the solar system, is no reason for Sagan to imply that no such relationships exist. The very axioms that Niels Bohr used to analyze the atom seem to be reflected by the planets of the solar system. In this case, it does seem that Sagan refused to make the 'quantum leap' which is his right, but he does ignore the evidence that appears to be found in atoms and solar systems.

HANNES ALFVEN

Hannes Alfven after many years of research postulates that the universe is actually dominated by electromagnetic forces, which he maintains is more significant in organizing and structuring the universe than is gravity. He has offered an electromagnetic theory in which this force can explain all the various types of galaxies, without seeking the "missing mass" to stop the universe's expansion. His theory accounts for the cosmic microwave background. Alfven's electromagnetic theories are supported by observations of solar system plasma and magnetic fields. Thus, Sagan's "we understand" is patently paternalistic; it is also arrogant.

Hannes Alfven, a Nobel prize winner in physics, claims the universe, as well as the solar system, is dominated by electrical forces. It is these forces which keep the atom together. Because of his view that electromagnetism plays a decisive role in structuring the universe, Alfven has also come under censure by the establishment. In the *American Scientist,* Vol. 76, (1988), p. 247, Alfven warns us that the scientific endeavor can be "petrified" by the political tactics employed by the establishment. He states,

"The (above) mentioned conditions and quite a few other factors have led to a disagreement between a very strong establishment (E) and a small group of dissidents (D) to which the present author belongs. This is nothing remarkable. What is more remarkable and regrettable is that it seems to be almost impossible to start a serious discussion between E and D. As a dissident is in a very unpleasant situation, I am sure that D would by very glad to change their views as soon as E gives convincing arguments. But the argument 'all knowledgeable people agree that...' (with the tacit addition that by not agreeing, you demonstrate that you are a crank) is not a valid argument in science. If science issues were decided by Gallup Polls, and not by scientific arguments, science will very soon be petrified forever."

Alfven has raised an issue similar to the one raised by the author of this book. He makes the point that scientists who oppose the major concepts accepted by the establishment will be called cranks or perhaps crackpots, that they may not be able to have their views published in scientific journals, that they may not be able to get funding or support for their research, and that they may be ostracized, that is, excommunicated by the scientific establishment. The cogent point, however, is that the debate will not be based on scientific evidence; rather it will be based on politics, namely through distortion of the evidence and censorship in the scientific journals.

THE NUB OF THE ISSUE

As a final note, Sagan stated,

"What then is all the furor about? It is about the time scale and the adequacy of the purported evidence. In the 4.6 billion year history of the solar system, many collisions must have occurred. But have there been major collisions in the last 3,500 years and can the study of the ancient writings demonstrate such collisions? That is the nub of the issue."[15]

Some 32 years after Velikovsky had published *Worlds in Collision*, another book appeared written by two noted British astronomers. The title of their book is *The Cosmic Serpent*, put out by Universe Books, (NY 1982). The authors, Victor Clube and Bill Napier are astronomers at the Royal Observatory, Edinburgh. They have published articles in many scientific journals, and are well respected establishment scientists. From the blurb of their book we read, "The cosmic serpent was a giant comet that terrorized mankind in prehistoric times. As a fiery dragon and hurler of thunderbolts, it wrought destruction and disaster upon the Earth. Comets have been the object of superstition and fascination since cavemen looked up at the night sky and saw 'shooting stars.' In the intervening millennia, much information about comets has been gathered, but even today we do not properly understand these cosmic phenomena.

"In *The Cosmic Serpent,* two internationally respected astronomers propose radical new theories about the origin and future of the Earth, focusing on worlds shaped by catastrophe. They claim that impacts from comets, asteroids, and meteoric bombardments caused devastation on a global scale and brought on massive evolutionary changes, including the demise of the dinosaurs. *And they say these cosmic events gave rise to ancient myths, inspired prophets and philosophers and affected the creation of astrological sciences and ancient calendars.* [emphasis added]

"Warning that 'there is something here to outrage everyone,' Clube and Napier *present impressive and persuasive arguments that challenge established scientific thought, mythological, and Biblical interpretation, and historical archaeological and anthropological evidence.* [emphasis added] Hardly an area of modern knowledge remains unaffected directly or indirectly by the ideas put forth in this book.

"...Perhaps most provocative of all are assertions concerning both the *Old* and *New Testaments. Here, Clube and Napier reveal that cometary collisions with the Earth were responsible for the Great Flood of Noah's time as well as the parting of the Red Sea during the Israelites' Exodus from Egypt.*" [emphasis added]

The only difference between Napier and Clube's thesis with that of Velikovsky's is that their comet called Venus is not the planet Venus. D.C. Stove remarked,

"Now from the start, Velikovsky's books, though officially despised by scientists, provided a rich source for some of them, especially younger ones, to borrow from. Many a decent list of publications, in respectable journals, have been built up in this way; but almost always, of course, without a word of acknowledgment to Velikovsky, since to mention him would be virtually to ensure the rejection of your article. In this anonymous and back-door way, Velikovsky's ideas, and ideas like his have been steadily gaining in respectability for over thirty years. This process has reached an unprecedented height in a book recently published. It is *The Cosmic Serpent*... Their book, considered strictly in itself, is by no means a bad one. But the theory it puts forward happens to be essentially Velikovsky's. Not only that, but most of the historical [and legendary] evidence that the authors offer for their theory is simply some warmed-over parts of Velikovsky's evidence. Of course their theory is not *exactly* the same as Velikovsky's. The main extra-terrestrial agent of destruction, according to Velikovsky, was Venus in a pre-planetary state; according to Clube and Napier, it was a certain asteroid stream." [16]

Carl Sagan's asks, "But have there been major collisions [between a comet and the Earth] in the last 3,500 years and can the study of ancient writings demonstrate such collisions? That is the nub of the issue." According to two internationally respected astronomers, Clube and Napier, the answer, is apparently "YES", and to make their case they essentially cite Velikovsky's evidence. So when we evaluate what Sagan termed the "nub of the issue" regarding Velikovsky's historical and legendary evidence, his own colleagues say Sagan is wrong!

With all the furor and name calling of Velikovsky, certain conclusions become obvious. One is that there is a double standard in the scientific community. The scientists readily give recognition and respect to certain pirated Velikovskian theses while at the same time vilifying these same views when presented by Velikovsky.

Patrick Moore is a highly respected British astronomer, internationally recognized for his researches on the Moon and planets, who in his books, resorts to name calling, labeling Velikovsky a "charlatan" and his hypothesis "crank science." But, on the back cover of *The Cosmic Serpent* Moore states: "This is one of the most extraordinary books I have ever read. Both the authors are professional astronomers, and both hold responsible scientific positions. Yet the views expressed are totally unorthodox... Not that I can accept many of the views put forward, but they are fascinating nonetheless... The book is extremely well written. It will create a stir. It will certainly be widely read and discussed."

Elsewhere, Moore writes of Napier and Clube's theory. "It is revolutionary and unorthodox, it is speculative, but it is worth taking seriously, and it is an honest effort to link astronomy and paleontology, geology, mythology and even history."[17] Patrick Moore in *Fireside Astronomy,* (NY 1992), p. 193, tells us, "support for...[Napier and Clube's] theories is said to come from records left by old civilizations—the Romans, Greeks, Mayas, Incas, Chinese, Babylonians, and so on, even the New Zealand Maoris and the aborigines of Australia."

When *astronomers* tell us that the study of ancient writings may demonstrate that there have been major collisions between the Earth and a comet in the last 3,500 years, they are neither "charlatans" nor do they practice "crank science." The nub of the issue comes down to this; it is not that Velikovsky's hypothesis lacks scientific foundation, but that establishment scientists have a double standard; one dirty little standard for Velikovsky and quite another standard when the same thesis and evidence is presented by *colleagues*.

In *The Cosmic Serpent,* it is, therefore, not in the least surprising that Carl Sagan is strangely unwilling to discuss the fact that Clube and Napier use Velikovsky's evidence to support their claims. Sagan, in his book *Cosmos,* p. 90 states, "Astronomers do not object to the idea of major collisions, only to major *recent* collisions." But Clube and Napier, apparently haven't read Sagan's injunction.

SCIENTIFIC ETHICS

Sagan ends his critique with the following,

"The attempt to rescue old time religion, in an age which seems to be seeking some religious roots, some cosmic significance for mankind, may or may not be creditable. I think there is much good and much evil in the old-time religions. But I do not understand the need for half-measures. If we are forced to choose between them—and *decidedly* we are not—is the evidence not better for the God of Moses, Jesus and Muhammed than for the comet of Velikovsky?"[18]

These religions about which Sagan writes have flourished because humanity seems to need, among other things, the assurance that honesty and truth are as fundamental as love and life. The scientific endeavor as an organized society of scholars also has the responsibility to tell the truth, and not betray this calling by substituting political tactics and misrepresentations of evidence. In this respect, Sagan's critique of Velikovsky has been destructive of this responsibility.

The findings in the several branches of science are significant not only because of the profound insights they may give, but also because of the benefits they can produce for mankind. While revolutionary new theories are of a most fundamental nature in fostering this process, they are also the views most subject to the antagonistic and unobjective opposition by organized science. In this respect, science itself becomes a religion, opposed to any great change in the doctrines (theories) it holds as truth. The power employed by the established scientific institutions to suppress or misrepresent and thus, subvert any revolutionary concept that would overthrow the *status quo,* is perhaps the greatest danger to science and humanity.

If new ideas are dismissed through the use of venal tactics, rather than through proofs, experiments and evidence, then science will cease to perform its function to search for objective knowledge and become just another religion which no longer seeks new knowledge. The evidence presented in this book clearly shows that there is a need for scientists to re-examine what the words "ethics", "truthfulness" and "open and honest debate" mean to them. If old time religion had much good, surely that good lay in its adherence to the ethic of truthfulness. If scientists wish to be respected for their search for objective knowledge, they must rid themselves of what is clearly "evil". The greatest "evil" for science is to lie through misrepresenta-tion and the use of yellow journalism. In this respect, there is much to be learned from the American Association for the Advancement of Science's symposium on Velikovsky and Sagan's critique.

Contrary to the view of how science operates presented by Sagan, a very different view is offered by Thomas S. Kuhn, the historian of science in his classic treatment of the subject, *The Scientific Revolution*. James Gleick's best-seller, *Chaos: Making a New Science,* (NY 1989), pp. 36–38 aptly condenses Kuhn's evaluation of how science operates.

"Kuhn's notions of how scientists work and how revolutions occur drew as much hostility as admiration when he first published them, in 1962, and the controversy has never ended. He pushes a sharp needle into the traditional view that science progresses by the accretion of knowledge, each discovery adding to the last, and that new theories emerge when new experimental facts require them. He deflated the view of science as an orderly process of asking questions and finding their answers. He emphasized a contrast between the bulk of what scientists do, working on legitimate, well-understood problems within their disciplines, and the exceptional, unorthodox work that creates revolutions. Not by accident, he made scientists seem less than perfect rationalists...

"Central to Kuhn's ideas is the vision of normal science as solving problems, the kinds of problems that students learn the first time they open their textbooks... 'Under normal conditions the research scientist is not an innovator but a solver of puzzles, and the puzzles upon which he concentrates are just those which he believes can be both stated and solved within the existing scientific tradition,' Kuhn wrote.

"Then there are revolutions. A new science arises out of one that has reached a dead end. Often a revolution has an interdisciplinary character—its central discoveries often come from people straying outside the normal bounds of their specialties. The problems that obsess these theorists are not recognized as legitimate lines of inquiry. Thesis proposals are turned down or articles are refused publication...

"Every scientist who turned to [study mathematical] chaos has a story to tell of discouragement or open hostility. Graduate students were warned that their careers could be jeopardized if they wrote theses in an untested discipline... Shallow ideas can be assimilated; ideas that require people to reorganize their picture of the world provoke hostility."

This is the picture of how science really operates. When Sagan suggested that Velikovsky submit his answers to attacks to recognized journals of science, he was implying that such articles would have a chance of being accepted. Velikovsky did try this route, but the door was rudely slammed in his face again and again. Sagan suggests that the AAAS symposium held to discuss Velikovsky's ideas is based on rational considerations when, in reality, it was organized to destroy Velikovsky's reputation and ideas. Count Leo Tolstoy understood long ago that educated men can be so narrow that new concepts bring forth hostility, meanness and ridicule. Gleick cites him, thus, on p. 38,

"I know that most men, including those at ease with problems of the greatest complexity, can seldom accept even the simplest and most obvious truth if it be such as would oblige them to admit the falsity of conclusions which they have delighted in explaining to colleagues, which they have proudly taught to others, and which they have woven, thread by thread into the fabric of their lives."

How modern science makes sophisticated warfare on the proponents of new ideas has been discussed in the conclusion of this work; for now it is clear that the approach to science that Sagan professes was in no manner, whatsoever, applied to Velikovsky. In fact, just the reverse was employed. It was political and dirty.

In Sagan's book *The Dragons Of Eden,* (NY 1977), soft cover edition, pp. 203-204, Sagan wrote,

> "...the future belongs to those societies that...enable the characteristically human components of our nature to flourish; to those societies that encourage diversity rather than conformity...*to those societies that treat new ideas as delicate fragile and immensely valuable pathways to the future.*" [emphasis added]

Mark Washburn, however, in *Mars at Last,* (NY 1977), p. 96 informs us, "Indeed the scientists [at the AAAS symposium on Velikovsky] seemed to relish the opportunity to dispose of Velikovsky once and for all." *That* was how Velikovsky's theory was greeted by the scientists. They did not come to examine the evidence, but to dispose of Velikovsky. The same year as the AAAS meeting we find Charles Fair in *The New Nonsense,* (NY 1974) stating, "Academia will never let him [Velikovsky] get away with it. No matter how many of his predictions prove correct, it will see him dead first." [19]

It is extremely difficult to believe that Sagan or the AAAS could ever treat Velikovsky's ideas with either honesty or respect. Like the governments that Sagan castigates that have shut its minds to new ideas, his mind, when it comes to Velikovsky's ideas, is of the same order. He has betrayed the assumption of openness fundamental to science. R*evolutionary ideas are fragile* and may be immensely valuable pathways to the future. But the gross animosity exhibited by Sagan and the AAAS to Velikovsky and his thesis are nothing but a public relations attempt to murder that thesis and cover up the evidence of the assassination. Such tactics can only grow out of deep seated prejudices. Such tactics are the American form of Lysenkoism.

At the beginning of this book we cited Anthony Aveni who ex- claimed that Carl Sagan had "driven the last nail" into the coffin of Velikovskyanism; that it was dead and buried. But, given the evidence presented here, Sagan did nothing of the sort. Rather the AAAS drove its nails into scientific integrity. The coffin buried at the AAAS Symposium contained not Velikovsky's theories, but rather the corpse of scientific ethics, mutilated by misrepresentation.

William James, the great American psychologist and philosopher made the following remark about a new discovery. He said that it was, "First...attacked as absurd; then it is admitted to be true, but obviously insignificant; finally it is seen to be so important that its adversaries claim that they themselves discovered it."[20] The work of B. Napier and V. Clube, which is built on the work of Velikovsky, is clearly a reflection of James' comment.

Consider current views about how the Earth evolved over the ages. Walter Sullivan of *The New York Times* writing in the November 1, 1988 issue discusses the emerging concept as it was presented under the title, "Global Catastrophes in Earth History" at Snowbird, Utah. On page C1 Sullivan, who has been strongly opposed to Velikovsky's theory, nevertheless reports,

"Catastrophic collisions with asteroids have played a major role on Earth in shaping geology, climate and evolution of life, a growing number of scientists assert.

"...New theories, some highly speculative, have linked past impacts to a wide variety of key events in the Earth's history. The theories suggest that impacts caused reversals of the Earth's magnetic field, the onset of ice ages, the splitting apart of continents 80 million years ago and great volcanic eruptions..."

On page C6 Sullivan continues,

"According to a theory advanced by...Berkeley scientists, the reversals [of the Earth's magnetic field] could have been triggered by extraterrestrial impacts that initiated a complicated series of climate and rotational changes..."

This new theory of the evolution of the Earth is, in almost all respects, the theory Velikovsky stated back in 1950. Almost every one of Velikovsky's concepts are being incorporated into the corpus of modern scientific thought, but without any acknowledgement of Velikovsky. As yet one concept of Velikovsky's work has not been accepted: most scientists still believe that the solar system is a stable arrangement that has not changed significantly over the aeons. Nevertheless, the theory of Nemesis—the death star—a companion star of the Sun, may well throw even this sacred cow of solar system stability out the window.

If the Sun does possess a small dark companion star on a highly elliptical orbit there can be no question that it may enter the the solar system every time it approaches the Sun. A body one-tenth the mass of the Sun would wreck havoc among the planetary bodies. There is little question that its gravitational influence would cause the planets to establish new orbits. Therefore, this theory itself points to an unstable solar system. The uniformitarian theory can only be maintained by divorcing the Earth from the heavens and the abundant evidence of solar system catastrophes. The house that Velikovsky built in 1950 is becoming the modern edifice of science. His contribution to this house should not be buried by propaganda masquerading as science.

But Velikovsky's theory is not so easily dismissed in view of all the supporting evidence. Thomas A. Mutch, et. al., in *The Geology of Mars,* (Princeton NJ 1976), p. 93 after looking at the evidence of Mars and its catastrophic topography were willing to admit...

> "The same seeds of doubt germinate when someone [Velikovsky] proposes that a large cometary body struck Earth many years ago, causing disruption of continents and oceans...
>
>   "It is interesting to note how [scientific] fashions change. Five hundred years ago the cosmological models that are now regarded with such disinterest would have been heresy. How might our prejudices change in the future? Today we refuse to believe that *catastrophic, interplanetary collisions have warped Earth's history.* a hundred years hence, when the large impact scars on the other planets are familiar landscapes, will we feel the same?" [emphasis added.]

In this respect, only time will tell.

NOTES & REFERENCES

[1] Raup, David; *The Nemesis Affair,* (NY 1986), pp. 148–149.

[2] Ibid.

[3] Ibid., p. 147.

[4] Ibid., p. 193.

[5] Ibid., p. 195.

[6] *Kronos* I, 4, p. 76.

[7] Velikovsky, I.; *Stargazers,* op. cit., pp. 287–295.

[8] *Pensée,* Vol. 4, No. 1, p. 7.

[9] Velikovsky, I.; *Stargazers,* op. cit., pp. 295–303.

[10] Burbidge, Geoffrey; "Anatomy of a Controversy", *Sky and Telescope.* Vol. 75, (Jan 1988), p. 40.

[11] Ibid., p. 42.

[12] *The New York Times,* (Jan. 19, 1988), p. 1C.

[13] SCV, p. 90; B.B., p. 125.

[14] Caswell, A.E.; "A Relation Between the Mean Distance of the Planets From the Sun", *Science,* Vol. 69, (1929) as summarized in *Mysterious Universe,* op. cit., pp. 543–544.

[15] B.B., p. 86.

[16] Stove, D.C.; "Velikovsky and the Cosmic Serpent", *Kronos* IX, 3, p. 41.

[17] Moore, Patrick, *Armchair Astronomy,* (NY 1984), p. 100.

[18] SCV, p. 93; B.B., p. 127.

[19] *Kronos,* V, 4, p. 61.

[20] Bartusiak, Marcia; *Thursday's Universe.* (NY 1986), pp. 170–171.

INDEX

439

441

FOR MORE INFORMATION

Velikovsky's legacy is being carried on by scientists and scholars in various countries. In the United States there is a journal that is devoted to continuing this research. *The Velikovskian* is published quarterly. For those interested it may be ordered by writing to:

Charles Ginenthal
65-35 108 Street
Forest Hills, NY 11375
U.S.A.

This work has been going on in the United States and elsewhere for several years.